Molecular Technology

Related Titles

Yamamoto, H., Kato, T. (eds.)

Molecular Technology

Volume 2: Life Innovation

2018

ISBN: 978-3-527-34162-7

Yamamoto, H., Kato, T. (eds.)

Molecular Technology

Volume 3: Materials Innovation

2019

ISBN: 978-3-527-34161-0

Yamamoto, H., Kato, T. (eds.)

Molecular Technology

Volume 4: Synthesis Innovation

2019

ISBN: 978-3-527-34588-5

Molecular Technology

Energy Innovation

Edited by Hisashi Yamamoto and Takashi Kato

Volume 1

Editors

Professor Hisashi Yamamoto
Chubu University
Molecular Catalyst Research Center
1200 Matsumoto
Kasugai
487-501 Aichi
Japan

Professor Takashi Kato
University of Tokyo
Department of Chemistry & Biotechnology
7-3-1 Hongo, Bunkyo-ku
113-8656 Tokyo
Japan

Cover
fotolia_VAlex and fotolia_Slanapotam

All books published by **Wiley-VCH** are carefully produced. Nevertheless, authors, editors, and publisher do not warrant the information contained in these books, including this book, to be free of errors. Readers are advised to keep in mind that statements, data, illustrations, procedural details or other items may inadvertently be inaccurate.

Library of Congress Card No.: applied for

British Library Cataloguing-in-Publication Data
A catalogue record for this book is available from the British Library.

Bibliographic information published by the Deutsche Nationalbibliothek
The Deutsche Nationalbibliothek lists this publication in the Deutsche Nationalbibliografie; detailed bibliographic data are available on the Internet at <http://dnb.d-nb.de>.

© 2018 Wiley-VCH Verlag GmbH & Co. KGaA, Boschstr. 12, 69469 Weinheim, Germany

All rights reserved (including those of translation into other languages). No part of this book may be reproduced in any form – by photoprinting, microfilm, or any other means – nor transmitted or translated into a machine language without written permission from the publishers. Registered names, trademarks, etc. used in this book, even when not specifically marked as such, are not to be considered unprotected by law.

Print ISBN: 978-3-527-34163-4
ePDF ISBN: 978-3-527-80279-1
ePub ISBN: 978-3-527-80278-4
Mobi ISBN: 978-3-527-80280-7

Cover Design Adam-Design, Weinheim, Germany

Typesetting SPi Global, Chennai, India

Printing and Binding C.O.S. Printers Pte Ltd Singapore

Printed on acid-free paper

10 9 8 7 6 5 4 3 2 1

Contents

Foreword by Dr Hamaguchi *xiii*
Foreword by Dr Noyori *xv*
Preface *xvii*

1 Charge Transport Simulations for Organic Semiconductors *1*
Hiroyuki Ishii
1.1 Introduction *1*
1.1.1 Historical Approach to Organic Semiconductors *1*
1.1.2 Recent Progress and Requirements to Computational "Molecular Technology" *4*
1.2 Theoretical Description of Charge Transport in Organic Semiconductors *4*
1.2.1 Incoherent Hopping Transport Model *6*
1.2.2 Coherent Band Transport Model *7*
1.2.3 Coherent Polaron Transport Model *9*
1.2.4 Trap Potentials *10*
1.2.5 Wave-packet Dynamics Approach Based on Density Functional Theory *11*
1.3 Charge Transport Properties of Organic Semiconductors *15*
1.3.1 Comparison of Polaron Formation Energy with Dynamic Disorder of Transfer Integrals due to Molecular Vibrations *15*
1.3.2 Temperature Dependence of Mobility *16*
1.3.3 Evaluation of Intrinsic Mobilities for Various Organic Semiconductors *17*
1.4 Summary *18*
1.4.1 Forthcoming Challenges in Theoretical Studies *19*
Acknowledgments *20*
References *20*

2 Liquid-Phase Interfacial Synthesis of Highly Oriented Crystalline Molecular Nanosheets *25*
Rie Makiura
2.1 Introduction *25*
2.2 Molecular Nanosheet Formation with Traditional Surfactants at Air/Liquid Interfaces *26*

2.2.1	History of Langmuir–Blodgett Film	*26*
2.2.2	Basics of Molecular Nanosheet Formation at Air/Liquid Interfaces	*27*
2.3	Application of Functional Organic Molecules for Nanosheet Formation at Air/Liquid Interfaces	*27*
2.3.1	Functional Organic Molecules with Long Alkyl Chains	*27*
2.3.2	Functional Organic Molecules without Long Alkyl Chains	*27*
2.3.3	Application of Functional Porphyrins on Metal Ion Solutions	*28*
2.4	Porphyrin-Based Metal–Organic Framework (MOF) Nanosheet Crystals Assembled at Air/Liquid Interfaces	*29*
2.4.1	Metal–Organic Frameworks	*29*
2.4.2	Method of MOF Nanosheet Creation at Air/Liquid Interfaces	*29*
2.4.3	Study of the Formation Process of MOF Nanosheets by *In Situ* X-Ray Diffraction and Brewster Angle Microscopy at Air/Liquid Interfaces	*32*
2.4.4	Application of a Postinjection Method Leading to Enlargement of the Uniform MOF Nanosheet Domain Size	*35*
2.4.5	Layer-by-Layer Sequential Growth of Nanosheets – Toward Three-Dimensionally Stacked Crystalline MOF Thin Films	*38*
2.4.6	Manipulation of the Layer Stacking Motif in MOF Nanosheets	*41*
2.4.7	Manipulation of In-Plane Molecular Arrangement in MOF Nanosheets	*46*
	References	*51*
3	**Molecular Technology for Organic Semiconductors Toward Printed and Flexible Electronics**	*57*
	Toshihiro Okamoto	
3.1	Introduction	*57*
3.2	Molecular Design and Favorable Aggregated Structure for Effective Charge Transport of Organic Semiconductors	*58*
3.3	Molecular Design of Linearly Fused Acene-Type Molecules	*59*
3.4	Molecular Technology of π-Conjugated Cores for p-Type Organic Semiconductors	*61*
3.5	Molecular Technology of Substituents for Organic Semiconductors	*64*
3.5.1	Bulky-Type Substituents	*64*
3.5.2	Linear Alkyl Chain Substituents	*65*
3.6	Molecular Technology of Conceptually-new Bent-shaped π-Conjugated Cores for p-Type Organic Semiconductors	*66*
3.6.1	Bent-Shaped Heteroacenes	*66*
3.7	Molecular Technology for n-Type Organic Semiconductors	*71*
3.7.1	Naphthalene Diimide and Perylene Diimide	*72*
	References	*77*

4	**Design of Multiproton-Responsive Metal Complexes as Molecular Technology for Transformation of Small Molecules** *81*
	Shigeki Kuwata
4.1	Introduction *81*
4.2	Cooperation of Metal and Functional Groups in Metalloenzymes *81*
4.2.1	[FeFe] Hydrogenase *82*
4.2.2	Peroxidase *82*
4.2.3	Nitrogenase *83*
4.3	Proton-Responsive Metal Complexes with Two Appended Protic Groups *84*
4.3.1	Pincer-Type Bis(azole) Complexes *84*
4.3.2	Bis(2-hydroxypyridine) Chelate Complexes *89*
4.4	Proton-Responsive Metal Complexes with Three Appended Protic Groups on Tripodal Scaffolds *94*
4.5	Summary and Outlook *98*
	Acknowledgments *98*
	References *98*
5	**Photo-Control of Molecular Alignment for Photonic and Mechanical Applications** *105*
	Miho Aizawa, Christopher J. Barrett, and Atsushi Shishido
5.1	Introduction *105*
5.2	Photo-Chemical Alignment *107*
5.3	Photo-Physical Alignment *112*
5.4	Photo-Physico-Chemical Alignment *115*
5.5	Application as Photo-Actuators *118*
5.6	Conclusions and Perspectives *123*
	References *123*
6	**Molecular Technology for Chirality Control: From Structure to Circular Polarization** *129*
	Yoshiaki Uchida, Tetsuya Narushima, and Junpei Yuasa
6.1	Chiral Lanthanide(III) Complexes as Circularly Polarized Luminescence Materials *130*
6.1.1	Circularly Polarized Luminescence (CPL) *130*
6.1.2	Theoretical Explanation for Large CPL Activity of Chiral Lanthanide(III) Complexes *131*
6.1.3	Optical Activity of Chiral Lanthanide(III) Complexes *132*
6.1.4	CPL of Chiral Lanthanide(III) Complexes for Frontier Applications *135*
6.2	Magnetic Circular Dichroism and Magnetic Circularly Polarized Luminescence *135*

6.2.1	Magnetic–Field-induced Symmetry Breaking on Light Absorption and Emission *136*	
6.2.2	Molecular Materials Showing MCD and MCPL and Applications *137*	
6.3	Molecular Self-assembled Helical Structures as Source of Circularly Polarized Light *138*	
6.3.1	Chiral Liquid Crystalline Phases with Self-assembled Helical Structures *139*	
6.3.2	Strong CPL of CLC Laser Action *139*	
6.4	Optical Activity Caused by Mesoscopic Chiral Structures and Microscopic Analysis of the Chiroptical Properties *140*	
6.4.1	Microscopic CD Measurements via Far-field Detection *142*	
6.4.2	Optical Activity Measurement Based on Improvement of a PEM Technique *143*	
6.4.3	Discrete Illumination of Pure Circularly Polarized Light *143*	
6.4.4	Complete Analysis of Contribution From All Polarization Components *145*	
6.4.5	Near-field CD Imaging *145*	
6.5	Conclusions *146*	
	References *147*	

7   Molecular Technology of Excited Triplet State *155*
Yuki Kurashige, Nobuhiro Yanai, Yong-Jin Pu, and So Kawata

7.1	Properties of the Triplet Exciton and Associated Phenomena for Molecular Technology *155*	
7.1.1	Introduction: The Triplet Exciton *155*	
7.1.2	Molecular Design for Long Diffusion Length *155*	
7.1.3	Theoretical Analysis for the Electronic Transition Processes Associated with Triplet *158*	
7.2	Near-infrared-to-visible Photon Upconversion: Chromophore Development and Triplet Energy Migration *162*	
7.2.1	Introduction *162*	
7.2.2	Evaluation of TTA-UC Properties *164*	
7.2.3	NIR-to-visible TTA-UC Sensitized by Metalated Macrocyclic Molecules *165*	
7.2.4	TTA-UC Sensitized by Metal Complexes with S–T Absorption *169*	
7.2.5	Conclusion and Outlook *171*	
7.3	Singlet Exciton Fission Molecules and Their Application to Organic Photovoltaics *171*	
7.3.1	Introduction *171*	
7.3.2	Polycyclic π-Conjugated Compounds *172*	
7.3.2.1	Pentacene *172*	
7.3.2.2	Tetracene *174*	
7.3.2.3	Hexacene *175*	
7.3.2.4	Heteroacene *175*	
7.3.2.5	Perylene and Terrylene *175*	

7.3.3	Nonpolycyclic π-Conjugated Compounds *177*
7.3.4	Polymers *178*
7.3.5	Perspectives *179*
	References *180*

8 Material Transfer and Spontaneous Motion in Mesoscopic Scale with Molecular Technology *187*

Yoshiyuki Kageyama, Yoshiko Takenaka, and Kenji Higashiguchi

8.1	Introduction *187*
8.1.1	Introduction of Chemical Actuators *187*
8.1.2	Composition of This Chapter *188*
8.2	Mechanism to Originate Mesoscale Motion *189*
8.2.1	Motion Generated by Molecular Power *189*
8.2.2	Gliding Motion of a Mesoscopic Object by the Gradient of Environmental Factors *189*
8.2.3	Mesoscopic Motion of an Object by Mechanical Motion of Molecules *191*
8.2.4	Toward the Implementation of a One-Dimensional Actuator: Artificial Muscle *191*
8.3	Generation of "Molecular Power" by a Stimuli-Responsive Molecule *193*
8.3.1	Structural Changes of Molecules and Supramolecular Structures *193*
8.3.2	Structural Changes of Photochromic Molecules *196*
8.3.3	Fundamentals of Kinetics of Photochromic Reaction *197*
8.3.4	Photoisomerization and Actuation *199*
8.4	Mesoscale Motion Generated by Cooperation of "Molecular Power" *199*
8.4.1	Motion in Gradient Fields *199*
8.4.2	Movement Triggered by Mobile Molecules *201*
8.4.3	Autonomous Motion with Self-Organization *203*
8.5	Summary and Outlook *204*
	References *205*

9 Molecular Technologies for Photocatalytic CO_2 Reduction *209*

Yusuke Tamaki, Hiroyuki Takeda, and Osamu Ishitani

9.1	Introduction *209*
9.2	Photocatalytic Systems Consisting of Mononuclear Metal Complexes *213*
9.2.1	Rhenium(I) Complexes *213*
9.2.2	Reaction Mechanism *216*
9.2.3	Multicomponent Systems *218*
9.2.4	Photocatalytic CO_2 Reduction Using Earth-Abundant Elements as the Central Metal of Metal Complexes *220*

9.3	Supramolecular Photocatalysts: Multinuclear Complexes	*223*
9.3.1	Ru(II)—Re(I) Systems	*224*
9.3.2	Ru(II)—Ru(II) Systems	*233*
9.3.3	Ir(III)—Re(I) and Os(II)—Re(I) Systems	*234*
9.4	Photocatalytic Reduction of Low Concentration of CO_2	*236*
9.5	Hybrid Systems Consisting of the Supramolecular Photocatalyst and Semiconductor Photocatalysts	*241*
9.6	Conclusion	*245*
	Acknowledgements	*245*
	References	*245*

10 Molecular Design of Photocathode Materials for Hydrogen Evolution and Carbon Dioxide Reduction *251*
Christopher D. Windle, Soundarrajan Chandrasekaran, Hiromu Kumagai, Go Sahara, Keiji Nagai, Toshiyuki Abe, Murielle Chavarot-Kerlidou, Osamu Ishitani, and Vincent Artero

10.1	Introduction	*251*
10.2	Photocathode Materials for H_2 Evolution	*253*
10.2.1	Molecular Photocathodes for H_2 Evolution Based on Low Bandgap Semiconductors	*253*
10.2.1.1	Molecular Catalysts Physisorbed on a Semiconductor Surface	*253*
10.2.1.2	Covalent Attachment of the Catalyst to the Surface of the Semiconductor	*256*
10.2.1.3	Covalent Attachment of the Catalyst Within an Oligomeric or Polymeric Material Coating the Semiconductor Surface	*258*
10.2.2	H_2-evolving Photocathodes Based on Organic Semiconductors	*260*
10.2.3	Dye-sensitised Photocathodes for H_2 Production	*263*
10.2.3.1	Dye-sensitised Photocathodes with Physisorbed or Diffusing Catalysts	*266*
10.2.3.2	Dye-sensitised Photocathodes Based on Covalent or Supramolecular Dye–Catalyst Assemblies	*268*
10.2.3.3	Dye-sensitised Photocathodes Based on Co-grafted Dyes and Catalysts	*270*
10.3	Photocathodes for CO_2 Reduction Based on Molecular Catalysts	*273*
10.3.1	Photocatalytic Systems Consisting of a Molecular Catalyst and a Semiconductor Photoelectrode	*274*
10.3.2	Dye-sensitised Photocathodes Based on Molecular Photocatalysts	*278*
	Acknowledgements	*281*
	References	*281*

11 Molecular Design of Glucose Biofuel Cell Electrodes *287*
Michael Holzinger, Yuta Nishina, Alan Le Goff, Masato Tominaga, Serge Cosnier, and Seiya Tsujimura

11.1	Introduction	*287*
11.2	Molecular Approaches for Enzymatic Electrocatalytic Oxidation of Glucose	*291*

11.3	Molecular Designs for Enhanced Electron Transfers with Oxygen-Reducing Enzymes *295*	
11.4	Conclusion and Future Perspectives *297*	
	References *300*	

Index *307*

Foreword by Dr Hamaguchi

Molecular Technology is a newly developed research field supported through Japan Science and Technology Agency (JST) research funding programs. These programs aim to establish an innovative research field that harnesses the characteristics of molecules to enable new scientific and commercial applications. It is our great pleasure to publish this book, with the ambition that it will develop both an understanding of and further support for this new research field within the research and student community.

Molecular Technology as introduced in this book began in 2012 as a research area within JST's Strategic Basic Research Programs. JST is an advanced network-based research institution that promotes state-of-the-art R&D projects and leads the way in the cocreation of future innovation in tandem with wider society. JST develops a wide range of funding programs related to the promotion of scientific and technological innovation, which include strategy planning, target-driven basic research, and promotion of research and development.

Various research projects focused on Molecular Technology are currently underway within JST's Strategic Basic Research Programs:

- The team-based research program "CREST (Core Research for Evolutionary Science and Technology)"
- The individual research program "PRESTO (Precursory Research for Embryonic Science and Technology)."

Dr Yamamoto (CREST) and Dr Kato (PREST) manage the Molecular Technology Research Area as research supervisors.

In addition, JST's Strategic International Collaborative Research Program promotes research projects in the area of Molecular Technology, including ongoing cooperation with *L'Agence nationale de la recherche* (The French National Research Agency, ANR).

A wide range of researchers from young to senior across the fields from green science, life science, and energy are participating in successful research aimed at establishing the new field of Molecular Technology. They are already producing excellent research results, and it is our hope that these will develop into technologies capable of initiating a new era in energy, green, and life sciences.

I encourage you to read not only researchers in related fields but also look more broadly to researchers working in other fields. Inspired by this book, I look forward to emerging new research fields and seeds toward future innovation.

Michinari Hamaguchi
President, Japan Science and Technology Agency

Foreword by Dr Noyori

As an affiliated institution of the Japan Science and Technology Agency (JST), the Center for Research and Development Strategy (CRDS) navigates the latest global trends in science, technology, and innovation to aid the Japanese government in formulating its national strategies. *Molecular Technology* is the outcome of a research project born of a CRDS Strategic Proposal realized under the excellent editorial supervision of Hisashi Yamamoto and Takashi Kato. To them and to the scientists who have made major advances in molecular technology through their uninhibited research, I extend my heartfelt congratulations and respect.

The significance of molecular science in all areas of scientific endeavor is certain to increase. Accurate understanding of molecular assemblies and molecular complexes is essential for comprehending the elaborate workings of natural phenomena and of the genesis and mechanisms of materials and life functions. Now, more than ever, science must be seen as a single entity, a comprehensive whole. Mathematical science and the most advanced technologies of observation and information help us to explore the essence of materials and substances in a way that brings together all fields of science. It is the nature of molecular science to continually advance and expand. Using the metaphor of light, we can say that molecules behave in the manner of both "waves and particles."

The traditional separation of science into physics, chemistry, and biology no longer applies. Neither does it make any sense to maintain those seemingly self-contained subdivisions of organic chemistry, inorganic chemistry, physical chemistry, or polymer chemistry. So long as specialized groups and rigid educational systems cling to outdated perceptions, the more important it is to encourage an "antidisciplinary" type of science in which diverse fields converge rather than conventional interdisciplinary or transdisciplinary attempts to link diverse fields.

Molecular Technology, while firmly grounded in fundamental scientific knowledge, aims for practical applications within contemporary society. Johann Wolfgang von Goethe once said, "Knowing is not enough; we must apply. Willing is not enough; we must do." Technology with no practical application is meaningless to society. Researchers should not hesitate to set their own themes and topics of exploration in academia where self-determination holds strong and creativity wins the highest respect. Researchers must show ingenuity in the pursuit of their chosen mission even as they fulfill their duty to pursue science-based technology for society. Never forget that it is by no means advisable to function purely as a support for activities that industry should actually undertake on its own.

The creative outcomes of the Molecular Technology Project launched in 2013 in conjunction with new collaborations are certain to lead to a wide range of innovations and to make significant contribution to achieving the Sustainable Development Goals (SDGs) of the United Nations' 2030 Agenda.

Science is one; and the world is one. Those who will follow us have a responsibility to the world after 2030, and it is my hope that new generations will pioneer revolutionary molecular technology that will bring science and humanity ever closer together. Brain circulation and international collaboration are essential to achieve these goals. V. S. Naipaul, winner of the 2001 Nobel Prize in Literature, once noted that knowing what you wanted to write was three-quarters of the task of writing. Humanity's future is to be found in the unbounded imagination of the young and in its ability to support the challenges they undertake.

December 2017

Ryoji Noyori
Tokyo, Japan

Preface

Chemical science enables us to qualitatively change exiting science and technology by purposefully designing and synthesizing molecules and creating the desired physical, chemical, and biological functions of materials and drugs at molecular level. In 2012, we started the big funding project in Japan, "Molecular Technology" (Establishment of Molecular Technology toward the Creation of New Functions (CREST) and Molecular Technology and New Functions (PRESTO)), and numerous research groups in Japan join the project of diverse research areas. All of these are typical transdisciplinary research projects between chemistry and various research areas of science and technology. In other words, Molecular Technology is the brand new scientific discipline. In principle, most of the proposed projects try to create the big bridge between chemistry and other basic science and technology. We thus propose a nice model for this bridge that is able to make valuable contribution for human welfares.

Between Japan Science and Technology Agency (JST) and French National Research Agency (ANR), we initiated a number of international collaboration projects of Molecular Technology in 2014. Since then 12 new collaboration projects started. The project provides quite unique collaboration opportunities between Japan and France, and quite active research groups involved in very close discussions of molecular technology between two countries. We are sure this project gave us close contacts between research groups of Japan and France for numerous discoveries. Overall, this international collaboration will be new entry for even more important discoveries in the future.

In 2016, we started the discussion for making a new and comprehensive book of molecular technology for the benefit of all researchers in the world to provide typical and leading examples of molecular technology. Overall, researchers of 15 CREST, 50 PRESTO, and 12 INTERNATIONAL groups have contributed to this book. Because of the wide areas of molecular technology, this book covers extremely diverse areas of science and technology from material to pharmaceuticals.

Hisashi Yamamoto
Chubu University, Supervisor of CREST

Takashi Kato
The University of Tokyo, Supervisor of PRESTO

1

Charge Transport Simulations for Organic Semiconductors

Hiroyuki Ishii

University of Tsukuba, Division of Applied Physics, 1-1-1 Tennodai, Tsukuba, Ibaraki, 305-8573, Japan

1.1 Introduction

1.1.1 Historical Approach to Organic Semiconductors

Organic semiconductors have the potential to be used in future electronic devices requiring structural flexibility and large-area coverage that can be fabricated by low-cost printing processes. Ordinary organic materials such as plastics (polyethylene) have primarily been regarded as typical electrical insulators. However, graphite exhibits the high electrical conductivity [1], which has been attributed to their molecular structures, which are made of network planes of the conjugated double bonds of carbon atoms with the π-electrons. There exist some organic molecules that have similar molecular structures, for example, aromatic compounds. Around 1950, Eley [2], Akamatu and Inokuchi [3], and Vartanyan [4] have reported that the phthalocyanines, violanthrones, and cyanine dyes have semiconductive characters, respectively. These characters are attributed to the intermolecular overlapping of the electron clouds of π-electrons in the condensed aromatic rings. These materials were named as *organic semiconductors* [5]. However, in general, these organic semiconductors were still recognized as the insulating materials because resistivity of these organic semiconductors is much higher than that of inorganic semiconductors such as silicon and gallium arsenide. The resistivity ρ is given as

$$\frac{1}{\rho} = nq\mu, \tag{1.1}$$

where n, q, and μ represent the carrier concentration, elementary charge of a carrier, and the electron (hole) mobility, respectively. The high resistivity of the organic materials originates from the low carrier concentration and the low mobility.

The carriers can be chemically doped by using the electron–donor–acceptor complexes. In 1954, Akamatu et al. found that the electron–donor–acceptor complex between perylene and bromine is relatively stable and has very good electrical conductance [6]. In 1973, Ferraris et al. have reported that

Molecular Technology: Energy Innovation, Volume 1, First Edition.
Edited by Hisashi Yamamoto and Takashi Kato.
© 2018 Wiley-VCH Verlag GmbH & Co. KGaA. Published 2018 by Wiley-VCH Verlag GmbH & Co. KGaA.

the complex between the electron donor tetrathiafulvalene (TTF) and the electron acceptor tetracyano-*p*-quinodimethane (TCNQ) has the very high conductivity comparable with the conductivities of metals such as copper [7]. Shirakawa et al. also showed that the organic polymer, polyacetylene, has a remarkably high conductivity at room temperature by chemical doping with iodine in 1977 [8]. These complexes are called organic conductors. The high electrical conductivity accelerated interest in organic conductors, not only because of their huge electrical conductivity but also by the possibility of superconductivity [9].

The multicomponent systems as mentioned above have some disadvantageous properties such as air and thermal instability in general. Therefore, semiconducting single-component organic compounds are likely to be much more suitable for use as molecular devices. From a viewpoint of the electronic device applications, mobility is very important to evaluate the device performance because it characterizes how quickly an electron can move in a semiconductor when an external electric field is applied. In 1960, Kepler [10] and LeBlanc [11] measured the mobility of an organic semiconductor by the time-of-flight (TOF) technique, where the flight time of carriers in a given electric field is determined by observing an arrival time kink in the current that is caused by a pulse-generated unipolar "charge carrier sheet" moving across a plane-parallel slice of a sample. They reported that the anthracenes have the mobility of $0.1–2.0$ cm^2 V^{-1} s^{-1} at room temperature and their mobilities increase as the temperature decreases. Friedman theoretically investigated the electrical transport properties of organic crystals using the Boltzmann equation treatment of narrow-band limit in the case of small polaron band motion [12]. Sumi also discussed the change from the band-type mobility of large polarons to the hopping type of small polarons, using the Kubo formula with the adiabatic treatment of lattice vibrations in the single-site approximation [13]. However, the mobility obtained by TOF technique is different from the mobility of actual devices such as field-effect transistors (FETs) because the charge carriers are induced at the interface between the organic semiconductor and the dielectric film by an applied gate voltage. Kudo et al. reported the field-effect phenomena of merocyanine dye films and their field-effect mobilities of $10^{-7}–10^{-5}$ cm^2 V^{-1} s^{-1} estimated from the measurements in 1984 [14]. Then, Koezuka et al. fabricated the actual FET utilizing polythiophene as a semiconducting material and reported the mobility of 10^{-5} cm^2 V^{-1} s^{-1} [15].

A major industrial breakthrough occurred in the application to electroluminescent (EL) devices. Tang and VanSlyke reported the first organic EL device based on a π-conjugated molecular material in 1987 [16]. After that, typical industrial applications spread to light-emitting diodes (LEDs) [17, 18] and solar cells [19–21]. Recently, organic semiconductors are expected as the future electronic device semiconducting materials requiring structural flexibility and large-area coverage that can be fabricated by low-cost printing processes [22, 23]. However, we have a massive task for the realization of the "printed electronics," for example, increasing the mobility, improvement of the solubility, and thermal durability, suppressing the variations of device characteristics, decreasing the threshold voltage, and so on.

1.1 Introduction

Although π-conjugated polymers with aromatic backbones have been widely investigated as soluble organic semiconductors, further improvement of mobility of polymer semiconductors has disadvantages owing to the statistical distribution of molecular size and structural defects caused by mislinkage of monomers, which act as carrier traps in the semiconducting channel. Therefore, small molecular materials, such as pentacene (see Figure 1.1a), have advantages in terms of their well-defined crystal structure and ease of purification. At first, the organic transistors were fabricated utilizing the organic polycrystals. For example, the field-effect mobility of polycrystal thin-film transistors (TFTs) increases in proportion to the grain size [63, 64]. The mobility in the polycrystals is mainly limited by the grain boundaries, and the typical highest value is generally below 1.0 cm^2 V^{-1} s^{-1} at room temperature. The temperature dependence with a thermally activated behavior indicates that the incoherent hopping process of spatially localized carriers between trap sites is dominated in the polycrystals [26]. In such a low-mobility regime, the charge transport mechanism has been investigated theoretically using the Marcus theory [65, 66] based on the small polaron model [67].

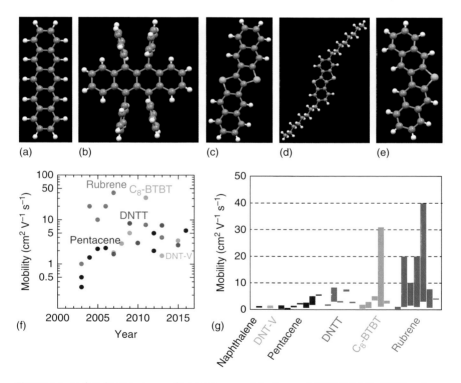

Figure 1.1 Molecular structures of (a) pentacene, (b) rubrene, (c) DNTT, (d) C$_8$-BTBT, and (e) DNT-V. (f) Annual change of the highest hole mobilities of different organic single-crystal field-effect transistors in the literature, and (g) the distribution of the reported mobilities for naphthalene [24], DNT-V [25], pentacene [26–35], DNTT [36–41], C$_8$-BTBT [42–49], and rubrene [29, 31], [50–62].

1.1.2 Recent Progress and Requirements to Computational "Molecular Technology"

Recent rapid progress in technology enables us to fabricate the very pure rubrene single-crystal FETs (see Figure 1.1b) with the high carrier mobility up to 40 cm^2 V^{-1} s^{-1} at room temperature [60], which exceeds the mobility of amorphous silicon [68]. The high mobility attributes the exclusion of trap sites such as grain boundary in organic semiconductors. The mobility monotonically decreases with increasing temperature, $\mu \propto T^{-n}$ [54]. The power-law temperature dependence is a typical characteristic of coherent band transport by spatially extended carriers, which is scattered by the molecular vibrations (phonons). The rubrene single crystals obtained by the physical vapor deposition method show the excellent high mobilities at room temperature, but their poor solubility is a serious problem for the printed electronics. In 2006, Takimiya et al. reported solution-processable organic semiconductors based on [1]benzothieno[3,2-b][1]benzothiophene (BTBT) core [69] and dinaphtho[2,3-b:2′,3′-f]thieno[3,2-b]thiophene (DNTT) core [36] with high mobility and stability, as shown in Figure 1.1c,d. Moreover, Okamoto et al. reported a new candidate semiconducting material based on V-shaped dinaphtho[2,3-b:2′,3′-d]thiophene (DNT-V) core (Figure 1.1e), with high mobility, solubility, and thermal durability [25]. Figure 1.1f,g shows annual change of the highest hole mobilities of different organic single-crystal FETs in the literature, and the distribution of the reported mobilities for naphthalene [24], DNT-V [25], pentacene [26–35], DNTT [36–41], C$_8$-BTBT [42–49], and rubrene [29, 31, 50–62].

As shown above, new organic semiconductors with higher mobilities have been required. It is a very important and an urgent issue for us to establish the method and system for finding new organic semiconductors with high mobility from among various kinds of the candidate materials. Computer simulation to predict the mobility of candidate organic semiconductors becomes a powerful tool to accelerate the material development.

1.2 Theoretical Description of Charge Transport in Organic Semiconductors

As shown in Figure 1.2a, different from covalent crystals such as silicon, organic semiconductors are formed with van der Waals interactions between molecules [70]. The very weak interactions give the organic semiconductors the property of mechanical flexibility and the solubility. Charge carriers in organic semiconductors strongly couple with the molecular vibrations, namely, phonons. The total Hamiltonian consists of that for the electron $\hat{H}_e$, the phonon $\hat{H}_{ph}$, and the interaction $\hat{H}_{e-ph}$,

$$\hat{H} = \hat{H}_e + \hat{H}_{ph} + \hat{H}_{e-ph}. \tag{1.2}$$

The general expression can be given on the basis of molecular orbitals and the eigenfunctions of phonon as follows:

$$\hat{H}_e = \sum_{m,n} h^0_{mn} \hat{a}^\dagger_m \hat{a}_n, \tag{1.3}$$

1.2 Theoretical Description of Charge Transport in Organic Semiconductors

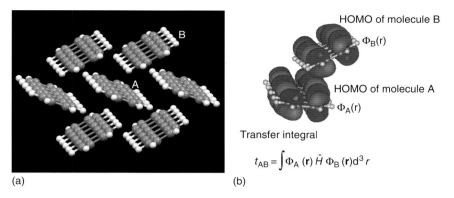

Figure 1.2 (a) Structure of a single crystal of pentacene. (b) HOMOs of the molecules labeled A and B in (a). The transfer integral between molecules t_{AB} is defined as the off-diagonal elements of the Hamiltonian matrix $\hat{H}$ on the molecular orbital basis set.

$$\hat{H}_{\text{ph}} = \sum_{\lambda,\mathbf{q}} \hbar\omega_{\lambda\mathbf{q}} \left(\hat{b}^{\dagger}_{\lambda\mathbf{q}} \hat{b}_{\lambda\mathbf{q}} + \frac{1}{2} \right), \tag{1.4}$$

$$\hat{H}_{\text{e-ph}} = \sum_{m,n} \sum_{\lambda,\mathbf{q}} \hbar\omega_{\lambda\mathbf{q}} g^{\lambda\mathbf{q}}_{mn} \hat{a}^{\dagger}_m \hat{a}_n (\hat{b}^{\dagger}_{\lambda\mathbf{q}} + \hat{b}_{\lambda,-\mathbf{q}}), \tag{1.5}$$

where $\hat{a}^{\dagger}_n$ and $\hat{b}^{\dagger}_{\lambda\mathbf{q}}$ represent the creation operator of electron at the nth orbital and the correlation operator of phonon with mode λ, wave-vector $\mathbf{q}$, and the vibration frequency $\omega_{\lambda\mathbf{q}}$. Here, $h^0_{mn(m\neq n)}$ and h^0_{nn} are the transfer integral t^0_{mn} between nth and mth molecular orbitals and the orbital energy ε^0_n at the equilibrium position, respectively. As an example, the highest occupied molecular orbitals (HOMOs) of the pentacene molecules are shown in Figure 1.2b [70]. The dimensionless electron–phonon coupling constant is defined by [71, 72]

$$g^{\lambda\mathbf{q}}_{nm} \equiv \sum_{k,s} \sqrt{\frac{1}{2\hbar MN\omega^3_{\lambda\mathbf{q}}}} e^{i\mathbf{q}\mathbf{R}_k} \left(\frac{\partial h_{nm}}{\partial \mathbf{R}_{ks}} \right) \mathbf{e}^{\lambda\mathbf{q}}_s, \tag{1.6}$$

where $\mathbf{e}^{\lambda\mathbf{q}}_s$ is the phonon eigenvector representing the direction of displacement of sth atom. M and N represent the mass of a single molecule and the number of unit cells. The position of sth atom in kth unit cell is given by $\mathbf{R}_{ks} = \mathbf{r}_s + \mathbf{R}_k$, where the relative position of sth atom in the unit cell is represented by $\mathbf{r}_s$, and $\mathbf{R}_k$ is the position vector of kth unit cell. The change in transfer integral due to molecular vibration $(\partial h/\partial \mathbf{R})$ is an intrinsic meaning of the electron–phonon interaction.

The relation between the mobility μ and the diffusion coefficient D is well used in theoretical studies of transport of carrier with charge q in organic semiconductors and known as the Einstein relation

$$\mu = \frac{qD}{k_B T}, \tag{1.7}$$

where $k_B T$ is the thermal energy defined as the product of the Boltzmann constant and the temperature. The problem remained is how we obtain the diffusion coefficient from the general Hamiltonian of Eqs. (1.3)–(1.5).

1.2.1 Incoherent Hopping Transport Model

The semiclassical Marcus theory [65, 66] based on the small polaron model [67] describes the hopping motion of charge carriers that are self-trapped in a single molecule by their induced intramolecular deformations. The schematic picture of hopping motion is shown in Figure 1.3a [70]. Before injection of a charge carrier, all molecules in the organic semiconductor are in the neutral state. As shown in Figure 1.3b, when a charge carrier is injected into a single molecule, the state is changed from the most stable neutral state (i) to the charged state (ii). Then, the state (ii) relaxes into the most stable charged state (iii) by their induced intramolecular distortion. If we assume that the transfer integrals are much smaller than the magnitude of electron–phonon couplings, the hopping rate to neighboring jth molecule can be calculated using the perturbation theory and takes the following thermally activated form:

$$\frac{1}{\tau_j^{\text{hop}}} = \frac{|t_j^0|^2}{\hbar} \sqrt{\frac{\pi}{\lambda k_B T}} e^{-\frac{\lambda}{4 k_B T}}, \tag{1.8}$$

where t_j^0 is the transfer integral at the equilibrium position. The quantity $\lambda \equiv \lambda^{(1)} + \lambda^{(2)}$ is the reorganization energy. Here, two components $\lambda^{(1)}$ and $\lambda^{(2)}$ correspond to going into a charged state and returning to a neutral state as shown in Figure 1.3b.

Moreover, we assume that the quantum coherence is lost after each hopping event, the diffusion coefficient is given by

$$D^{\text{hop}} = \sum_j a_j^2 \frac{P_j}{\tau_j^{\text{hop}}}, \tag{1.9}$$

where a_j represents the intermolecular distance and the hopping probability is defined by $P_j \equiv (\tau_j^{\text{hop}})^{-1} / \sum_k (\tau_k^{\text{hop}})^{-1}$. Finally, we can obtain the thermally activated form of hopping mobility using the Einstein relation of Eq. (1.7). In the case of simple one-dimensional molecular crystals with the intermolecular transfer integral t and the intermolecular distance a, the hopping mobility is written by

$$\mu^{\text{hop}} = \frac{1}{2} \frac{a^2 |t|^2}{\hbar} \frac{q}{k_B T} \sqrt{\frac{\pi}{\lambda k_B T}} e^{-\frac{\lambda}{4 k_B T}}. \tag{1.10}$$

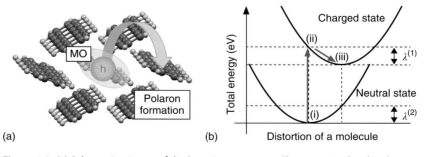

Figure 1.3 (a) Schematic picture of the hopping transport. Charge carrier localized at a molecule moves to the neighboring molecules by the thermally activated hopping process. (b) Potential energy surface for the neutral state and the charged state of the single molecule.

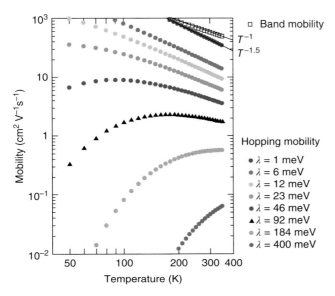

Figure 1.4 Temperature-dependent behavior of the hopping and band mobilities of pentacene single crystal with the transfer integrals between adjacent molecules are 43.6, 71.5, and −111.3 meV [73]. The reorganization energy of a pentacene molecule is $\lambda = 92$ meV and the calculated hopping mobility is shown by the black triangles. For the reference, the hopping mobilities are calculated for the several reorganization energies from 1 to 400 meV. The calculated band mobility is represented by the white squares. Source: Ishii et al. 2017 [73]. Reproduced with permission of American Physical Society.

Figure 1.4 shows the temperature-dependent behavior of the hopping mobility of the two-dimensional pentacene single crystal by the black triangles. The calculated intermolecular transfer integrals 43.6, 71.5, and −111.3 meV [73] are comparable to the reorganization energy $\lambda = 92$ meV. Interestingly, the calculated hopping mobility of pentacene single crystal exhibits temperature-independent mobility around room temperature with a mobility of ∼ 1 cm² V⁻¹ s⁻¹. The calculated results seem to well explain the experimentally observed temperature-independent mobility [27]. However, the Marcus theory is generally applicable for $t \ll \lambda$. It indicates that we should not employ the Marcus theory for the analysis of charge transport of high-mobility organic semiconductors.

1.2.2 Coherent Band Transport Model

The band transport model starts from the solution of the electronic problem in an unperturbed, perfect lattice (perfect periodicity). In this limit, the electrons form the Bloch waves identified by a well-defined momentum **k** and the energy band dispersion $E(\mathbf{k})$. As an example, the HOMO band dispersion calculated by the density functional theory (DFT) using the plane-wave basis set is shown in Figure 1.5b [70]. The existence of HOMO band dispersion of pentacene crystal [74–77] is experimentally demonstrated using angle-resolved photoelectron spectroscopy (ARPES). Quantum mechanics tells us that a charge carrier having the effective mass $\mathbf{m}(\mathbf{k})$ propagates at the group velocity $\mathbf{v}(\mathbf{k})$ without

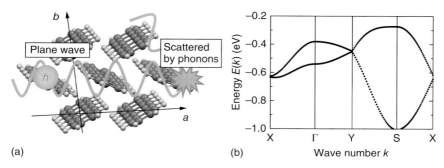

Figure 1.5 (a) Schematic picture of the band transport. Extended charge carriers are described by the Bloch states with the wave-vector **k** and scattered by the molecular vibrations (phonons). (b) HOMO bands obtained from DFT using the plane-wave basis set with symmetry points of Γ(0, 0, 0), X(1/2, 0, 0), Y(0, 1/2, 0), and S(1/2, 1/2, 0). The Fermi energy is located at $E = 0$ eV.

any scattering in the perfect lattice. The effective mass and the group velocity are obtained from the band dispersion as follows: $\mathbf{v}(\mathbf{k}) \equiv \partial E(\mathbf{k})/\hbar \partial \mathbf{k}$ and $\mathbf{m}(\mathbf{k}) \equiv (\partial^2 E(\mathbf{k})/\hbar^2 \partial \mathbf{k}^2)^{-1}$. However, as shown in Figure 1.5a, even if the perfect single crystal can be made, the molecular vibrations disturb the periodicity and become a dominant origin of electric resistance at room temperature. Scattering of the Bloch states by the molecular vibrations is included as a perturbation in the band transport model. The momentum relaxation time (scattering time) of τ^{band} is given by Fermi's golden rule,

$$\frac{1}{\tau^{\text{band}}(\mathbf{k})} = \frac{2\pi}{\hbar} \sum_{\mathbf{k}'} \sum_{\lambda \mathbf{q}} |\langle \mathbf{k}'|\hat{H}|\mathbf{k}\rangle|^2 \times \delta(E(\mathbf{k}') - E(\mathbf{k}) \pm \hbar\omega_{\lambda\mathbf{q}})(1 - \cos\theta_{\mathbf{k}'\mathbf{k}}), \quad (1.11)$$

where $|\mathbf{k}\rangle$ represents the eigenfunction of $\hat{H}_e$ with the eigenenergy $E(\mathbf{k})$ in the momentum representation, and $\theta_{\mathbf{k}'\mathbf{k}}$ is the angle between **k** and **k**′. In the acoustic deformation potential model for two-dimensional transport [78], the relaxation time can be calculated as

$$\frac{1}{\tau^{\text{band}}(T)} = \frac{\varepsilon_{\text{ac}}^2 m_d k_B T}{\hbar^3 B L_{\text{eff}}}. \quad (1.12)$$

Here, ε_{ac} is the acoustic deformation potential defined by $\varepsilon_{\text{ac}} \equiv \Omega dE_{\text{hbm}}/dV$, where E_{hbm} represents the value to the HOMO band maximum. The acoustic deformation potential should be the quantity as a function of the electron–phonon coupling constant g of Eq. (1.6). B is the elastic modulus, L_{eff} is the effective channel width of carrier confinement layer, and m_d is the density of states mass, which is equal to $\sqrt{m_a m_b}$. m_a and m_b represent the effective mass at the HOMO band maximum along a and b axes, respectively.

Different from the hopping model, the diffusion coefficient along x axis is defined as $D_x^{\text{band}} \equiv \int_0^{+\infty} v_x(s) v_x(0) ds$, using the velocity correlation functions, where $v_x(s)$ is the x component of the group velocity **v** at time s. When the velocity correlation is disappeared by the molecular vibrations, the diffusion coefficient is calculated as

$$D_x^{\text{band}} = \int_0^{+\infty} v_x^2 \exp\left(\frac{-s}{\tau^{\text{band}}}\right) ds, \qquad (1.13)$$
$$= v_x^2 \tau^{\text{band}}.$$

The band mobility is obtained from the Einstein relation of Eq. (1.7) as follows:

$$\begin{aligned}\mu_x^{\text{band}} &= \frac{q}{k_B T} v_x^2 \tau^{\text{band}}, \\ &= \frac{q \tau^{\text{band}}(T)}{m_x}, \\ &= \frac{q}{m_x} \frac{\hbar^3 B L_{\text{eff}}}{\varepsilon_{ac}^2 m_d k_B T}.\end{aligned} \qquad (1.14)$$

Here, the relation between the kinetic energy and the temperature, $\frac{3}{2}mv^2 = \frac{3}{2}k_B T$, has been used. The calculated band mobility decreases with increasing temperature, according to $\mu \propto T^{-1}$ shown by the white squares in Figure 1.4. Such power-law temperature dependence is a typical character of coherent band transport and has been observed in some experiments for organic single crystals with high mobility [24, 45, 54, 55]. Moreover, recent experiments of Hall effects on the organic FETs provide us with an evidence of possible coherent charge transport in the organic semiconductors [55, 56, 59, 79]. On the other hand, a difficult problem is still remained in the coherent band picture. That is, the estimated mean free path is comparable to or shorter than the distance between adjacent molecules [27], which implies a breakdown of the coherent band transport.

1.2.3 Coherent Polaron Transport Model

Some experiments reported that the width of HOMO bands observed by the ARPES is narrowing with increasing temperature [76, 77]. This phenomenon is known as the band narrowing and can be rationalized by means of the concept of polaron, as described in the following. When there exist the electron–phonon interactions, the bare electrons get dressed by phonons (molecular vibrations) and form quasiparticles called polarons. The charge carriers have to carry the phonon cloud as well, which always accompanies the carrier. With increasing temperature, the effective mass of polaron becomes larger since much more phonons are available that coupled to the charge carrier.

Hannewald et al. have derived analytically the total Hamiltonian of coherent polaron $\tilde{H}$ from $\hat{H}$ of Eq. (1.2), using the method of the Lang-Firsov canonical transformation [71, 72, 80],

$$\tilde{H} = e^{\hat{S}} \hat{H} e^{\hat{S}^\dagger}, \qquad (1.15)$$
$$\simeq \sum_{m,n} \tilde{t}_{mn} \hat{a}_m^\dagger \hat{a}_n + \hat{H}_{\text{ph}}, \qquad (1.16)$$

where $\hat{S} \equiv \sum_{m,n} \hat{C}_{mn} \hat{a}_m^\dagger \hat{a}_n$ and $\hat{C}_{mn} \equiv \sum_{\lambda,\mathbf{q}} g_{mn}^{\lambda\mathbf{q}}(\hat{b}_{\lambda\mathbf{q}}^\dagger - \hat{b}_{\lambda-\mathbf{q}})$. To get Eq. (1.16) from Eq. (1.15), they have assumed that the electron–intramolecular vibration coupling is much stronger than the electron–intermolecular vibration coupling ($|g_{mm}| \gg |g_{mn}|$) and the coherent part is dominant than the incoherent hopping

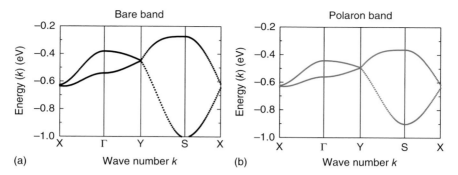

Figure 1.6 (a) HOMO band dispersion of bare electron of the pentacene single crystal. (b) Schematic picture of band narrowing by polaron formation.

part described by the Marcus theory. The transfer integrals of coherent polaron $\tilde{t}$ are expressed using the bare transfer integral t^0, the electron–phonon couplings g and the number of phonons n as follows:

$$\tilde{t}_{mn} \simeq t^0_{mn} \exp\left\{-\sum_{\lambda,\mathbf{q}} \left(n_{\lambda\mathbf{q}} + \frac{1}{2}\right) |g^{\lambda\mathbf{q}}_{mm} - g^{\lambda\mathbf{q}}_{nn}|^2\right\}, \quad (1.17)$$

where $n_{\lambda\mathbf{q}} = (\exp(\hbar\omega_{\lambda\mathbf{q}}/k_B T) - 1)^{-1}$. The band narrowing is included in this expression because $\tilde{t}$ is always smaller than t^0. The schematic pictures of HOMO band dispersion for bare electrons and for polarons are shown in Figure 1.6a,b, respectively.

In analogy with the band mobility of Eq. (1.14), the coherent polaron mobility is obtained using the effective mass of polaron band,

$$\mu_x^{\text{polaron}} = \frac{q\tau}{\tilde{m}_x(T)}. \quad (1.18)$$

Here, τ is the scattering time by static disorders such as impurities and defects. Although the temperature dependence of polaron mobility exhibits the power-law behavior around room temperature, different from the band mobility, the effective mass of polaron, not the scattering time τ, decides the temperature dependence of mobility.

Band narrowing is a prominent feature of the polaron model. However, as already discussed in Section 1.2.1, the polaron concept is strictly valid only when the transfer integrals are much smaller than the reorganization energy. This condition is actually hardly fulfilled in a number of organic semiconductors. Recently, Brédas and coworkers theoretically demonstrated that the thermal expansion of the crystal structures, rather than the polaron formations, is the main factor responsible for the thermal bandwidth narrowing in organic semiconductors [81].

1.2.4 Trap Potentials

Most theoretical studies try to understand the intrinsic transport nature, namely, the thermally activated hopping behavior for low mobility and the power-law temperature-dependent band-like behavior for high mobility, in

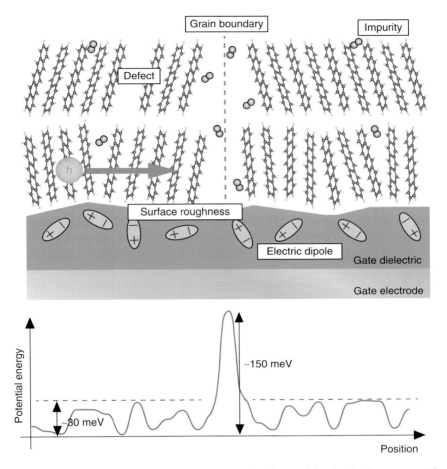

Figure 1.7 Schematic picture of various extrinsic disorders and the depth of trap potential.

organic semiconductors by taking the electron–phonon couplings into account. However, experimental data obtained on high-quality single crystals indicate that the appearance of an activated transport is in many instances more likely due to the presence of extrinsic disorder effects such as structural disorder, chemical defects [82], interaction with the substrate [57] and so on (see Figure 1.7) [70]. Such extrinsic disorders inevitably exist in actual devices and trap the carriers, resulting in decreasing the mobility. The existence of carrier-trap potentials has been confirmed by atomic force microscope potentiometry [63] and electron spin resonance spectra [83]. The depth of trap potentials W is in the range of $10–10^2$ meV, which is comparable to the magnitude of transfer integrals [84].

1.2.5 Wave-packet Dynamics Approach Based on Density Functional Theory

As discussed in Sections 1.2.1–1.2.4, the polaron concept is valid in the case of $t \ll \lambda$, whereas the band transport, where charge carriers are scattered by phonons, is applicable in the case of $t \gg \lambda$. However, for typical organic semiconductors the transfer integrals t are in the range $10-10^2$ meV, which has similar energetic orders of reorganization energy λ. Furthermore, the transfer integrals t are comparable to the depth of carrier-trap potentials W. It is important for us to understand the carrier transport mechanism in competition among the electron–phonon scatterings, the polaron formations, and the trap potentials. Especially, a unified theoretical description from the thermally activated hopping transport behavior to the band-like transport behavior represents a very challenging problem.

Therefore, I have developed the methodology named the time-dependent wave-packet diffusion (TD-WPD) method [73, 85–88], which enables us to carry out the transport calculations including the strong electron–phonon couplings and the trap potentials on equal footing without perturbative treatment. The mobility of a charge q along the x direction for an organic semiconductor with volume Ω is calculated using the following Kubo formula:

$$\mu_x = \lim_{t \to +\infty} \frac{q}{n} \int_{-\infty}^{+\infty} dE \left(-\frac{df}{dE}\right) \left\langle \frac{\delta(E - \hat{H}_e)}{\Omega} \frac{\{\hat{x}(t) - \hat{x}(0)\}^2}{t} \right\rangle, \quad (1.19)$$

where the concentration of charge carriers is obtained by $n = \int dE f(E) \langle \delta(E - \hat{H}_e) \rangle / \Omega$. The Heisenberg picture of the position operator is defined by $\hat{x}(t) = \hat{U}^\dagger(t) \hat{x} \hat{U}(t)$, where $\hat{U}(t = N_t \Delta t) \equiv \Pi_{n=0}^{N_t-1} \exp\{i\hat{H}_e(n\Delta t)\Delta t/\hbar\}$ is the time evolution operator. The dynamical change of electronic states induced by molecular vibrations and distortions is included in the time-dependent expression of Hamiltonian $\hat{H}_e(t)$. The quantity $\langle \cdots \rangle$ is evaluated as $\sum_{m=1}^{N_{wp}} \langle \Psi_m(0) | \cdots | \Psi_m(0) \rangle / N_{wp}$, where N_{wp} is the number of random-phase wave-packets $|\Psi_m(0)\rangle$. Note that when the Fermi distribution function is approximated as $f(E) \simeq e^{-\beta(E-E_F)}$, the Einstein relation $\mu_x = qD_x/k_B T$ of Eq. (1.7) can be reproduced from Eq. (1.19), where the diffusion coefficient D_x is defined as $D_x \equiv \lim_{t \to +\infty} (1/t)[\int dE f(E) \langle \delta(E - \hat{H}_e) \{\hat{x}(t) - \hat{x}(0)\}^2 \rangle]/[\int dE f(E) \langle \delta(E - \hat{H}_e) \rangle]$.

From Eqs. (1.3) and (1.5), the transfer integrals including the electron–phonon couplings are written by

$$t_{mn} = t_{mn}^0 + \sum_{\lambda, \mathbf{q}} \hbar \omega_{\lambda \mathbf{q}} g_{mn}^{\lambda \mathbf{q}} (\hat{b}_{\lambda \mathbf{q}}^\dagger + \hat{b}_{\lambda -\mathbf{q}}). \quad (1.20)$$

Then, to reduce the calculation cost, I adopt the semiclassical approximation to evaluate the molecular vibrations. The phonon operators are replaced by the displacements of molecules, where the displacement of the sth molecule in the kth unit cell is defined as $\Delta \mathbf{R}_{ks} = \sum_{\lambda, \mathbf{q}} X_{\lambda \mathbf{q}} e^{i\mathbf{q}\mathbf{R}_k} \mathbf{e}_s^{\lambda \mathbf{q}}$ with $X_{\lambda \mathbf{q}} = \sqrt{\hbar/2MN\omega_{\lambda \mathbf{q}}} (\hat{b}_{\lambda \mathbf{q}}^\dagger + \hat{b}_{\lambda -\mathbf{q}})$. Furthermore, I assume that the transfer integrals t_{mn} depend solely on the relative coordinate $\mathbf{R}_{mn} \equiv \mathbf{R}_m - \mathbf{R}_n$, then the semiclassical expression of transfer integrals of Eq. (1.20) is obtained as

$$t_{mn}(t) \simeq t_{mn}^0 + \frac{\partial h_{mn}}{\partial \mathbf{R}_{mn}} \Delta \mathbf{R}_{mn}(t), \quad (1.21)$$

where $\Delta \mathbf{R}_{mn}(t)$ represents the change in intermolecular distance at time t due to molecular vibrations. The equation of motion for the nth molecule with mass M is

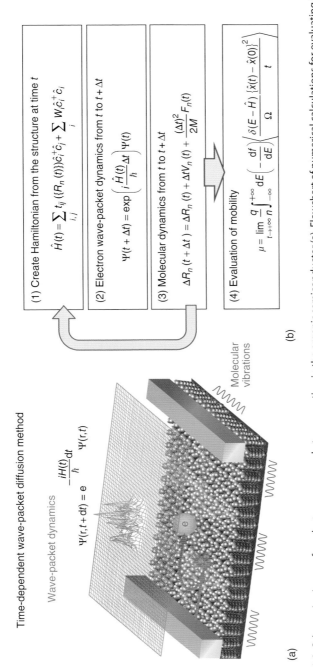

Figure 1.8 Schematic picture of an electron wave-packet propagation in the organic semiconductor (a). Flowchart of numerical calculations for evaluating the mobility using the wave-packet dynamics combined with the molecular dynamics (b).

derived from the canonical equation $M\Delta\ddot{\mathbf{R}}_n = -\partial E_{\text{tot}}(\{\Delta\mathbf{R}_{ij}\})/\partial\Delta\mathbf{R}_n$, where E_{tot} is the total energy defined by the summation of electron energy and the molecular vibration energy including these interactions [87]. By extracting $\Delta\mathbf{R}_{mn}(t)$ at each time step of the molecular dynamics calculations, I can introduce the effects of strong electron–phonon couplings as the ever-changing transfer integral and obtain the mobility from Eq. (1.19). Flowchart of numerical calculations for evaluating the mobility using the wave-packet dynamics combined with the molecular dynamics and the schematic picture are shown in Figure 1.8 [70].

To reduce the computational cost, I employ the Chebyshev polynomial expansion of the time evolution operator [85, 89],

$$e^{i\frac{\hat{H}_e(t)}{\hbar}\Delta t} = \sum_{n=0}^{+\infty} e^{-i\frac{a\Delta t}{\hbar}} h_n i^n J_n\left(-\frac{b\Delta t}{\hbar}\right) T_n\left(\frac{\hat{H}_e(t) - a}{b}\right), \tag{1.22}$$

where the HOMO band is included within the energy interval $[a - b, a + b]$, and $h_0 = 1$ and $h_n = 2$ $(n \geq 1)$. The Chebyshev polynomials obey the following recursive relation: $T_{n+1}(x) = 2xT_n(x) - T_{n-1}(x)$ with $T_0(x) = 1$ and $T_1(x) = x$. As a result, the approach enables us to perform the order-N computation. Figure 1.9 shows the computing time and memory usage as a function of the number of molecules N. I confirm that the order-N calculations with respect to both the computing time and the memory usage are realized for the system of up to 10^8 molecules. The maximum system size corresponds to the two-dimensional monolayer organic semiconductor with each side length of a few micro meters. This shows that I can directly compare the transport properties calculated from atomistic treatments with the experimentally observed one.

Using the dimer approach [90, 91], I evaluate the transfer integrals t, the elastic constants K, and the electron–phonon couplings $(\partial t/\partial\Delta\mathbf{R})$ from the DFT

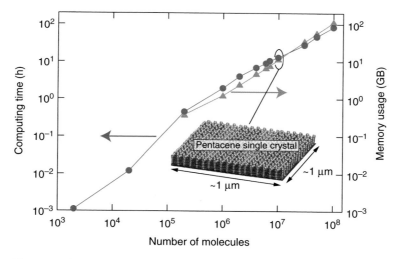

Figure 1.9 Computing time and memory usage per one wave-packet as a function of number of molecules on the TD-WPD method. The number of molecule of two-dimensional monolayer pentacene single crystal with each side length of 1 μm is shown as an example. The number of time step is set to 1000.

calculations [92] including the van der Waals interactions at the DFT-D level [93] with the Becke three-parameter Lee–Yang–Parr (B3LYP) functional in conjunction with the 6-31G(d) basis set. The all material parameters can be obtained from the DFT calculations, and the TD-WPD method enables us to evaluate the mobility of any organic semiconductors without fitting parameters.

1.3 Charge Transport Properties of Organic Semiconductors

1.3.1 Comparison of Polaron Formation Energy with Dynamic Disorder of Transfer Integrals due to Molecular Vibrations

First, the polaron formation energy of pentacene single crystal was investigated. For the simplicity, the one-dimensional crystal was employed [87]. The polaron state is obtained by self-consistent calculations to minimize the total energy E_{tot} with respect to the molecular displacements $\Delta \mathbf{R}$, namely, $\partial E_{\text{tot}}/\partial \Delta \mathbf{R} = 0$ [94, 95]. When evaluating the reorganization energy used in the Marcus theory, one assumes the small polaron, thus the calculation is done for an isolated single molecule in general. As shown in Figure 1.10a, the evaluated binding energy of small polaron is 93 meV, which is enough larger than the thermal energy at room temperature [70]. However, there exist large transfer integrals of a few 10 meV between molecules in the organic semiconductor. It is considered that the polaron state is spatially extended over the crystal. Therefore, the binding energy of polaron state in the crystal having transfer integrals of 75 meV was calculated. The polaron binding energy decreases to only 14 meV, which indicates that the polaron is unstable in the crystal around room temperature.

Then, the dynamic disorder of the transfer integrals, induced by the molecular vibrations, was investigated. Figure 1.10b shows the time-dependent transfer integrals $t_{mn}(t)$ defined by Eq. (1.21) for several intermolecular bonds at 300 K

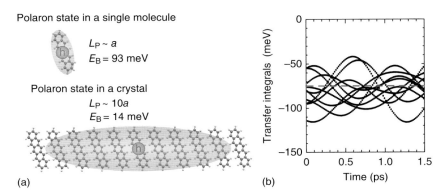

Figure 1.10 (a) Schematic picture of a polaron state in an isolated pentacene molecule and in a pentacene single crystal. The calculated binding energies of polaron state and their size are also shown. (b) Fluctuation of some transfer integrals between molecules induced by the intermolecular vibrations.

[70]. The transfer integral without any molecular vibration t^0_{mn} is 75 meV and shown by the red dashed line for comparison. The amplitude of the thermally fluctuating transfer integrals corresponding to the second term of Eq. (1.21) reaches 80 meV, which is comparable with t^0_{mn}. This calculated result indicates that the electron–phonon scattering cannot be treated by the perturbation theory such as band transport theory discussed in Section 1.2.2. Furthermore, it can be concluded that the polaron state with the binding energy 14 meV is completely destroyed by the strong dynamic disorder of transfer integrals.

1.3.2 Temperature Dependence of Mobility

Before discussing extrinsic effects of trap potentials on the charge transport, the intrinsic transport of organic semiconductors without the trap potential will be investigated. Figure 1.11a shows the logarithmic plot of mobility μ as a function of temperature [70]. As shown by the white circles, the calculated mobility μ decreases monotonically with increasing temperature approximately by the power-law dependence, which shows apparent evidence of the band-like transport. Similar power-law dependence has been reported in other theoretical works [96, 97]. The mean free path is one of the important quantities to understand the transport mechanism, since if the mean free path is shorter than the intermolecular distance, then the concept of band transport is break down. The mean free path defined by $l_{\text{mfp}} \equiv v_x \tau$ can be obtained as $l_{\text{mfp}} \equiv \lim_{t \to +\infty} D_x(t)/v_x$ in the TD-WPD formalism. White circles in Figure 1.11b show that the calculated mean free path is approximately 10 times longer than the intermolecular distance at room temperature [70]. It supports that the band-like transport can be realized when the trap potential is absent. However, the ideal coherent band transport is not realized in organic semiconductors because the HOMO band-edge states are spatially localized owing to the strong electron-phonon scatterings [96–99].

Next, how the mobilities are affected by the extrinsic trap potentials will be investigated, which are caused by chemical impurities, defects, randomly oriented dipoles in gate dielectric, and so on. To take the trap potentials into account, the author introduces the Anderson-type static-disorder potentials, which modulate the on-site orbital energies randomly within the energy width $[-W/2, +W/2]$ as shown in Figure 1.11c [70]. Some experiments show that the depth of trap potentials are estimated as about 50 meV [84]; thus, W is changed from 50 to 200 meV in this study. By the introduction of the trap potentials W, the magnitude of mobility is significantly decreased from 10^2 to 10^{-1} cm^2 V^{-1} s^{-1}. Furthermore, the author obtained a change in the temperature dependence from power-law dependence to thermally activated behavior via temperature-independent behavior. This behavior is experimentally observed in pentacene and rubrene devices [26, 27, 57]. When $W = 200$ meV, the mean free path is shorter than the intermolecular distance. It implies that the concept of band transport is break down, which is consistent with the hopping transport behavior of mobility. These calculated results indicate that competition between the electron–phonon scattering and the trap potential provides important clues to understand the transport nature of organic semiconductor devices.

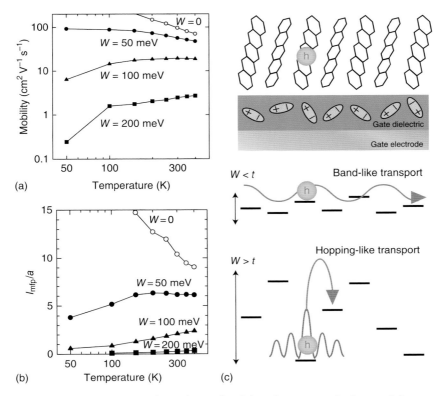

Figure 1.11 (a) Temperature dependence of mobility of pentacene single crystals for several magnitudes of trap potential W. (b) Mean free path normalized by the intermolecular distance a vs temperature characteristics for several W. (c) Schematic picture of electron (hole) transport of organic semiconductor on the gate dielectric. Randomly oriented dipoles in gate dielectric are possible origin of trap potentials. If the transfer integral t^0 is larger than the trap potential W, the electron–phonon scattering is dominated, thus the mobility decreases as increasing temperature. On the other hand, if $W > t^0$, the charge carrier is trapped tightly by the potential W; Therefore, the transport properties are close to typical thermally activated behaviors.

1.3.3 Evaluation of Intrinsic Mobilities for Various Organic Semiconductors

Finally, the author evaluated the intrinsic mobilities of representative organic semiconductors. Numerical evaluation of the intrinsic mobilities for various materials becomes a useful technique to find promising high-mobility materials suitable for organic electronics from among a number of candidate materials. Especially, the author focuses on the magnitude relation of mobilities among the various materials. As a demonstration, the TD-WPD method to the naphthalene, DNT-V, pentaene, DNTT, C_8-BTBT, and rubrene single crystals was applied. Figure 1.12b shows the mobilities obtained by the TD-WPD method. The distributions of experimentally observed mobilities in Figure 1.1g are shown by vertical bars. It can be confirmed that the calculation results well reproduce

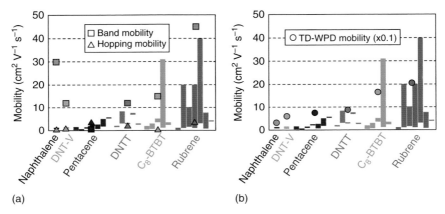

Figure 1.12 (a) Band and hopping mobilities at 300 K are plotted by squares and triangles, respectively, for naphthalene, DNT-V, pentacene, DNTT, C_8-BTBT, and rubrene. (b) Mobilities at 300 K calculated by TD-WPD method for naphthalene, DNT-V, pentacene, DNTT, C_8-BTBT, and rubrene. Note that the mobility divided by 10 for the results of TD-WPD method was plotted. Here, same material parameters, such as the transfer integrals, are employed even in above different theories. For comparison, the distributions of mobilities observed at room temperature are drawn by vertical bars.

the magnitude relation of experimentally observed mobilities for these materials. Note that the mobility divided by 10 for the results of TD-WPD method was plotted, since the wave-packet approach overestimates the magnitude of intrinsic mobility. The possible origin of the overestimation is suggested in some papers [99, 100], but still under consideration. For comparison, the band and hopping mobilities are shown in Figure 1.12a. The band and hopping mobilities cannot reproduce the magnitude relation of mobilities for some materials discussed here. For example, although the C_8-BTBT exhibits high mobility in experiments, the calculated hopping mobility is the lowest among the materials discussed here. The calculated band mobility of naphthalene is very high, whereas the mobility observed in experiment is quite low. In comparison with the conventional band and hopping models, the TD-WPD method is expected to become a useful tool to estimate and predict the magnitude relation of mobilities for candidate materials.

1.4 Summary

Organic semiconductors are expected to become key materials for realizing the printed electronics. New organic semiconductors with higher mobility have been strongly desired. However, in general, it requires time-consuming processes to synthesize these molecules, fabricate devices using the molecules, and evaluate the device performance. It is a very important and an urgent issue for us to establish the numerical simulation method for finding new organic semiconductors with high mobility from among various kinds of the candidate materials. Evaluating the intrinsic charge transport properties of organic semiconductors requires a new theory that is able to describe the electron–phonon coupling using nonperturbative manner.

The author introduced some recent topics in the field of charge transport of organic semiconductors and presented the fundamental transport theories from an atomistic viewpoint. Then, the author's theoretical study using the TD-WPD method was presented, which enables us to evaluate the transport properties taking into account the electron–phonon couplings and the trap potentials on an equal footing without any perturbative treatment. Using this method, it was shown that the calculated temperature dependence of mobility of pentacene single crystal agrees well with experimentally observed characteristics. Furthermore, the calculated mobilities of representative organic semiconductors well reproduce the magnitude relation of experimentally observed mobilities. These calculated results indicate that, in comparison with the conventional band and hopping models, the TD-WPD method is expected to become a useful tool to estimate and predict the magnitude relation of mobilities for candidate materials.

1.4.1 Forthcoming Challenges in Theoretical Studies

As far as I know, the theoretical studies on the charge transport of organic semiconductors are divided into two approaches. One is the coherent transport approach based on the band transport, including my present study, and induces decreases in mobility with increasing temperature. Thermally activated behavior is seen only if the trap potentials are introduced into the crystal. Another one is the incoherent transport approach based on the Marcus theory, which induces the thermally activated behavior even if there are no trap potentials in crystals. Recently, a new incoherent transport approach based on a flexible surface hopping scheme was applied to a model Hamiltonian and showed the crossover from hopping transport at low electronic couplings to a band-like transport at high couplings [101]. The two approaches mentioned above give similar temperature dependences of mobility, but the physical origins are considerably different from each other. Systematic understanding of these different approaches remains an important issue.

When we try to predict the transport properties of new organic semiconductors, the precise crystal structure is the required information. Therefore, the theoretical prediction of packing structure of molecular crystal is very important. But it is known as a difficult problem in general because there exists a number of crystalline polymorphs reflecting the weak intermolecular interactions.

Furthermore, one of the main scientific challenges is to identify the microscopic origin of trap potentials in realistic organic devices. In this study, the Anderson-type static disorder potentials using W as a parameter were introduced. The inclusion of more realistic trap potentials is crucial for understanding and improving the device performance of organic semiconductors.

Acknowledgments

The author thanks K. Hirose, N. Kobayashi, J. Takeya, and S. Fratini for their valuable comments and suggestions. This work was supported by the JST-PRESTO program "Molecular technology and creation of new functions." I also acknowledge JSPS KAKENHI Grants No. 15H05418.

References

1 Wallace, P.R. (1947). *Phys. Rev.* 71: 622.
2 Eley, D.D. (1948). *Nature* 162: 819.
3 Akamatu, H. and Inokuchi, H. (1950). *J. Chem. Phys.* 18: 810.
4 Vartanyan, A.T. (1950). *Zhru. Fhim. Khim.* 24: 1361.
5 Inokuchi, H. (1954). *Bull. Chem. Soc. Jpn.* 27: 22.
6 Akamatu, H., Inokuchi, H., and Matsunaga, Y. (1954). *Nature* 173: 168.
7 Ferraris, J., Cowan, D.O., Walatka, V. Jr., and Perlstein, J.H. (1973). *J. Am. Chem. Soc.* 95: 948.
8 Shirakawa, H., Louis, E.J., MacDiarmid, A.G. et al. (1977). *J. Chem. Soc., Chem. Commun.* 578.
9 Ishiguro, T., Yamaji, K., and Saito, G. (1998). *Organic Superconductors*. Berlin, Heidelberg: Springer-Verlag.
10 Kepler, R.G. (1960). *Phys. Rev.* 119: 1226.
11 LeBlanc, O.H. (1960). *J. Chem. Phys.* 33: 626.
12 Friedman, L. (1964). *Phys. Rev.* 133: A1668.
13 Sumi, H. (1972). *J. Phys. Soc. Jpn.* 33: 327.
14 Kudo, K., Yamashita, M., and Moriizumi, T. (1984). *Jpn. J. Appl. Phys.* 23: 130.
15 Koezuka, H., Tsumura, A., and Ando, T. (1987). *Synth. Met.* 18: 699.
16 Tang, C.W. and VanSlyke, S.A. (1987). *Appl. Phys. Lett.* 51: 913.
17 Burroughes, J.H., Bradley, D.D.C., Brown, A.R. et al. (1990). *Nature* 347: 539.
18 Ohmori, Y., Uchida, M., Muro, K., and Yoshino, K. (1991). *Jpn. J. Appl. Phys.* 30: L1941.
19 Tang, C.W. (1986). *Appl. Phys. Lett.* 48: 183.
20 Xue, J., Uchida, S., Rand, B.P., and Forrest, S.R. (2004). *Appl. Phys. Lett.* 85: 5757.
21 Kim, J.Y., Lee, K., Coates, N.E. et al. (2007). *Science* 317: 222.
22 Yan, H., Chen, Z., Zheng, Y. et al. (2009). *Nature* 457: 679.
23 Rivnay, J., Jimison, L.H., Northrup, J.E. et al. (2009). *Nat. Mater.* 8: 952.
24 Karl, N. (2003). *Synth. Met.* 133–134: 649.
25 Okamoto, T., Mitsui, C., Yamagishi, M. et al. (2013). *Adv. Mater.* 25: 6392.
26 Nelson, S.F., Lin, Y.-Y., Gundlach, D.J., and Jackson, T.N. (1998). *Appl. Phys. Lett.* 72: 1854.
27 Takeya, J., Goldmann, C., Haas, S. et al. (2003). *J. Appl. Phys.* 94: 5800.
28 Butko, V.Y., Chi, X., Lang, D.V., and Ramirez, A.P. (2003). *Appl. Phys. Lett.* 83: 4773.

29 Goldmann, C., Haas, S., Krellner, C. et al. (2004). *J. Appl. Phys.* 96: 2080.
30 Roberson, L.B., Kowalik, J., Tolbert, L.M. et al. (2005). *J. Am. Chem. Soc.* 127: 3069.
31 Reese, C., Chung, W.-J., Ling, M.-M. et al. (2006). *Appl. Phys. Lett.* 89: 202108.
32 Lee, J.Y., Roth, S., and Park, Y.W. (2006). *Appl. Phys. Lett.* 88: 252106.
33 Uemura, T., Yamagishi, M., Soeda, J. et al. (2012). *Phys. Rev. B* 85: 035313.
34 Tateyama, Y., Ono, S., and Matsumoto, Y. (2012). *Appl. Phys. Lett.* 101: 083303.
35 Arabi, S.A., Dong, J., Mirza, M. et al. (2016). *Cryst. Growth Des.* 16: 2624.
36 Yamamoto, T. and Takimiya, K. (2007). *J. Am. Chem. Soc.* 129: 2224.
37 Uno, M., Tominari, Y., Yamagishi, M. et al. (2009). *Appl. Phys. Lett.* 94: 223308.
38 Haas, S., Takahashi, Y., Takimiya, K., and Hasegawa, T. (2009). *Appl. Phys. Lett.* 95: 022111.
39 Yamagishi, M., Soeda, J., Uemura, T. et al. (2010). *Phys. Rev. B* 81: 161306(R).
40 Xie, W., Willa, K., Wu, Y. et al. (2013). *Adv. Mater.* 25: 3478.
41 Kraft, U., Sejfić, M., Kang, M.J. et al. (2015). *Adv. Mater.* 27: 207.
42 Ebata, H., Izawa, T., Miyazaki, E. et al. (2007). *J. Am. Chem. Soc.* 129: 15732.
43 Izawa, T., Miyazaki, E., and Takimiya, K. (2008). *Adv. Mater.* 20: 3388.
44 Uemura, T., Hirose, Y., Uno, M. et al. (2009). *Appl. Phys. Express* 2: 111501.
45 Liu, C., Minari, T., Lu, X. et al. (2011). *Adv. Mater.* 23: 523.
46 Tanaka, H., Kozuka, M., Watanabe, S.-I. et al. (2011). *Phys. Rev. B* 84: 081306(R).
47 Soeda, J., Hirose, Y., Yamagishi, M. et al. (2011). *Adv. Mater.* 23: 3309.
48 Minemawari, H., Yamada, T., Matsui, H. et al. (2011). *Nature* 475: 364.
49 Kwon, S., Kim, J., Kim, G. et al. (2015). *Adv. Mater.* 27: 6870.
50 Podzorov, V., Pudalov, V.M., and Gershenson, M.E. (2003). *Appl. Phys. Lett.* 82: 1739.
51 Sundar, V.C., Zaumseil, J., Podzorov, V. et al. (2004). *Science* 303: 1644.
52 Stassen, A.F., de Boer, R.W.I., Iosad, N.N., and Morpurgo, A.F. (2004). *Appl. Phys. Lett.* 85: 3899.
53 Menard, E., Podzorov, V., Hur, S.-H. et al. (2004). *Adv. Mater.* 16: 2097.
54 Podzorov, V., Menard, E., Borissov, A. et al. (2004). *Phys. Rev. Lett.* 93: 086602.
55 Podzorov, V., Menard, E., Rogers, J.A., and Gershenson, M.E. (2005). *Phys. Rev. Lett.* 95: 226601.
56 Takeya, J., Tsukagoshi, K., Aoyagi, Y. et al. (2005). *Jpn. J. Appl. Phys.* 44: L1393.
57 Hulea, I.N., Fratini, S., Xie, H. et al. (2006). *Nat. Mater.* 5: 982.
58 Reese, C. and Bao, Z. (2007). *Adv. Mater.* 19: 4535.
59 Takeya, J., Kato, J., Hara, K. et al. (2007). *Phys. Rev. Lett.* 98: 196804.
60 Takeya, J., Yamagishi, M., Tominari, Y. et al. (2007). *Appl. Phys. Lett.* 90: 102120.
61 Marumoto, K., Arai, N., Goto, H. et al. (2011). *Phys. Rev. B* 83: 075302.
62 Lee, B., Chen, Y., Fu, D. et al. (2013). *Nat. Mater.* 12: 1125.

63 Ohashi, N., Tomii, H., Matsubara, R. et al. (2007). *Appl. Phys. Lett.* 91: 162105.
64 Jung, M.-C., Leyden, M.R., Nikiforov, G.O. et al. (2015). *ACS Appl. Mater. Interfaces* 7: 1833.
65 Marcus, R.A. (1956). *J. Chem. Phys.* 24: 966.
66 Hush, N.S. (1958). *J. Chem. Phys.* 28: 962.
67 Holstein, T. (1959). *Ann. Phys.* 8: 325; (1959). *Ann. Phys.* 8: 343.
68 Dimitrakopoulos, C.D. and Malenfant, P.R.L. (2002). *Adv. Mater.* 14: 99.
69 Takimiya, K., Ebata, H., Sakamoto, K. et al. (2006). *J. Am. Chem. Soc.* 128: 12604.
70 Ishii, H. (2016). *J. Inst. Elect. Eng. Jpn.* 136: 434.
71 Hannewald, K. and Bobbert, P.A. (2004). *Phys. Rev. B* 69: 075212.
72 Ortmann, F., Bechstedt, F., and Hannewald, K. (2009). *Phys. Rev. B* 79: 235206.
73 Ishii, H., Kobayashi, N., and Hirose, K. (2017). *Phys. Rev. B* 95: 035433.
74 Fukagawa, H., Yamane, H., Kataoka, T. et al. (2006). *Phys. Rev. B* 73: 245310.
75 Yamane, H., Kawabe, E., Yoshimura, D. et al. (2008). *Phys. Status Solidi B* 245: 793.
76 Koch, N., Vollmer, A., Salzmann, I. et al. (2006). *Phys. Rev. Lett.* 96: 156803.
77 Nakayama, Y., Mizuno, Y., Hikasa, M. et al. (2017). *J. Phys. Chem. Lett.* 8: 1259.
78 Northrup, J.E. (2011). *Appl. Phys. Lett.* 99: 062111.
79 Okada, Y., Sakai, K., Uemura, T. et al. (2011). *Phys. Rev. B* 84: 245308.
80 Ortmann, F., Bechstedt, F., and Hannewald, K. (2011). *Phys. Status Solidi B* 248: 511.
81 Li, Y., Coropceanu, V., and Brédas, J.-L. (2012). *J. Phys. Chem. Lett.* 3: 3325.
82 Bussolotti, F., Kera, S., Kudo, K. et al. (2013). *Phys. Rev. Lett.* 110: 267602.
83 Mishchenko, A.S., Matsui, H., and Hasegawa, T. (2012). *Phys. Rev. B* 85: 085211.
84 Kalb, W.L. and Batlogg, B. (2010). *Phys. Rev. B* 81: 035327.
85 Ishii, H., Kobayashi, N., and Hirose, K. (2008). *Appl. Phys. Express* 1: 123002.
86 Ishii, H., Kobayashi, N., and Hirose, K. (2010). *Phys. Rev. B* 82: 085435.
87 Ishii, H., Honma, K., Kobayashi, N., and Hirose, K. (2012). *Phys. Rev. B* 85: 245206.
88 Ishii, H., Tamura, H., Tsukada, M. et al. (2014). *Phys. Rev. B* 90: 155458.
89 Roche, S., Jiang, J., Triozon, F., and Saito, R. (2005). *Phys. Rev. Lett.* 95: 076803.
90 Valeev, E.F., Coropceanu, V., da Silva Filho et al. (2006). *J. Am. Chem. Soc.* 128: 9882.
91 Ishii, H., Kobayashi, N., and Hirose, K. (2013). *Phys. Rev. B* 88: 205208.
92 I used the GAMESS program at the DFT-D/B3LYP-D3/6-31G(d) level: Schmidt, M.W., Baldridge, K.K., Boatz, J.A. et al. (1993). *J. Comput. Chem.* 14: 1347.
93 Grimme, S. (2004). *J. Comput. Chem.* 25: 1463.
94 Su, W.P., Schrieffer, J.R., and Heeger, A.J. (1979). *Phys. Rev. Lett.* 42: 1698.
95 Su, W.P., Schrieffer, J.R., and Heeger, A.J. (1980). *Phys. Rev. B* 22: 2099.
96 Troisi, A. and Orlandi, G. (2006). *Phys. Rev. Lett.* 96: 086601.

- 97 Fratini, S. and Ciuchi, S. (2009). *Phys. Rev. Lett.* 103: 266601.
- 98 Picon, J.-D., Bussac, M.N., and Zuppiroli, L. (2007). *Phys. Rev. B* 75: 235106.
- 99 Ciuchi, S., Fratini, S., and Mayou, D. (2011). *Phys. Rev. B* 83: 081202(R).
- 100 Parandekar, P.V. and Tully, J.C. (2005). *J. Chem. Phys.* 122: 094102.
- 101 Wang, L. and Beljonne, D. (2013). *J. Phys. Chem. Lett.* 4: 1888.

2

Liquid-Phase Interfacial Synthesis of Highly Oriented Crystalline Molecular Nanosheets

Rie Makiura

Osaka Prefecture University, Department of Materials Science, Osaka 599-8570, Japan

2.1 Introduction

Over the past decade, significant progress has been made in the field of functional molecular-based materials, resulting in an increased understanding of the relationship between molecular structure/ordering and performance in various applications in diverse fields such as electronics, energy creation/storage, and biomaterials. For most potential applications, organizing organic molecules into two-dimensional (2D) thin film state is necessary. In order to create thin films with organic molecules, two experimental approaches have been followed: gas-phase and liquid-phase protocols [1]. Representative gas-phase processes include vacuum deposition, sputtering, and molecular beam epitaxy. In these methods, molecules are directly deposited on the substrate surface during the fabrication procedure. For liquid-phase processes, fabrication techniques are further divided into two categories – with or without solid substrates. Representative methods, which include deposition on solid substrates, are dip casting, drop casting, and spin casting, whereas in the absence of solid substrates, the methods employed utilize gas/liquid or liquid/liquid interfaces. In this chapter, we focus primarily on the formation of highly crystalline molecular nanosheets at such liquid interfaces. Compared with the direct deposition methods on solid substrates, nanosheets formed on liquid surfaces can be transferred to various types of substrates, thereby increasing the options for potential applications. Especially, we discuss the assembly of new types of crystalline nanosheets – metal–organic framework (MOF) nanosheets – rationally built using molecular components, which do not contain alkyl chains and can be connected via coordination bonds with metal ions in a directionally controllable manner (Figure 2.1).

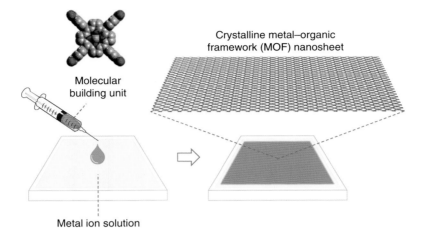

Figure 2.1 Schematic illustration of the assembly of crystalline metal–organic framework (MOF) nanosheets at the air/liquid interface.

2.2 Molecular Nanosheet Formation with Traditional Surfactants at Air/Liquid Interfaces

2.2.1 History of Langmuir–Blodgett Film

Langmuir–Blodgett (LB) film is the most well-known molecular thin film formed at the gas/liquid interface. History of LB film begins in 1774 [2]. Benjamin Franklin observed that when he dropped a teaspoon of oil in a pond, waves became calmer and almost flat like a mirror over several square yards. Almost a century later, Lord Rayleigh considered the report by Franklin and provided an explanation of the phenomenon involved – oil is composed of oleic acids, which spread evenly on the water surface. Rayleigh evaluated the thickness of the thin film to be less than 2 nm from the mount of spread oil. Around the same time, Agnes Pockels had accomplished an analysis of monolayer formation at gas/liquid interfaces. She was interested in oil dirt spread on sink water and created a primitive instrument for measuring surface pressure. Her letter on the observations to Rayleigh was elaborated with Rayleigh's help and published in Nature in 1891 and 1892 [3, 4]. Inspired by Pockels' work, Irving Langmuir developed the Langmuir trough and conducted systematic works on monolayers assembled at air/liquid interfaces [5]. Langmuir discovered together with Katherine Blodgett that the spread organic layer can be uniformly deposited on a solid substrate by insertion of the substrate vertically into the water subphase [6]. In accord with the history of the discoveries, the name "Langmuir film" refers to a monolayer floating on the liquid surface, while a "LB film" represents a multistacked film deposited on a substrate by the repetitive process of substrate immersion/lift-up into/from the liquid subphase.

2.2.2 Basics of Molecular Nanosheet Formation at Air/Liquid Interfaces

Typical organic molecules used for monolayer creation at air/liquid interfaces are surfactants containing both hydrophobic and hydrophilic parts [2]. The hydrophobic parts are mainly hydrocarbon or fluorocarbon chains, whereas the hydrophilic parts are polar groups such as —COOH, —OH, and —NH$_2$. Formation of monolayer at the air/liquid interface depends on (i) miscibility between polar groups and subphase (commonly water), (ii) van der Waals interactions between hydrophobic parts, and (iii) relative importance of factors (i) and (ii).

Amphiphilic surfactant molecules are dissolved in highly volatile organic solvents such as hexane and chloroform and the solution is dropped onto the water surface. As soon as the droplets reach the water surface, they spread over the surface and insoluble monolayers form at the interface between air and liquid. The surfactant molecules arrange themselves in such a way so that their hydrophilic parts are down into the water subphase and the hydrophobic parts stay on the air side. The monolayer formed at the air/liquid interface is called a Langmuir film or L film.

In this chapter, we use the term "nanosheet" to refer both thin films floating on liquid surfaces and free-standing films. Therefore, nanosheets necessarily include Langmuir films. Following the LB film history, however, we use the term "Langmuir film" only for monolayers composed of surfactants with long hydrocarbon chains.

2.3 Application of Functional Organic Molecules for Nanosheet Formation at Air/Liquid Interfaces

2.3.1 Functional Organic Molecules with Long Alkyl Chains

Following the initial discovery of the LB method, components are no longer restricted to simple surfactants and various types of molecules have been utilized for creating functional thin films at the air/liquid interface. Fujihira et al. reported an electrochemical photodiode fabricated by depositing complex LB films, which consisted of an electron acceptor, a sensitizer, and an electron donor surfactant derivative [7]. The organization of an amphiphilic azobenzene derivative in monolayers at the air/water interface is strongly controlled by association phenomena due to intermolecular interactions of the azobenzene moiety [8, 9]. Ariga et al. discussed progress in hydrogen-bond-based molecular recognition at the air/water interface [10]. Nonetheless, all the molecules reported in these works incorporate long alkyl chains, a feature that sometimes restricts potential functionalities because of the associated highly insulating behavior.

2.3.2 Functional Organic Molecules without Long Alkyl Chains

Application of the LB method to functional molecules without any alkyl chain has also been reported. Floating layers of C$_{60}$ at the air/water interface were first observed by Obeng et al. [11] Mixed Langmuir films of C_{60} with icosanoic acid were then reported by Nakamura et al. [12]. After that, the lattice image of C_{60} films prepared on the pure water subphase was shown [13]. A high-performance photo-responsive molecular field-effect transistor was formed from Langmuir monolayers of copper phthalocyanine (CuPc) [14]. High carrier mobility and low threshold voltage for both holes and electrons in an organic field-effect transistor were realized with ordered layers composed of a europium triple-decker complex assembled at the air/liquid interface [15]. In most cases where functional molecules without alkyl chains are used, the molecules pack in the layered structure at the air/liquid interface via weak interactions such as van der Waals and π–π interactions. On the other hand, much stronger interactions including covalent, coordination, and multiple hydrogen bonds can lead to layers with increased stability, and molecular arrangements can be more finely controlled by applying such directly guided interactions. Schlüter and King groups reported the synthesis of a large area nanoporous two-dimensional polymer at the air/liquid interface [16]. Compression followed by photopolymerization at the interface provides organic nanosheets in which the molecular building units are covalently connected with their long-range structural order confirmed by scanning tunneling microscopy.

In this chapter, we focus on nanosheets formed via coordination bonds between metal ions and organic ligands, so-called MOFs or porous coordination polymers (PCPs), for their rich variety of materials design, as described in Section 2.4. Especially, we discuss the progress in the assembly of MOF nanosheets composed of porphyrins at the air/liquid interface. Porphyrins are important functional molecules, which have generated significant research interest from the viewpoint of topological design of ordered networks as well as for their rich chemical/physical properties without long alkyl chains.

2.3.3 Application of Functional Porphyrins on Metal Ion Solutions

Porphyrins and metalloporphyrins are very stable π-conjugated macrocyclic molecules with square-planar geometry. As such, they are highly suitable building components for the formation of 2D supramolecular architectures. There is a large family of porphyrins and metalloporphyrins with or without incorporation of different center-occupying metal ions and with a rich variety of peripheral functional substituents, such as alkyl chains, (hetero)aromatic groups, hydroxy groups, carboxylic acid, and amino groups [17–23]. Combination of a hydrophobic macrocyclic core and peripheral polar substituents in porphyrins allows us to apply them to air/liquid interfacial synthesis for nanosheet formation. Qian et al. employed the functional porphyrin, 5,10,15,20-tetrakis(4-pyridyl)-21H,23H-porphyrin (H_2TPyP), and its metal–ion derivatives (MTPyP, M = Zn^{II}, Mn^{III}, $Ti^{IV}O$) on palladium salt aqueous solutions [24–29]. Drawing from the structural results on bulk crystals synthesized in solution of Drain et al., they conjectured that a similar check-patterned structure was also formed in their films by coordination of Pd^{2+} ions and the peripheral pyridine units in the porphyrin building blocks [30–32]. However, no direct experimental evidence for such structure formation in the films was provided.

2.4 Porphyrin-Based Metal–Organic Framework (MOF) Nanosheet Crystals Assembled at Air/Liquid Interfaces

2.4.1 Metal–Organic Frameworks

The current explosion of interest in new MOFs, PCPs, or related porous hybrid solids arising from the virtually unlimited structural possibilities involving combinations of inorganic and organic building components has implications for general fundamental crystal growth techniques and for a number of potential applications ranging from the materials science to life science fields [33–37]. In many cases of targeted synthesis of polycrystalline or single crystalline bulk MOFs using solvothermal reactions, the desired structures that possessed well-defined pores were generated following rational molecular design routes, strategic choice of combination of molecular components based on coordination chemistry knowledge, and exploration of the reaction conditions. On the other hand, controlling the size of porous objects at the nanoscale and arranging such objects on/with various substances in certain desired ways (e.g. sequentially layered on top of each other, patterned on surfaces) remains as a challenge that needs to be addressed. In addition, understanding the roles of nanoscale size and interfacial conditions in integrated systems in defining the properties of such porous solids can lead to optimization of their potential functionalities.

In fact, reports on various attempts to process bulk polycrystalline MOFs and to fabricate nanoscaled ones by developing new methodologies such as microwave synthesis, inverse emulsion technique, and liquid-phase epitaxy have been increasing in recent years [38–48]. Especially, 2D sheet assemblies are necessary when considering the use of such coordination materials, which frequently incorporate functional π-electron components, in nanotechnological thin film devices.

2.4.2 Method of MOF Nanosheet Creation at Air/Liquid Interfaces

We have recently succeeded to create highly crystalline nanosheets with a porphyrin derivative at the air/liquid interface. In our first report on porphyrin nanosheet creation, we used 5,10,15,20-tetrakis(4-carboxyphenyl)-porphyrinato-cobalt(II) (CoTCPP) and a copper (II) ion aqueous solution as a subphase (NAFS-1) [49]. At that time, we established the nanosheet structure only after multiple transfer/deposition processes of the nanosheets formed at the air/liquid interface onto a solid substrate [49–51]. In subsequent work, we were able to follow the nanosheet formation directly by in situ synchrotron X-ray diffraction (XRD) at the air/liquid interface [52]. Therefore, here we start by considering the creation of the porphyrin nanosheet, NAFS-13, that was followed by in situ XRD

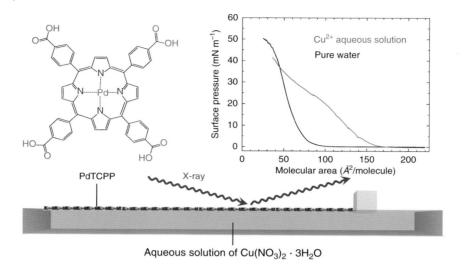

Figure 2.2 Formation of NAFS-13 (PdTCPP-Cu nanosheet) was monitored by in situ grazing incident XRD measurements. The solution of PdTCPP molecular building units is spread onto the copper ion aqueous solution in a Langmuir trough. The surface pressure, π, is controlled by the movement of a single barrier and is kept constant during the collection of each GIXRD profile at the air/liquid interface. The right top figure shows surface pressure – mean molecular area ($\pi - A$) isotherms for NAFS-13 (red) and for PdTCPP solution spread onto pure water subphase (black). *Source*: From Makiura and Konovalov 2013 [52]. Reproduced with permission of Nature Publishing Group.

measurements. In this work, we used 5,10,15,20-tetrakis(4-carboxyphenyl)-porphyrinato-palladium(II) (PdTCPP) as a building unit (Figure 2.2, left top). The procedure for preparing crystalline nanosheets was initiated with the spreading of the PdTCPP solution onto $Cu(NO_3)_2 \cdot 3H_2O$ aqueous solution (Figure 2.2, bottom). The formation of the NAFS-13 nanosheets where PdTCPP units were connected into a 2D nano-architecture via copper bridging ions was monitored by measuring the surface pressure – mean molecular area ($\pi - A$) isotherm. The surface pressure was increased to $40\,mN\,m^{-1}$ by compression with the barrier walls of the trough that moved at a constant speed. When the same PdTCPP solution was spread onto a purified water subphase without any metal ions in the Langmuir trough and compressed at the same barrier speed, the mean molecular area, A, recorded at the same surface pressure was significantly smaller (Figure 2.2, right top). This implies that the PdTCPP units stand vertically at some angle to the liquid surface (Figure 2.3, left top) or remain in the horizontal orientation but pack very closely (Figure 2.3, left bottom). In contrast, when copper ion solution is used, they lie flat on the solution surface as a result of the coordinative linking of copper ions with the tetratopic PdTCPP molecules (Figure 2.3, right). The nanosheets were transferred from the liquid subphase to a quartz substrate, and ultraviolet–visible (UV–Vis) absorption spectroscopic measurements were conducted. The absorbance of the Soret band

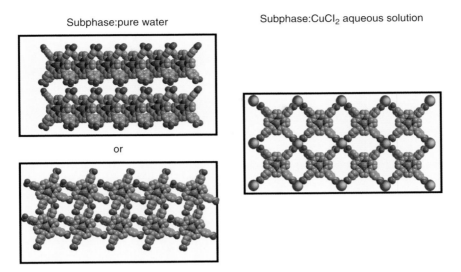

Figure 2.3 Proposed molecular arrangements of PdTCPP nanosheets in two different subphases as deduced by concentration of the molecular areas derived by the $\pi - A$ isotherms. *Source*: From Makiura and Kitagawa 2010 [50]. Reproduced with permission of John Wiley and Sons.

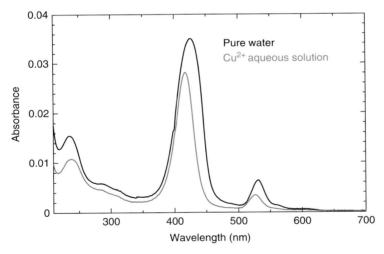

Figure 2.4 UV–Vis absorption spectra for PdTCPP nanosheets (monolayer) fabricated on copper ion solution (NAFS-13, red solid line) or pure water (black solid line) subphase.

of PdTCPP in nanosheets fabricated on purified water is larger than that of NAFS-13 nanosheets formed on copper ion aqueous solution (Figure 2.4). This is consistent with a larger number of PdTCPP units packing in the same area of the monolayer in agreement with the $(\pi - A)$ isotherm measurements (Figure 2.2, right top).

2.4.3 Study of the Formation Process of MOF Nanosheets by *In Situ* X-Ray Diffraction and Brewster Angle Microscopy at Air/Liquid Interfaces

Detailed insights into the formation of the NAFS-13 nanosheets were obtained by in situ synchrotron XRD measurements [52]. These were conducted directly at the air/liquid interface in grazing incidence (GI) mode with the incident X-ray beam almost parallel to the liquid surface, as illustrated in Figure 2.2. Figure 2.5a shows the evolution of the in-plane grazing incidence X-ray diffraction (GIXRD) profiles measured for NAFS-13 with increasing surface pressure. Many sharp Bragg peaks were observed, providing the signature of the formation of a highly crystalline organization. Importantly, such a profile with many sharp peaks was seen at the low surface pressure point ($\pi \approx 0$) after spreading the PdTCPP solution onto the copper ion solution. The result suggests that the formation of NAFS-13 with long-range order occurs in a self-assembling manner without the need for surface compression. The assemblies were induced by the interfacial reaction between the peripheral carboxylic acid groups of the PdTCPP molecules and the copper ions in the subphase.

A homologous series of porphyrin-based bulk crystalline MOFs (porphyrin paddle-wheel frameworks, PPFs) has been reported [53–56]. Stacking sequences were controlled by utilizing different metal-centered porphyrins, MTCPP (M = Co(III), Zn(II), Pd(II)), in which the coordination number of the central metal site was varied between six (octahedral, Co(III)), five (square-pyramidal, Zn(II)), and four (square-planner, Pd(II)), whereas the in-plane linkage and the checkerboard pattern remained identical for all the three crystalline materials. Returning to the NAFS-13 nanosheets, we find that all observed Bragg peaks in their GIXRD profile index as ($hk0$), as shown in Figure 2.5a. They correspond to a 2D square unit cell with lattice parameters, $a = b \approx 16.6$ Å. The lattice metrics are extremely close to those of the bulk crystalline analogues composed of MTCPP porphyrin building units [53–56]. The comparable lattice sizes unambiguously confirm that the molecular arrangement parallel to the sheet plane in NAFS-13 is that of PdTCPP molecules and dimeric paddle-wheel $Cu_2(COO)_4$ secondary building units, which adopt a checkerboard structural motif (Figure 2.5b and c). The diffraction profile comprising more sharp peaks does not change as the surface pressure is progressively increased, implying that the molecular arrangement of the NAFS-13 nanosheet remains intact upon the surface compression. On the other hand, the peak width and intensity of the Bragg reflections gradually change with surface compression (Figure 2.5d–g). In the low surface pressure region ($\pi = 0$–1 mN m^{-1}), the unit cell size of the square lattice and the average crystalline domain size of the NAFS-13 nanosheets do not change, indicating that surface compression has little influence on the sheet formation process. On the other hand, the peak intensity increases sharply with the surface compression in the same surface pressure range. These results suggest that the effect of surface compression is first to increase the surface coverage by gathering the

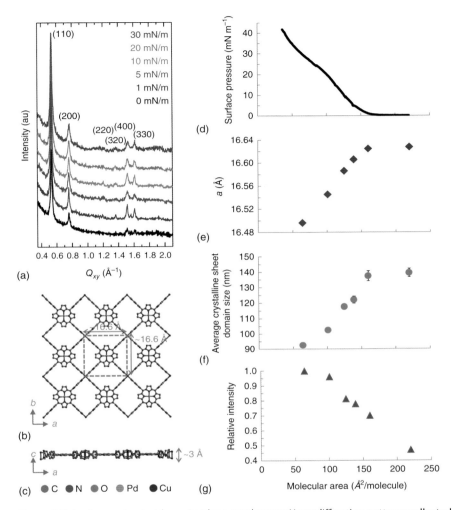

Figure 2.5 In situ grazing incidence in-plane synchrotron X-ray diffraction patterns collected at the air/liquid interface for NAFS-13 nanosheets. (a) Observed GIXRD ($\lambda = 1.549$ Å, incidence angle, $\alpha = 0.12°$) profiles at surface pressures, $\pi = 0, 1, 5, 10, 20$, and 30 mN m^{-1}. (b) Basal plane projection of the crystalline structure of NAFS-13, which consists of a 2D "checkerboard" motif of PdTCPP units linked by binuclear Cu$_2$(COO)$_4$ paddle wheels. (c) Schematic diagram of the crystalline structure of NAFS-13, which consists of 2D sheets of thickness ~3 Å viewed along the b axis. (d–g) Evolution of the crystalline structure and morphology of the molecularly thin NAFS-13 nanofilms with change in surface compression. (d) Surface pressure – mean molecular area ($\pi - A$) isotherm. (e) Unit cell basal plane dimension, a. (f) Average crystalline sheet domain size estimated from the full width at half maximum of the intense (110) Bragg reflection. (g) Relative intensity of the (110) reflection normalized to the value at the highest surface compression versus mean molecular area. *Source*: From Makiura and Konovalov 2013 [52]. Reproduced with permission of Nature Publishing Group.

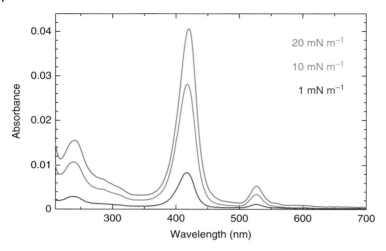

Figure 2.6 UV–Vis absorption spectra for NAFS-13 nanosheets. The films formed at a surface pressure of 1, 10, and 20 mN m^{-1} were deposited onto quartz substrates and the spectra were measured after a rinsing/solvent immersion/drying process.

preassembled NAFS-13 nanosheet domains to a smaller area without affecting the crystalline domain size. This interpretation was also supported by the growth in the intensity of the Soret band of PdTCPP with increasing surface pressure observed by UV–Vis spectroscopy for NAFS-13 formed at different surface pressure points and transferred onto quartz substrates (Figure 2.6).

However, further compression of the surface influences in a pronounced way both the crystalline structure and the morphology of the nanosheets. Firstly, the in-plane lattice parameter, a, decreases monotonically as π increases (Figure 2.5e). This is accompanied by a contraction in the nanosheet domain size and a continuous growth of the intensity of the (110) Bragg reflection (Figure 2.5f). In order to understand this, we recall that as the surface pressure increases, the coverage of the surface increases until it is fully occupied by the NAFS-13 nanosheets. Upon further compression, the nanosheets gather into a smaller area until their peripheral parts start to squeeze neighboring sheets, thereby leading to increased surface roughness through deformation or multiple sheet stacking. The assembling process of NAFS-13 at the air/liquid interface is schematically illustrated in Figure 2.7. Brewster angle microscopy (BAM) measurements support this phenomenological interpretation (Figure 2.8). The BAM images show first a progressive increase of surface coverage by the nanosheets upon compression, followed by the appearance of cracks and surface deformation at high surface pressure points. In addition, white spots are observed in the BAM images regardless of the surface pressure, implying the existence of molecular aggregation in the nanosheets.

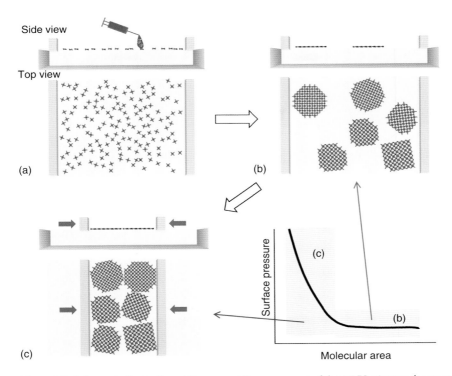

Figure 2.7 Schematic illustration of the assembling processes of the NAFS-13 nanosheets at the air/liquid interface. Spreading molecular components PdTCPP on the copper ion aqueous solution subphase (a) drives immediate formation of 2D arrays where PdTCPP molecules are highly ordered because of the coordinative interaction between their carboxylic parts and copper ions. (b) The 2D domain arrays are distributed inhomogeneously when the surface area is relatively large in comparison with the number of spread molecules – the low surface pressure condition. (c) Further pressing of the surface gathers the 2D arrays to a smaller area, resulting in the high coverage of the sheets after deposition onto the solid substrates and the size of the crystalline domains becomes smaller by squeezing neighboring domains each other. *Source*: From Makiura et al. 2011 [51]. Reproduced with permission of Royal Society of Chemistry.

2.4.4 Application of a Postinjection Method Leading to Enlargement of the Uniform MOF Nanosheet Domain Size

The in situ XRD measurements at the air/liquid interface on the NAFS-13 nanosheet together with the complementary BAM measurements have established that the important process in the nanosheet creation procedure is the interfacial coordination reaction, which occurs immediately after spreading the molecular building unit solution onto the subphase containing metal ions [52]. Therefore, this is the most critical step in determining the sheet domain size. However, it is unavoidable that droplets of the spread solution produce surface

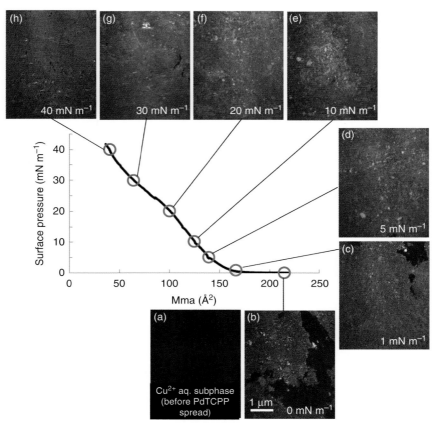

Figure 2.8 Brewster angle microscopy (BAM) images of the NAFS-13 nanosheet at the air/water interface captured during compression. (a) Surface of the aqueous Cu^{2+} ion aqueous solution subphase before spreading PdTCPP molecules, (b) at 0 mN m^{-1}, after spreading PdTCPP molecules, (c) at 1 mN m^{-1}, (d) at 5 mN m^{-1}, (e) at 10 mN m^{-1}, (f) at 20 mN m^{-1}, (g) at 33 mN m^{-1}, and (h) at 40 mN m^{-1}.

ripples in the conventional protocol of the nanosheet creation at the air/liquid interface. By recalling the procedure for obtaining large single crystals of coordination compounds with slow diffusion protocols of reactant solutions, we have attempted to integrate comparable diffusion protocols into the nanosheet growth strategy. As illustrated in Figure 2.9, we first spread the solution of the molecular building units of PdTCPP on the purified water subphase and then gently inject the concentrated copper ion aqueous solution into the subphase from the corner of the Langmuir trough.

After spreading the PdTCPP solution but before the injection of the copper ion solution, the in-plane XRD pattern observed at the air/liquid interface does not contain any Bragg peaks (Figure 2.10 black). This confirms that the PdTCPP molecules do not self-assemble to form a crystalline nanosheet in the absence of copper ions. After injection of the concentrated copper ion solution into the purified water subphase, more number of very sharp peaks appear in

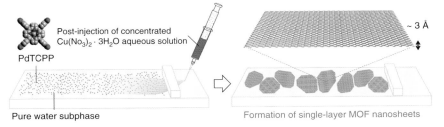

Figure 2.9 Schematic illustration of the postinjection methodology employed in the fabrication of NAFS-13 nanosheets. A PdTCPP solution is first spread directly on the pure water subphase. A concentrated copper ion aqueous solution is then slowly injected into the water subphase from the side surface, which is separated from the Langmuir trough by the compression barrier. Source: From Makiura and Konovalov 2013 [52]. Reproduced with permission of Nature Publishing Group.

the GI in-plane XRD profile (Figure 2.10 red). The positions of the observed Bragg peaks coincide with those recorded for the NAFS-13 nanosheet formed by our conventional method – the PdTCPP solution is spread on the copper ion solution subphase (Figure 2.10 inset). This confirms the formation of the nanosheets with the same crystalline structure and identical lattice constants. However, the peak widths of the Bragg peaks in the GI in-plane XRD pattern of the nanosheet formed by the postinjection method are considerably smaller than those observed for the nanosheet by the conventional method. This implies a significant increase in the lateral size of crystalline domains in the nanosheets.

In addition, the recorded BAM images of the NAFS-13 nanosheets formed by the postinjection method reveal much smoother surface morphology than that of those assembled by the conventional method (Figure 2.11). Furthermore, white spots associated with the presence of molecular aggregations are now completely absent. Therefore, surface rippling and instant reaction between the spread molecular building units and copper ions from the subphase both limit nanosheet domain growth and lead to the creation of molecular aggregates restricting the formation of smooth uniform nanosheets. On the other hand, the postinjection method, which effectively realizes gentle interfacial reaction under slow diffusion condition, leads to the enlargement of the sheet domain size and provides smooth surface morphology of the monomolecularly thin nanosheets.

We also successfully achieved further increase of the domain size of the nanosheet by changing the molecular building unit to 5,10,15,20-tetra(4-pyridyl)-porphyrinato-zinc (II) (ZnTPyP), while keeping the same subphase with copper ions (Figure 2.12) [57]. The MOF nanosheets (NAFS-21) composed of ZnTPyP are of monomolecular thickness with a crystalline domain size in the direction parallel to the liquid surface of submicron scale, larger than 410 nm – as the XRD peak widths were limited by the resolution of the synchrotron XRD experimental setup, the precise sheet size is expected to be larger approaching the micrometer size. We consider that the change of the molecular building units from anionic MTCPP to neutral ZnTPyP slows down the interfacial coordination reaction with the copper ions, leading to the formation of large nanosheet domains without the need to apply the postinjection method.

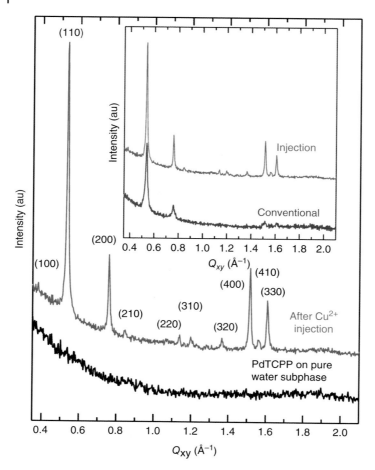

Figure 2.10 Observed grazing incidence in-plane synchrotron GIXRD profiles ($\lambda = 1.549$ Å, incidence angle, $\alpha = 0.12°$) for PdTCPP spread on the pure water subphase before (black line) and after (red line) injection of the copper ion aqueous solution in the absence of barrier compression. The *inset* shows a comparison of the in-plane GIXRD profiles for the NAFS-13 nanosheets formed by the conventional method of Figure 2.2 (blue line) and the postinjection method (red line) at $\pi = 0$ mN m^{-1}. Source: From Makiura and Konovalov 2013 [52]. Reproduced with permission of Nature Publishing Group.

2.4.5 Layer-by-Layer Sequential Growth of Nanosheets – Toward Three-Dimensionally Stacked Crystalline MOF Thin Films

Layer-by-layer (LbL) deposition is a representative technique of thin film growth and the notation LbL has usually been applied to describe alternate adsorption of electrostatically charged polymers or colloids [58, 59]. Nowadays, the term LbL is widely used to describe various stepwise bottom-up protocols of thin film growth [60–64]. Typically, two solutions consisting of different components are prepared and a solid substrate is immersed into the solutions in an alternating manner.

2.4 Porphyrin-Based MOF Nanosheet Crystals Assembled at Air/Liquid Interfaces | 39

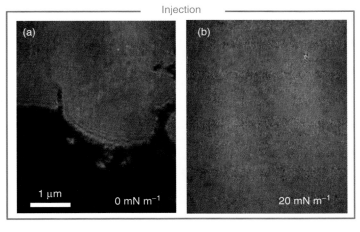

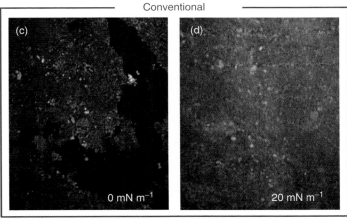

Figure 2.11 Brewster angle microscopy (BAM) images taken during the formation of the NAFS-13 nanofilms fabricated by the postinjection (a and b) and the conventional (c and d) methods at compressions, $\pi = 0$ and 20 mN m^{-1}. Source: From Makiura and Konovalov 2013 [52]. Reproduced with permission of Nature Publishing Group.

Focusing on the coordinative reaction between metal ions and organic ligands, such LbL protocols have been widely used to grow supramolecular networks on solid surfaces [65–68]. However, the LbL method was used for the fabrication of thin films of MOFs with high crystallinity only 10 years ago [69–75]. Shekhah et al. reported a step-by-step route of the synthesis of crystalline MOFs for understanding the MOF formation process in a rational way rather than to fabricate MOF thin films [70, 71]. Although the reports show promising results in obtaining partially ordered systems or with some ordering preference, the assembly of MOF thin films with completely controlled size and growth direction was not achieved, and the structural details, especially in the molecular arrangement parallel to the substrate surface, remain unknown.

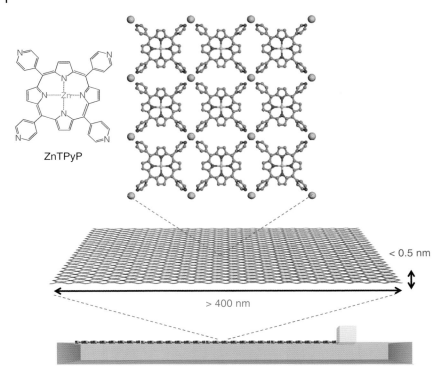

Figure 2.12 Schematic illustration of the assembly of the ZnTPyP-Cu nanosheets (NAFS-21) at the air/liquid interface.

A distinct advantage of the LbL growth mode is that it provides good control of the film orientation and the number of layers in the vertical direction to the substrate (out-of-plane). However, comparable control is not built into the technique for the horizontal direction parallel to the substrate surface (in-plane). This key aspect was addressed by the incorporation of the air/liquid interface into the film growth methodology allowing exquisite control in both lateral and transverse directions for the first time. Application of the air/liquid interfacial reaction mentioned above together with the LbL protocol can be integrated in a modular manner to fabricate perfectly crystalline MOF thin films on a solid surface under mild conditions. Here, we describe the formation of a porphyrin-based MOF thin film (NAFS-2) endowed with highly crystalline order both in the out-of-plane and in-plane orientations to the substrates by applying the combination of the air/liquid interfacial reaction and the LbL growth [76].

The NAFS-2 nanosheet composed of 5,10,15,20-tetrakis(4-carboxyphenyl) porphyrin (H_2TCPP) formed on the copper ion aqueous solution was transferred to the solid substrate by the horizontal dipping method (the substrate surface is parallel to the subphase surface) (Figure 2.13). Excess copper ions were removed by following immersion of the substrate into purified water. To stack further layers, the substrate with the predeposited NAFS-2 nanosheet was again dipped on the subphase surface. The number of the layers deposited is

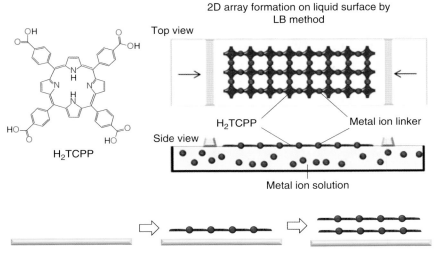

Figure 2.13 Schematic illustration of the assembly process of the H$_2$TCPP-Cu nanosheets (NAFS-2) at the air/liquid interface and their layer-by-layer (LbL) growth. *Source*: From Motoyama et al. 2011 [76]. Reproduced with permission of American Chemical Society.

precisely controlled by the number of cycles of sheet transfer. The successive LbL growth of the NAFS-2 nanosheet was monitored by UV–Vis spectroscopy. The linear increase in absorbance of the H$_2$TCPP Soret band indicates that roughly the same amount of H$_2$TCPP molecules is deposited in each step of the layer stacking protocol (Figure 2.14a and b). Similarly, a linear increase in the IR absorbance attributed to the COO symmetric stretch band of NAFS-2 with increasing number of cycles was also confirmed (Figure 2.14c and d). This further supports that to a good approximation, each cycle leads to the transfer of the same amount of H$_2$TCPP molecules. The XRD profile measured in out-of-plane scans shows two peaks, which can be indexed as (001) and (003). This reveals that NAFS-2 has a highly ordered growth perpendicular to the film growth direction. The value of the interlayer spacing was estimated to be 7.026 Å from the Bragg reflection positions. Detailed explanations of the structure will be described in the next section.

2.4.6 Manipulation of the Layer Stacking Motif in MOF Nanosheets

Linkages between the neighboring layers can be tuned by introducing pillaring ligands (Figure 2.15). The nanosheets can also be layered by adjacent sheet adherence via weak van der Waals interactions when no pillaring ligands are applied (Figure 2.15a). On the other hand, introducing pillaring ligands between the layers that can coordinate to the available axial sites of metal ion linkers and project from the 2D network plane leads to the enlargement of the interlayer spacing (Figure 2.15b and c). Applying monodentate ligands expands the interlayer distance, but the neighboring sheets are not strongly bound (Figure 2.15b), while

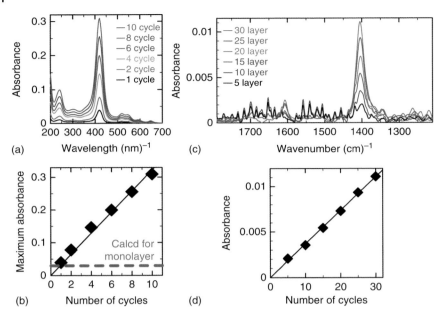

Figure 2.14 (a and b) Evolution of the UV–Vis absorption spectra (a) and IR absorption spectra (b) of NAFS-2 on a SiO$_2$/Cr/Au substrate with successive cycles of sheet deposition, rinsing, and drying. (c and d) Maximum absorbance of the H$_2$TCPP Soret band (c) and the COO symmetric stretch band (d) of NAFS-2 as a function of the number of film growth cycles.

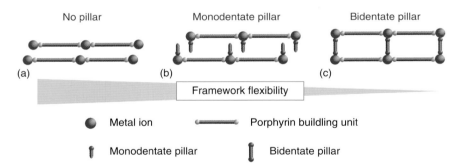

Figure 2.15 Schematic illustration of possible structural variations of MOF nanosheets by LbL growth. *Source*: From Yamada et al. 2013 [77]. Reproduced with permission of Royal Society of Chemistry.

introducing bidentate ligands binds the neighboring sheets in an ordered manner – the lattice size of the attached sheet networks matches (Figure 2.15c) [77].

To investigate the assembly of a nonpillared structure, a free-base porphyrin, H$_2$TCPP was selected (Figure 2.16, left top). Applying pyridine (py) molecules as monodentate pillaring ligands together with a metalloporphyrin, CoTCPP (Figure 2.16, right top), was another combination of building components that could lead to the formation of an interdigitated stacking motif. Copper(II) ions were selected as the metal ion linkers for both molecular systems.

2.4 Porphyrin-Based MOF Nanosheet Crystals Assembled at Air/Liquid Interfaces | 43

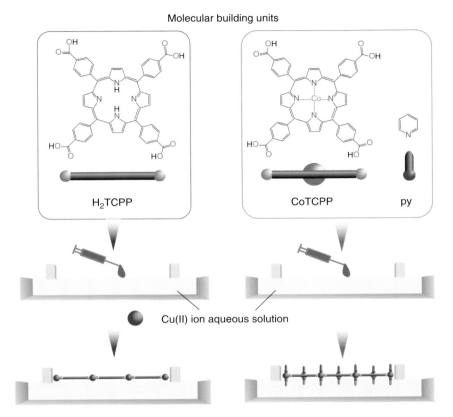

Figure 2.16 Schematic illustration of the MOF nanosheet fabrication protocol at the air/liquid interface and the molecular structures of the building units used. The combination of H_2TCPP and copper ion leads to the formation of the nonpillar type of sheet described in Figure 2.15a, whereas the combination of CoTCPP, pyridine (py), and copper ion provides the sheet that includes the monodentate pillar projecting from the sheet plane, as shown in Figure 2.15b. Source: From Yamada et al. 2013 [77]. Reproduced with permission of Royal Society of Chemistry.

The GI in-plane XRD profiles of H_2TCPP-Cu (NAFS-2) [76] and CoTCPP-py -Cu (NAFS-1) [49] are shown in Figure 2.17. The number of sharp diffraction peaks indicates that both NAFS-2 and NAFS-1 are highly crystalline in the substrate-plane direction. Very importantly, all of the reflections in both films could be indexed as ($hk0$) by using a pseudo-2D tetragonal unit cell with basal plane dimensions, $a = b = 16.477(2)$ Å in NAFS-2 and 16.460(3) Å in NAFS-1, supporting the adoption of a common 2D network, in which porphyrin units are linked by binuclear $Cu_2(COO)_4$ paddle wheels. This results in the checkerboard pattern shown in Figure 2.17a and b inset.

XRD measurements using the out-of-plane scattering geometry were also carried out. Figure 2.18 shows the wide-range out-of-plane XRD patterns of the two nanofilms, NAFS-2 and NAFS-1. Observation of Bragg reflections reveals the highly oriented nature of both nanofilms. They can be indexed as (001) and

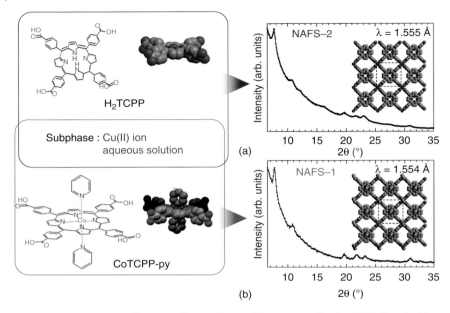

Figure 2.17 Grazing incidence synchrotron X-ray diffraction profiles for NAFS films. (a, b) Observed wide-range in-plane XRD profiles for NAFS-2 (a) and NAFS-1 (b). The inset shows a schematic diagram of the proposed in-plane crystalline structures and basal plane dimensions by using a pseudo-2D tetragonal unit cell. The left image shows the combination of the molecular components and metal ion linkers – H_2TCPP-Cu for NAFS-2 (top) and CoTCPP-py-Cu for NAFS-1 (bottom). *Source*: From Yamada et al. 2013 [77]. Reproduced with permission of Royal Society of Chemistry.

(002) for NAFS-2 (Figure 2.18a), and (001) and (003) for NAFS-1 (Figure 2.18b), leading to values for the size of the interlayer spacing, c of 7.026(3) Å for NAFS-2 and 9.380(3) Å for NAFS-1; thus, the interlayer distance, c, in NAFS-2 is significantly smaller than that found in NAFS-1. This can be explained by the absence of pyridine molecules (Figure 2.18c). In NAFS-1, pyridine molecules coordinate axially to both the copper binuclear units and the central metal ions of CoTCPP and protrude from each 2D sheet (Figure 2.18d). On the other hand, in NAFS-2, water molecules are available to coordinate to the axial sites of the binuclear paddle-wheel units to complete the coordination sphere. When we consider the sheets in NAFS-2 with no water molecules coordinating to the axial sites, their thickness is estimated as ≈3.9 Å (i.e. the distance between hydrogen atoms of the phenyl rings connected to the carboxylic part, where the phenyl ring plane stands perpendicular to the porphyrin flat plane), which is much smaller than the observed interlayer spacing. The evaluated H_2TCPP-Cu sheet thickness taking into account such axially coordinated water molecules (i.e. the distance between hydrogen atoms of the coordinated water molecules across each sheet) is ≈7.6 Å. This value is slightly larger than the NAFS-2 interlayer spacing obtained from the XRD measurements. This supports a sheet stacking pattern in NAFS-2, in which adjacent sheets do not stack directly on top of each other along c where their in-plane lattice matches but are arbitrarily shifted, arranging neighboring axially

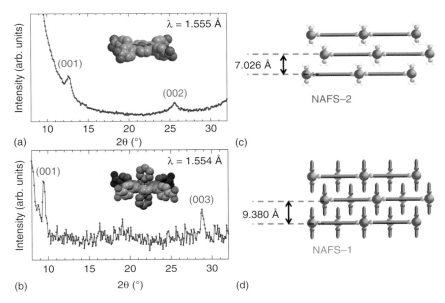

Figure 2.18 Out-of-plane synchrotron X-ray diffraction profiles for NAFS films and schematic images of the sheet stacking motifs. (a and b) Observed wide-range out-of-plane XRD profiles for NAFS-2 (a) and NAFS-1 (b). (c and d) Schematic images of the layering manner in NAFS-2 (c) and NAFS-1 (d). *Source*: From Yamada et al. 2013 [77]. Reproduced with permission of Royal Society of Chemistry.

coordinated water molecules in the interlayer space away from each other and leading to a slightly smaller interlayer distance (Figure 2.18c) than the magnitude of the film thickness estimated by considering the attached water molecules.

In NAFS-1, the observed interlayer spacing of 9.4 Å is significantly smaller than the calculated layer thickness (i.e. the distance between nitrogen atoms of the coordinated pyridine molecules across each sheet) of $\approx$12.4 Å. Therefore, an interlayer stacking order in which neighboring sheets are shifted along the a-axis by 1/4 of the unit cell was considered. The interlayer stacking model includes the pyridine molecules, bound to both the cobalt ion at the center of the porphyrin and the $Cu_2(COO)_4$ paddle-wheel units (Figure 2.18d). The simulated in-plane XRD profile of NAFS-1 taking into account such layer stacking corresponds well to the observed GI in-plane pattern. Therefore, the results suggest that the π–π interaction plays an important role in controlling the direction of growth of NAFS-1 as the axially coordinated pyridine molecules projecting from the 2D sheets allow each further layer to attach in a highly ordered interdigitated manner.

To obtain further structural information on the ordering of the stacking layers, we also carried out rocking curve (θ-scan) measurements at the (001) peak position in the out-of-plane orientation (Figure 2.19). Rocking curves provide information on the degree of crystallinity. The full width at half maximum (FWHM) of a peak observed in the rocking curve of a layer-structured material provides a measure of the average tilting angle of the stacking layers (Figure 2.19a). This was

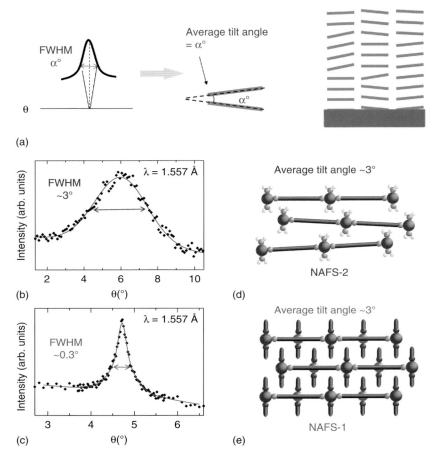

Figure 2.19 Rocking curve θ scan at the (001) reflection position in the out-of-plane orientation. (a) Rocking curves provide information on the degree of crystallinity. Full width at the half maximum (FWHM) of a peak observed in the rocking curve of a layer-structured material represents the average tilting angle of the stacking layers. (b and c) Rocking curves observed for the NAFS-2 (b) and NAFS-1 (c) films. (d and e) Schematic images of the layer stacking motif for NAFS-2 (d) and NAFS-1 (e). *Source:* From Yamada et al. 2013 [77]. Reproduced with permission of Royal Society of Chemistry.

estimated to be ≈3° in NAFS-2 (Figure 2.19b). The larger tilting angle than that in NAFS-1 (0.3°) (Figure 2.19c) is attributed to the weaker interaction between the layers (Figure 2.19d), as in NAFS-2 there are no "layer-locking" molecules projecting from the 2D sheets to exercise stereoelectronic control.

2.4.7 Manipulation of In-Plane Molecular Arrangement in MOF Nanosheets

We start with a tetratopic porphyrin MOF nanofilm, NAFS-2, the fabrication of which was described in the previous section and proceed to modify the size

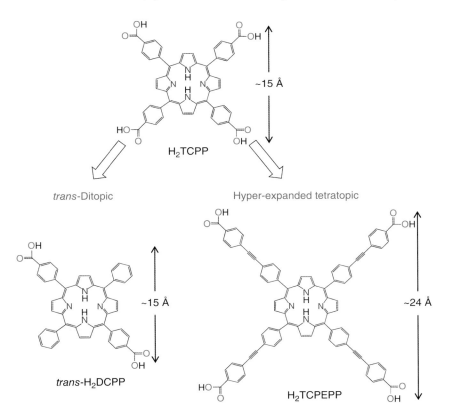

Figure 2.20 Molecular structures of the porphyrin building units, 5,10,15,20-tetrakis(4-carboxyphenyl)-porphyrin (H$_2$TCPP), *trans*-5,15-diphenyl-10,20-di(4-carboxyphenyl)porphine (*trans*-H$_2$DCPP), and 5,10,15,20-tetrakis[4-(4-carboxyphenylethynyl)phenyl]porphine (H$_2$TCPEPP). *Source*: From Makiura et al. 2014 [78]. Reproduced with permission of John Wiley and Sons.

and shape of the framework. In order to achieve this, we employ two specially designed building units, namely a *trans*-ditopic (*trans*-5,15-diphenyl-10,20-di(4-carboxyphenyl)porphine, *trans*-H$_2$DCPP) and an expanded tetratopic porphyrin (5,10,15,20-tetrakis[4-(4-carboxyphenylethynyl)phenyl]porphine, H$_2$TCPEPP), and while we retain the copper(II) ions as linkers (Figure 2.20) [78]. The *trans*-H$_2$DCPP molecule is basically of the same size as the MTCPP porphyrins but it differs in its denticity – it incorporates only two carboxylic substituents in the trans peripheral phenyl parts and can act as a linear ditopic ligand. On the other hand, H$_2$TCPEPP is a tetratopic ligand, but its size is significantly enlarged – the molecular size is now ~24 Å (the distance between the two oxygen atoms of neighboring carboxylic acids), which represents a substantial increase relative to the corresponding size of MTCPP of ~15 Å.

Grazing incidence in-plane GIXRD patterns for *trans*-H$_2$DCPP-Cu nanosheets (NAFS-31) and H$_2$TCPEP-Cu nanosheets (NAFS-41) fabricated on Si(100) substrates are shown in Figure 2.21. The observation of a number of sharp peaks

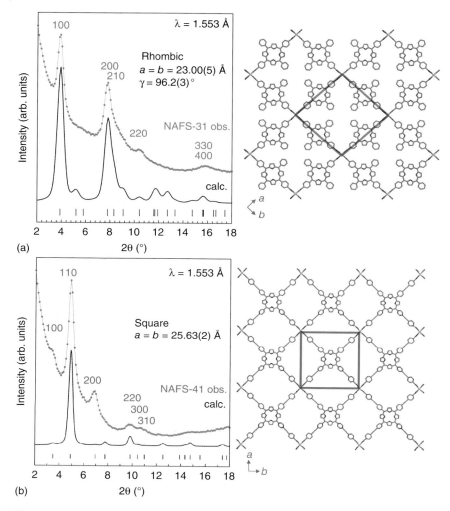

Figure 2.21 Grazing incidence (GI) in-plane synchrotron X-ray diffraction (XRD) patterns (green dots) for NAFS-31 (a) and NAFS-41 (b) nanosheets (left panels). In-plane lattice parameters and indexing of the Bragg peaks are shown in the plots. Derived structural models for NAFS-31 (a) and NAFS-41 (b) are shown in the right panels. Calculated GIXRD patterns (black lines) and reflection positions (black bars) using the structural models are also shown. *Source*: From Makiura et al. 2014 [78]. Reproduced with permission of John Wiley and Sons.

in the profiles indicates that both NAFS-31 and NAFS-41 are highly crystalline in the substrate-plane direction. The NAFS films composed of the tetratopic TCPP units adopt metrically tetragonal structures with in-plane lattice parameters, $a = b = 16.4$–16.6 Å, which coincide with those of comparable bulk crystals composed of TCPP units. For NAFS-31, the observed peak positions of the Bragg reflections cannot be described with a tetragonal unit cell but can be indexed on a metrically rhombic unit cell with basal plane dimensions, $a = b = 23.00(5)$ Å and $\gamma = 96.2(3)°$ (Figure 2.21a). The lattice size is approximately equal to the value obtained by multiplying the lattice dimensions of the NAFS-1 sheets (16.46 Å) by $\sqrt{2}$, corresponding to the basal plane diagonal of the NAFS-1 unit cell. Then the rhombic unit cell can be accounted for by the molecular arrangement shown in the right panel of Figure 2.21a. Four *trans*-H$_2$DCPP molecules are connected via binuclear copper paddle-wheel units, forming a slightly distorted checkerboard motif. The distortion from square to rhombic geometry may be attributed to the reduced number of binding parts in *trans*-H$_2$DCPP. The tetratopic TCPP leads to a symmetric square grid assembly, which reflects the shape of the molecule and the square disposition of its connecting ligands. On the other hand, the ditopic *trans*-H$_2$DCPP in which the connecting carboxylate units are linearly arranged cannot form an undistorted square grid, leading to the development of a small in-plane rhombic distortion.

On the other hand, all observed peaks in the GIXRD profile of NAFS-41 composed of the hyperexpanded tetratopic H$_2$TCPEPP building units index on a metrically tetragonal unit cell with basal plane dimensions, $a = b = 25.63(2)$ Å (Figure 2.21b). This is significantly enlarged compared with the lattice size of NAFS films with MTCPP building units ($a = b = 16.4–16.6$ Å). The symmetry and size of the unit cell can be well accounted for by a structural model (Figure 2.21b, right panel), which incorporates an in-plane structural motif comprising H$_2$TCPEPP units linked by binuclear Cu$_2$(COO)$_4$ paddle-wheel secondary building units and resulting in a similar 2D checkerboard pattern to that encountered in NAFS films composed of MTCPP.

Figure 2.22 summarizes the in-plane molecular arrangements of NAFS-31 and NAFS-41 together with that of NAFS-2 for comparison [78]. Calculated Connolly surfaces – graphical representations of the available empty space obtained with a 1.0 Å Connolly radius – are also included. The size of the in-plane unit cell of NAFS-31 is approximately twice that of NAFS-2 – there are two porphyrin molecular units in the NAFS-31 unit cell. The checkerboard framework of NAFS-2 accommodates a single type of distorted hexagonal pore with an aperture of $\sim$10.5 Å. On the other hand, the absence of half of the binuclear Cu$_2$(COO)$_4$ paddle-wheel connectors in NAFS-31 partially open the framework walls and impose a different shape on the void space – in addition to the distorted hexagonal pores found in NAFS-2, there are elongated narrow-shaped pores, which result from the merging of the smaller pores following the framework wall removal. On the other hand, NAFS-41 and NAFS-2 are isoreticular with the unit cell of NAFS-41 retaining the same square planar topology as in NAFS-2. However, the linear extension of the four organic struts in NAFS-41 is accompanied by a distinct expansion of the unit cell by $\sim$9 Å (25.63 Å in NAFS-41, 16.48 Å in NAFS-2) together with a simultaneous increase in the pore apertures. The dimension of the long diagonal of the pore apertures (the distance between the facing Connolly surfaces) estimated from the in-plane structural model of NAFS-41 is about 19.6 Å – double the size of that in NAFS-2 ($\sim$10.5 Å) – with a void area ($\sim$196 Å^2) four times as large as that of NAFS-2 ($\sim$43 Å^2).

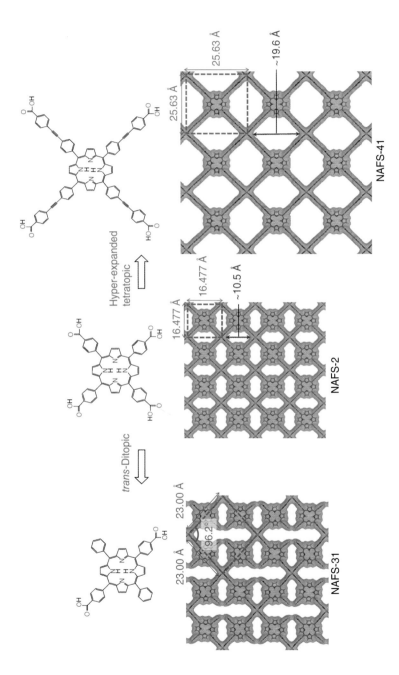

Figure 2.22 Summary of the in-plane structural arrangements in porphyrin-based MOF nanosheets assembled at the air/liquid interface. Molecular building units (top) and corresponding in-plane crystalline structures (bottom) are shown together with the unit cells obtained from synchrotron GIXRD measurements and calculated Connolly surfaces. *Source*: From Makiura et al. 2014 [78]. Reproduced with permission of John Wiley and Sons.

References

1 Ozin, G.A., Arsenault, A., and Cademartiri, L. (2008). *Nanochemistry: A Chemical Approach to Nanomaterials: Edition 2*. RSC Publishing.
2 Roberts, G. (1990). *Langmuir–Blodgett Films*. Springer.
3 Rayleigh, L. (1891). Surface tension. *Nature* 43: 437–439.
4 Pockels, A. (1892). On the relative contamination of the water-surface by equal quantities of different substances. *Nature* 46: 418–419.
5 Langmuir, I. (1917). The constitution and fundamental properties of solids and liquids. II. Liquids. *J. Am. Chem. Soc.* 39: 1848–1906.
6 Blodgett, K.B. (1935). Films built by depositing successive monomolecular layers on a solid surface. *J. Am. Chem. Soc.* 57: 1007–1022.
7 Fujihira, M., Nishiyama, K., and Yamada, H. (1985). Photoelectrochemical responses of optically transparent electrodes modified with Langmuir–Blodgett films consisting of surfactant derivatives of electron donor, acceptor and sensitizer molecules. *Thin Solid Films* 132: 77–82.
8 Pedrosa, J.-M., Romero, M.T.M., and Camacho, L. (2002). Organization of an amphiphilic azobenzene derivative in monolayers at the air-water interface. *J. Phys. Chem. B* 106: 2583–2591.
9 Wang, R., Jiang, L., Iyoda, T. et al. (1996). Investigation of the surface morphology and photoisomerization of an azobenzene-containing ultrathin film. *Langmuir* 12: 2052–2057.
10 Ariga, K. and Kunitake, T. (1998). Molecular recognition at air–water and related interfaces: complementary hydrogen bonding and multisite interaction. *Acc. Chem. Res.* 31: 371–378.
11 Obeng, Y.S. and Bard, A.J. (1991). Langmuir films of C60 at the air–water interface. *J. Am. Chem. Soc.* 113: 6279–6280.
12 Nakamura, T., Tachibana, H., Yumura, M. et al. (1992). Formation of Langmuir-Blodgett films of a fullerene. *Langmuir* 8: 4–6.
13 Long, C.-F., Xu, Y., Guo, F.-X. et al. (1992). Lattice imaging of C60 Langmuir-Blodgett films. *Solid State Commun.* 82: 381–383.
14 Cao, Y., Wei, Z., Liu, S. et al. (2010). High-performance Langmuir–Blodgett monolayer transistors with high responsivity. *Angew. Chem. Int. Ed.* 49: 6319–6323.
15 Kan, J., Chen, Y., Qi, D. et al. (2012). High-performance air-stable ambipolar organic field-effect transistor based on tris(phthalocyaninato) europium(III). *Adv. Mater.* 24: 1755–1758.
16 Murray, D.J., Patterson, D.D., Payamyar, P. et al. (2015). Large area synthesis of a nanoporous two-dimensional polymer at the air/water interface. *J. Am. Chem. Soc.* 137: 3450–3453.
17 Smith, K.M. (1972). *Porphyrins and Metaloporphyrins*. Elsevier.
18 Sheldon, R.A. (1994). *Metalloporphyrins in Catalytic Oxidations*. Marcel Dekker.
19 Kadish, K., Smith, K.M., and Guiard, R. (2003). *The Porphyrin Handbook*, vol. 1999. Academic Press.

20 Lee, S.J. and Hupp, J.T. (2006). Porphyrin-containing molecular squares: design and applications. *Coord. Chem. Rev.* 250: 1710–1723.

21 Suslick, K.S., Bhyrappa, P., Chou, J.-H. et al. (2005). Microporous porphyrin solids. *Acc. Chem. Res.* 38: 283–291.

22 Goldberg, I. (2008). Crystal engineering of nanoporous architectures and chiral porphyrin assemblies. *CrystEngComm* 10: 637–645.

23 Drain, C.M., Varotto, A., and Radivojevic, I. (2009). Self-organized porphyrinic materials. *Chem. Rev.* 109: 1630–1658.

24 Qian, D.-J., Nakamura, C., and Miyake, J. (2001). Spectroscopic studies of the multiporphyrin arrays at the air–water interface and in Langmuir–Blodgett films. *Thin Solid Films* 397: 266–275.

25 Qian, D.-J., Nakamura, C., and Miyake, J. (2000). Multiporphyrin array from interfacial metal-mediated assembly and its Langmuir-Blodgett films. *Langmuir* 16: 9615–9619.

26 Zhang, C.-F., Chen, M., Nakamura, C. et al. (2008). Electrochemically driven generation of manganese(IV,V)-oxo multiporphyrin arrays and their redox properties with manganese(III) species in Langmuir-Blodgett films. *Langmuir* 24: 13490–13495.

27 Qian, D.-J., Nakamura, C., and Miyake, J. (2001). Layer-by-layer assembly of metal-mediated multiporphyrin arrays. *Chem. Commun.* 2312–2313.

28 Liu, B., Qian, D.-J., Chen, M. et al. (2006). Metal-mediated coordination polymer nanotubes of 5,10,15,20-tetrapyridylporphine and tris(4-pyridyl)-1,3,5-triazine at the water–chloroform interface. *Chem. Commun.* 3175–3177.

29 Liu, B., Qian, D.-J., Huang, H.-X. et al. (2005). Controllable growth of well-defined regular multiporphyrin array nanocrystals at the water–chloroform interface. *Langmuir* 21: 5079–5084.

30 Drain, C.M., Nifiatis, F., Vasenko, A., and Batteas, J.D. (1998). Porphyrin tessellation by design: metal-mediated self-assembly of large arrays and tapes. *Angew. Chem. Int. Ed.* 37: 2344–2347.

31 Milic, T.N., Chi, N., Yablon, D.G. et al. (2002). Controlled hierarchical self-assembly and deposition of nanoscale photonic materials. *Angew. Chem. Int. Ed.* 41: 2117–2119.

32 Milic, T.N., Garno, J.C., Batteas, J.D. et al. (2004). Self-organization of self-assembled tetrameric porphyrin arrays on surfaces. *Langmuir* 20: 3974–3983.

33 Yaghi, O.M., O'Keeffe, M., Ockwig, N.W. et al. (2003). Reticular synthesis and the design of new materials. *Nature* 423: 705–714.

34 Férey, G. (2008). Hybrid porous solids: past, present, future. *Chem. Soc. Rev.* 37: 191–214.

35 Kitagawa, S., Kitaura, R., and Noro, S. (2004). Functional porous coordination polymers. *Angew. Chem. Int. Ed.* 43: 2334–2375.

36 Li, Q., Zhang, W., Miljanić, O.Š. et al. (2009). Docking in metal–organic frameworks. *Science* 325: 855–859.

37 Li, J.-R., Sculley, J., and Zhou, H.-C. (2012). Metal–organic frameworks for separations. *Chem. Rev.* 112: 869–932.

38 Yerushalmi, R., Scherz, A., and Van Der Boom, M.E. (2004). Enhancement of molecular properties in thin films by controlled orientation of molecular building blocks. *J. Am. Chem. Soc.* 126: 2700–2701.

39 Hermes, S., Schröder, F., Chelmowski, R. et al. (2005). Selective nucleation and growth of metal–organic open framework thin films on patterned COOH/CF$_3$-terminated self-assembled monolayers on au (111). *J. Am. Chem. Soc.* 127: 13744–13745.

40 Biemmi, E., Scherb, C., and Bein, T. (2007). Oriented growth of the metal organic framework Cu$_3$(BTC)$_2$(H$_2$O)$_3 \cdot x$H$_2$O Tunable with functionalized self-assembled monolayers. *J. Am. Chem. Soc.* 129: 8054–8055.

41 Haruki, R., Sakata, O., Yamada, T. et al. (2008). Structural evaluation of an iron oxalate complex layer grown on an ultra-smooth sapphire (0001) surface by a wet method. *Transactions of the Materials Research Society of Japan* 33: 629–631.

42 Zacher, D., Schmid, R., Wöll, C., and Fischer, R.A. (2010). Surface chemistry of metal–organic frameworks at the liquid–solid interface. *Angew. Chem. Int. Ed.* 50: 176–199.

43 Liu, B., Tu, M., Zacher, D., and Fischer, R.A. (2013). Multi variant surface mounted metal–organic frameworks. *Adv. Funct. Mater.* 23: 3790–3798.

44 Sakata, Y., Furukawa, S., Kondo, M. et al. (2013). Shape-memory nanopores induced in coordination frameworks by crystal downsizing. *Science* 339: 193–196.

45 Xu, G., Yamada, T., Otsubo, K. et al. (2012). Facile "modular assembly" for fast construction of a highly oriented crystalline MOF nanofilm. *J. Am. Chem. Soc.* 134: 16524–16527.

46 Lee, H.J., Cho, Y.J., Cho, W., and Oh, M. (2013). Controlled isotropic or anisotropic nanoscale growth of coordination polymers: formation of hybrid coordination polymer particles. *ACS Nano* 7: 491–499.

47 Falcaro, P., Buso, D., Hill, A.J., and Doherty, C.M. (2012). Patterning techniques for metal organic frameworks. *Adv. Mater.* 24: 3153–3168.

48 Falcaro, P., Ricco, R., Doherty, C.M. et al. (2014). MOF positioning technology and device fabrication. *Chem. Soc. Rev.* 43: 5513–5560.

49 Makiura, R., Motoyama, S., Umemura, Y. et al. (2010). Surface nano-architecture of a metal–organic framework. *Nat. Mater.* 9: 565–571.

50 Makiura, R. and Kitagawa, H. (2010). Porous porphyrin nanoarchitectures on surfaces. *Eur. J. Inorg. Chem.* (24): 3715–3724.

51 Makiura, R., Tsuchiyama, K., and Sakata, O. (2011). Self-assembly of highly crystalline two-dimensional MOF sheets on liquid surfaces. *Cryst. Eng. Comm.* 13: 5538–5541.

52 Makiura, R. and Konovalov, O. (2013). Interfacial growth of large-area single-layer metal–organic framework nanosheets. *Sci. Rep.* 3: 2506–2513.

53 Choi, E.-Y., Barron, P.M., Novotny, R.W. et al. (2009). Pillared porphyrin homologous series: intergrowth in metal–organic frameworks. *Inorg. Chem.* 48: 426–428.

54 Chung, H., Barron, P.M., Novotny, R.W. et al. (2009). Structural variation in porphyrin pillared homologous series: influence of distinct coordination centers for pillars on framework topology. *Cryst. Growth Des.* 9: 3327–3332.

55 Choi, E.-Y., Wray, C.A., Hu, C., and Choe, W. (2009). Highly tunable metal–organic frameworks with open metal centers. *CrystEngComm* 11: 553–555.

56 Makiura, R., Usui, R., Pohl, E., and Prassides, K. (2014). Porphyrin-based coordination polymer composed of layered pillarless two-dimensional networks. *Chem. Lett.* 43: 1161–1163.

57 Makiura, R. and Konovalov, O. (2013). Bottom-up assembly of ultrathin sub-micron size metal–organic framework sheets. *Dalton Trans.* 42: 15931–15936.

58 Decher, G., Hong, J.D., and Schmitt, J. (1992). Buildup of ultrathin multilayer films by a self-assembly process: III. Consecutively alternating adsorption of anionic and cationic polyelectrolytes on charged surfaces. *Thin Solid Films* 210–211: 831–835.

59 Lvov, Y., Decher, G., and Möhwald, H. (1993). Assembly, structural characterization, and thermal behavior of layer-by-layer deposited ultrathin films of poly(vinyl sulfate) and poly(allylamine). *Langmuir* 9: 481–486.

60 Ariga, K., Yamauchi, Y., Rydzek, G. et al. (2014). Layer-by-layer nanoarchitectonics: invention, innovation, and evolution. *Chem. Lett.* 43: 36–68.

61 Richardson, J.J., Cui, J., Björnmalm, M. et al. (2016). Innovation in layer-by-layer assembly. *Chem. Rev.* 116: 14828–14867.

62 Sinha Ray, S. and Okamoto, M. (2003). Polymer/layered silicate nanocomposites: a review from preparation to processing. *Prog. Polym. Sci.* 28: 1539–1641.

63 Kovtyukhova, N.I. (1999). Layer-by-layer assembly of ultrathin composite films from micron-sized graphite oxide sheets and polycations. *Chem. Mater.* 11: 771–778.

64 Lvov, Y., Ariga, K., Ichinose, I., and Kunitake, T. (1995). Assembly of multicomponent protein films by means of electrostatic layer-by-layer adsorption. *J. Am. Chem. Soc.* 117: 6117–6123.

65 Yang, H.C., Aoki, K., Hong, H.-G. et al. (1993). Growth and characterization of metal(II) alkanebisphosphonate multilayer thin films on gold surfaces. *J. Am. Chem. Soc.* 115: 11855–11862.

66 Altman, M., Shukla, A.D., Zubkov, T. et al. (2006). Controlling structure from the bottom-up: structural and optical properties of layer-by-layer assembled palladium coordination-based multilayers. *J. Am. Chem. Soc.* 128: 7374–7382.

67 Haga, M., Kobayashi, K., and Terada, K. (2007). Fabrication and functions of surface nanomaterials based on multilayered or nanoarrayed assembly of metal complexes. *Coord. Chem. Rev.* 251: 2688–2701.

68 Nishihara, H., Kanaizuka, K., Nishimori, Y., and Yamanoi, Y. (2007). Construction of redox- and photo-functional molecular systems on electrode surface for application to molecular devices. *Coord. Chem. Rev.* 251: 2674–2687.

69 Zacher, D., Shekhah, O., Wöll, C., and Fischer, R.A. (2009). Thin films of metal–organic frameworks. *Chem. Soc. Rev.* 38: 1418–1429.

70 Shekhah, O., Wang, H., Kowarik, S. et al. (2007). Step-by-step route for the synthesis of metal–organic frameworks. *J. Am. Chem. Soc.* 129: 15118–15119.

71 Shekhah, O., Wang, H., Strunskus, T. et al. (2007). Layer-by-layer growth of oriented metal organic polymers on a functionalized organic surface. *Langmuir* 23: 7440–7442.

72 Munuera, C., Shekhah, O., Wang, H. et al. (2008). The controlled growth of oriented metal–organic frameworks on functionalized surfaces as followed by scanning force microscopy. *Phys. Chem. Chem. Phys.* 10: 7257–7261.

73 Shekhah, O., Wang, H., Paradinas, M. et al. (2009). Controlling interpenetration in metal–organic frameworks by liquid-phase epitaxy. *Nat. Mater.* 8: 481–484.

74 Shekhah, O., Wang, H., Zacher, D. et al. (2009). Growth mechanism of metal–organic frameworks: insights into the nucleation by employing a step-by-step route. *Angew. Chem. Int. Ed.* 48: 5038–5041.

75 Kanaizuka, K., Haruki, R., Sakata, O. et al. (2008). Construction of highly oriented crystalline surface coordination polymers composed of copper dithiooxamide complexes. *J. Am. Chem. Soc.* 130: 15778–15779.

76 Motoyama, S., Makiura, R., Sakata, O., and Kitagawa, H. (2011). Highly crystalline nanofilm by layering of porphyrin metal–organic framework sheets. *J. Am. Chem. Soc.* 133: 5640–5643.

77 Yamada, T., Otsubo, K., Makiura, R., and Kitagawa, H. (2013). Designer coordination polymers: dimensional crossover architectures and proton conduction. *Chem. Soc. Rev.* 42: 6655–6669.

78 Makiura, R., Usui, R., Sakai, Y. et al. (2014). Towards rational modulation of in-plane molecular arrangements in metal–organic framework nanosheets. *ChemPlusChem* 79: 1352–1360.

3

Molecular Technology for Organic Semiconductors Toward Printed and Flexible Electronics

Toshihiro Okamoto[1,2]

[1] *Graduate School of Frontier Sciences, Department of Advanced Materials Science, The University of Tokyo, 5-1-5 Kashiwanoha, Kashiwa, Chiba, 277-8561, Japan*
[2] *Precursory Research for Embryonic Science and Technology (PRESTO), Japan Science and Technology Agency (JST), 4-1-8 Honcho, Kawaguchi, Saitama, 332-0012, Japan*

3.1 Introduction

Organic semiconductors are of scientific and technological interest because of their potential utilization in printable and flexible electronic applications such as organic field-effect transistors (OFETs) [1], organic light-emitting diodes (OLEDs) [2], and organic photovoltaic cells (OPVs) [3]. The functions of organic semiconductors are generated by intermolecular charge transport between their constituent neighboring π-conjugated molecules and the geometric and energetic features of which essentially determine the device performance [4–6]. Therefore, versatility in material design offers enormous possibilities in building the molecules desired for organic semiconductors, which would facilitate the initiation of practical industrialization.

The general requirements for printed and flexible electronic applications are as follows: (i) high chemical stability under ambient condition (toward oxygen and water), (ii) high carrier mobility (exceeding the performance of conventional amorphous silicon semiconductors (0.5–1.0 $cm^2 V^{-1} s^{-1}$) and to compete with metal-oxide semiconductors), (iii) enough solubility in common organic solvents for the achievement of low-cost solution process technology, and (iv) high thermal durability (highly stabilized crystal phase above at least 150 °C to survive the device fabrication processes, such as curing metallic paints or sealing, for completing such products as display panels) [7]. Essentially, the stability of aggregated form against thermal stress and molecular thermal motion is an indispensable requirement for intermolecular charge transport because the highly fluctuating position of each molecule is similar to the situation in disordered molecular systems [8], where fundamental electronic coherence can be destroyed [9]. Furthermore, in the viewpoint of plant scalability for organic semiconductors, a low-cost and environmentally friendly synthetic route is also required in the future (Figure 3.1).

To develop applicable organic semiconductors, many researchers with various approaches based on their own molecular designs focus on the aforementioned

Molecular Technology: Energy Innovation, Volume 1, First Edition.
Edited by Hisashi Yamamoto and Takashi Kato.
© 2018 Wiley-VCH Verlag GmbH & Co. KGaA. Published 2018 by Wiley-VCH Verlag GmbH & Co. KGaA.

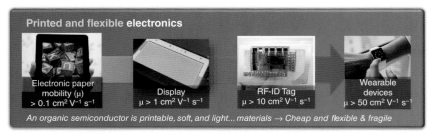

Requirement of next generation organic **semiconductor**

(1) High chemical stability under ambient condition

(2) High carrier mobility (>0.5–1.0 cm^2 V^{-1} s^{-1})

(3) Enough solubility in common organic solvents

(4) High thermal durability (>150 °C)

(5) Plant-scalability

Figure 3.1 Requirements of the next-generation organic semiconductor.

requirements. A great number of organic semiconducting materials based on π-electron-conjugated molecules such as linear and quasilinear acenes and heteroacenes composed of benzene- and/or heterole π-electronic cores (π-cores) have been extensively studied and developed for such applications. Additional features such as luminescence [10] and sensing [1, 11] are fascinating characteristics of organic semiconductors and have been the focus areas in recent research.

This chapter focuses on the historically important small molecular-type semiconducting π-cores, such as acenes, heteroacenes, and the other promising and high-impact organic semiconducting molecules of hole- (p-type) and electron- (n-type) transport characteristics based on *"molecular technology."* Furthermore, the chapter also describes the achievement to the transistor performance of organic semiconductors in polycrystalline- and single-crystalline-based devices along with recently developed solution process technique.

3.2 Molecular Design and Favorable Aggregated Structure for Effective Charge Transport of Organic Semiconductors

In an organic semiconductor consisting of the weak intermolecular interactions such as van der Waals force, charge carriers travel through the overlap of π-orbitals of neighboring molecules, whereas through covalent bonds in an inorganic semiconductor. Therefore, the aggregated structure is crucial to effective charge transport. Considering that an organic semiconducting molecule has π-conjugated skeletons such as benzene and heterole rings, an organic material shows aggregated structures classified to one-dimensional (1D) and two-dimensional (2D) structures, as shown in Figure 3.2. From the theoretical investigation, charges would transport more effectively in 2D aggregated structures such as herringbone and brick-work packing than in 1D structures

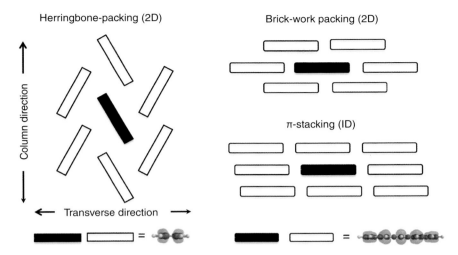

Figure 3.2 Herringbone packing (2D), brick-work packing (2D), and π-stacking (1D).

such as a π-stacking. In the case of herringbone and brick-work packings, one molecule is surrounded by other six and four molecules, which correspond to the effective orbital overlap (transfer integral, t) among the six and four molecules, respectively. On the other hand, in the case of the π-stacking, one molecule is surrounded by only two molecules. It is considered that 2D structure is preferable to 1D structure because a real material solid includes unavoidable defects or contaminations preventing orbital overlaps. Thus, it is important to understand how to construct two-dimensionally ordered aggregated structures, achieving organic semiconductors of high carrier mobility.

3.3 Molecular Design of Linearly Fused Acene-Type Molecules

As organic semiconductors, various aromatic compounds have been investigated. Among them, linearly fused benzene ring systems, called "acenes", such as naphthalene, anthracene, tetracene, and pentacene, are one of the most important and well-studied organic semiconducting materials. As shown in Figure 3.3, acenes have shallow highest occupied molecular orbital (HOMO) levels depending on the number of fused rings, resulting in p-type semiconducting behavior (hole transporting). As the HOMO levels of tetracene and pentacene are closer to the Fermi level of gold electrodes (−5.1 to −4.9 eV), holes can be smoothly injected into their HOMOs.

The aggregated structures of naphthalene [12], anthracene [13], tetracene [14], and pentacene [15] are summarized in Figure 3.4. Notably, all acenes form herringbone packing. These packings might be induced by the difference between the inner and outer electrostatic potential derived from the difference of the electronegativity of carbon and hydrogen atoms, called "C—H··· π interaction" [16]. As is well known, benzene itself is aggregated to the face-to-edge forms

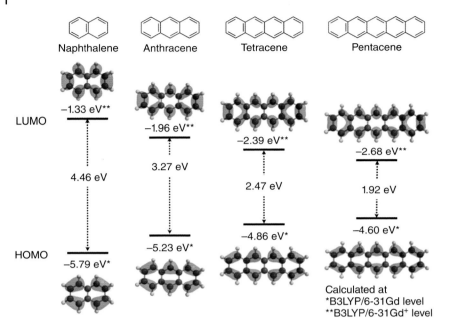

Figure 3.3 HOMO and LUMO levels of a series of acenes (HOMO and LUMO levels are calculated at B3LYP/6-31Gd and B3LYP/6-31Gd$^+$ levels, respectively).

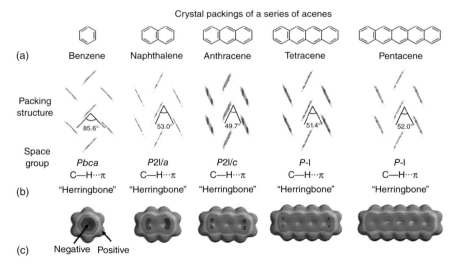

Figure 3.4 (a) Chemical structures of benzene, naphthalene, anthracene, tetracene, and pentacene. (b) Packing structures of these acenes along with their space group in single-crystal structures. (c) Electrostatic potential maps of all acenes in the property range from −100 to +100 kJ mol^{-1} (calculated by SPARTAN 16 program).

by C—H···π interaction in single-crystal state [17, 18]. Acenes, more than naphthalene, show two-plane angles of 49.7° to 53.0° between neighboring molecules in crystal phase. The trend indicates that the balances between the inner and outer electrostatic potentials in the π-core should play a role to form the 2D herringbone packing (according to detailed theoretical investigations, the dispersion interactions are major source of the attraction [19, 20]).

Although early research on conductivity of naphthalene crystals were reported [21], tetracene and pentacene have been well studied as organic semiconductors in chemical, physical, or device engineering fields. In a polycrystalline thin film fabricated by vacuum deposition, tetracene and pentacene exhibited hole mobilities of 0.1 and 3 $cm^2 V^{-1} s^{-1}$, respectively [22, 23]. Compared with polycrystalline films having grain boundaries at which molecules misalign, single crystals could neglect such extrinsic factors to evaluate the intrinsic carrier-conduction in organic semiconductors. Using single crystals grown by physical vapor transport technique [24, 25], the hole mobility of tetracene and pentacene in single-crystal-field-effect transistors (SC-FETs) is 1.4 [26] and over 10 $cm^2 V^{-1} s^{-1}$ [27, 28], which are apparently higher than those in polycrystalline field-effect transistors (FETs). As the most promising tetracene derivatives, 5,6,11,12-tetraphenyltetracene, called "rubrene", whose aggregated structure is also in herringbone packing, exhibited an extremely high hole mobility of up to 40 $cm^2 V^{-1} s^{-1}$ in SC-FETs [29], whereas the polycrystalline thin film of rubrene showed very low performance due to the possible difficulty of formation of oriented polycrystalline films.

3.4 Molecular Technology of π-Conjugated Cores for p-Type Organic Semiconductors

Although tetracene, rubrene, and pentacene are the promising organic semiconducting materials, they are gradually decomposed under ambient condition [30]. Among them, pentacene has a shallow HOMO level of −4.60 eV and a highly reactive site at central 6,13 positions to easily convert the oxidized pentacene such as pentacene quinone. One of the solutions for overcoming this instability is a new π-conjugated core based on *molecular technology*, as shown in Figure 3.5. First, Katz and Bao developed anthra[2,3-*b*:6,7-*b'*]dithiophene (ADT) and anthra[2,3-*b*]thiophene (AT), and tetraceno[2,3-*b*]thiophene (TT) as new stable π-cores [31, 32], which were simply replaced with thiophene ring at the terminal benzene ring. ADT, AT, and TT have slightly deeper HOMO levels of −4.80, −4.98, and −4.70 eV, respectively, and show a substantial reduction of HOMO coefficient at certain reactive sites.

Their aggregated structures form the herringbone packing similar to that of acenes, which results in a high hole mobility of up to 0.5 $cm^2 V^{-1} s^{-1}$ in vacuum-deposited polycrystalline films [32]. These derivatives triggered the synthesis of thiophene and other heterole-containing acenes, namely heteroacene [33]. As typical heteroacenes, the chemical structures and HOMO of bisthieno[2',3':4,5]thieno[3,2-*b*:2',3'-*d*]thiophene ([5]TAc) and thieno[3,2-*b*:4,5-

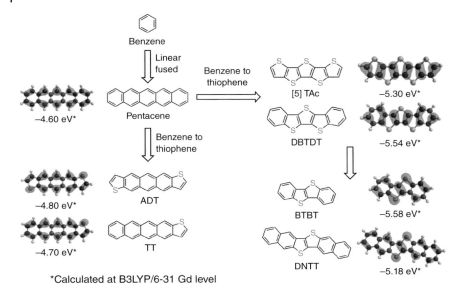

Figure 3.5 Historical molecular design of p-type semiconductors: Chemical structures and HOMO of a series of acenes and heteroacenes.

b']bis[1]benzothiophene (DBTDT) are illustrated in Figure 3.5. The introduction of thiophene rings in the whole or central sites of molecules lead to the deeper HOMO levels of −5.30 and −5.54 eV for [5]TAc and DBTDT, respectively. Both molecules exhibit high chemical stability under ambient condition as well as thermal stability to easily purify them by sublimation. Their HOMO coefficients are located at the double bonds and benzene rings, are not located at the sulfur atoms. This trend is totally different from that of ADT, AT, and TT as described earlier. The packing structures of [5]TAc and DBTDT are shown in Figure 3.6. DBTDT has typical herringbone packing with the tile angle of 56.5° in C—H···π and S···π short contact. On the other hand, [5]TAc has 1D-like herringbone packing (π-stacking) with the large tile angle of 130° and S···S short contact. As described in the section of acenes, since [5]TAc does not have any hydrogen atoms at the molecular longitudinal site and good balance of electrostatic potential map, it prefers π-stacking to herringbone packing. Furthermore, although [5]TAc has the attractive short contacts between sulfur atoms, no HOMO coefficient at sulfur atoms results in the small intermolecular electronic coupling, indicating that the column direction is well overlapped to give a large absolute transfer integral of 172 meV compared with the value of transverse direction of 2 meV. Thus, the behavior of DBTDT single crystals is 1D-type charge transport characteristics. In the case of DBTDT, the absolute transfer integrals (11 and 17 meV) are somewhat small as shown in Figure 3.6 due to the same reason as the case of [5]TAc with no HOMO coefficient at sulfur atoms. The reported hole mobility of [5]TAc is 0.045 cm^2 V^{-1} s^{-1} in polycrystalline thin films [34].

3.4 Molecular Technology of π-Conjugated Cores for p-Type Organic Semiconductors

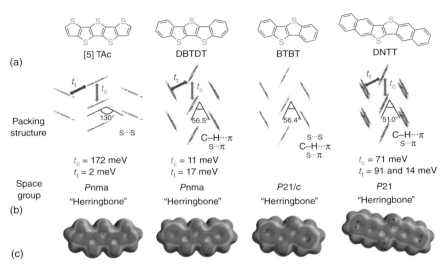

Figure 3.6 (a) Chemical structures of [5]TAc, DBTDT, BTBT, and DNTT. (b) Packing structures of [5]TAc, DBTDT, BTBT, and DNTT along with the absolute transfer integral values in the direction of column and transverse and their space group in single-crystal structures. (c) Electrostatic potential maps of [5]TAc, DBTDT, BTBT, and DNTT in the property range from −100 to +150 kJ mol^{-1} (calculated by SPARTAN 16 program).

To increase the intermolecular orbital overlap between neighboring molecules in the heteroacene system, the molecule having effective HOMO coefficients on sulfur atoms should be constructed based on molecular technology. Takimiya et al. have developed two terminal benzene-fused thienothiophene, [1]benzothieno[1]benzothiophene (BTBT). BTBT exhibits the same electronic structure of the chrysene π-core. Notably, unlike [5]TAc and DBTDT described above, the sulfur atoms have effective HOMO coefficient in BTBT (Figure 3.5). The crystal structure of BTBT [35] forms the typical herringbone packing exhibiting C—H···π, S···π, and S···S short contact at the same time. Other promising molecular technological approaches to enhance orbital coupling is to apply highly π-extended molecules. The phenyl-substituted BTBT (DPh–BTBT) [36] and the π-extended naphthalene-terminated thienothiophene, dinaphtho[2,3-b:2′,3′-f]thieno[3,2-b]thiophene (DNTT), and others were also developed [37, 38]. BTBT and DNTT exhibit excellent chemical stability under ambient condition due to the low-lying HOMO levels of −5.58 and −5.18 eV and form the typical herringbone packing with the tilted angles of 56.4° and 51.0°, respectively. Intermolecular electronic couplings of HOMO (transfer integral, t_{HOMO}) based on the packing structure of DNTT had been evaluated using the Amsterdam Density Functional (ADF) program package [39–41]. Although the absolute transfer integral of DNTT in the column direction is 71 meV, which is fairly large compared with that of DBTDT, the values of DNTT in the transverse direction are 91 and 14 meV, which is an anisotropic transport character. Among them, DNTT have high carrier mobilities of 2.9 and 8.3 cm² V⁻¹ s⁻¹ in polycrystalline and single-crystalline thin films, respectively [37].

3.5 Molecular Technology of Substituents for Organic Semiconductors

3.5.1 Bulky-Type Substituents

As described earlier, although pentacene exhibits hole mobilities of 1–3 cm² V⁻¹ s⁻¹ in vacuum-deposited polycrystalline films [42–46], it is instable under ambient condition due to the existence of the highly reactive site. Moreover, pentacene itself is insoluble in commonly used organic solvents, which makes it difficult to apply solution process to pentacene. To overcome its instability and insolubility issues, Anthony et al. developed the molecular design of the introduction of bulky-type substituents [47–50] such as triisopropylsilylethynyl groups to the pentacene core at the central reactive position [47]. Furthermore, the electron-withdrawing nature of the ethynyl substituent helps electronic delocalization, that is, stability of pentacene π-core. The compound, called triisopropylsilylethynyl-pentacene (TIPS-Pen), was easily synthesized. TIPS-Pen shows high chemical stability even in solution state due to the lower lying HOMO level ($E_{HOMO} = -5.1$ eV, determined by cyclic voltammetry). It has high solubility in common organic solvents, resulting in purification by using column chromatography. Interestingly, TIPS-Pen forms "brick-work" packing as shown in Figure 3.7. As described in the previous section, this

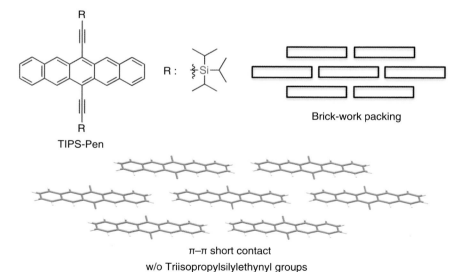

Figure 3.7 Chemical structures and packing structure of TIPS-Pen. (Triisopropylsilylethynyl groups are omitted for clarify.)

packing is expected as one of the two-dimensional charge-transporting systems because four molecules are located surrounding one molecule via effective π-orbital overlaps. In fact, after Anthony's first paper has been published, the mobility of TIPS-Pen was reported to be 0.18–11 cm^2 V^{-1} s^{-1} using several solution processes. Very recently, a new solution process technique "shearing method" developed by Bao et al. induced the more favorable brick-work packing, resulting in a much higher mobility of 11 cm^2 V^{-1} s^{-1} [51, 52]. By means of the molecular technological approach, unexplored more π-extended acene systems, hexacene and heptacene, bearing TIPSE substituents were studied as new field of chemistry [4, 53, 54]. Of course, the molecular technological strategy has been applied to use other heteroacenes such as ADT and TT. For example, triethylsilylethynyl-substituted anthra[2,3-b:6,7-b']dithiophene (TES-ADT) was synthesized and also formed brick-work packing, resulting that it shows more than 1 cm^2 V^{-1} s^{-1} by a simple solution process technique [48].

3.5.2 Linear Alkyl Chain Substituents

The molecular design to produce solution-processable organic semiconductors is the introduction of flexible and solubilizing long alkyl chains into the main π-cores such as a small molecule and a polymer. In addition, such a long alkyl group can act as the driving force for molecular ordering in the solid state with the aid of the van der Waals intermolecular interaction between the alkyl chains. Thus, the long alkyl chains render the semiconducting core to pack tightly, enhancing charge-transporting properties. Katz and coworkers first reported alkyl-substituted ADT, which is soluble in toluene at room temperature, whereas ADT itself is not soluble in the same condition [31]. As other solution-processable organic semiconductors, alkyl-substituted BTBT (C_n-BTBT) with different alkyl chains, from n-C_5H_{11} to n-$C_{14}H_{29}$, were developed with a facile synthesis via two reaction steps using parent BTBT as a starting material. The aggregated structures of C_8-BTBT also form the typical herringbone packing with the tiled angle of 56.4°, C—H···π, and S···π short contacts in single crystal. C_n-BTBTs exhibit excellent solubility in chloroform at room temperature and construct homogenous thin films by spin coating, whose mobility was higher than 10^{-1} cm^2 V^{-1} s^{-1} in polycrystalline thin films [55]. To improve their charge carrier mobility and evaluate their intrinsic mobilities, the single-crystalline films of C_n–BTBT grown by several solution techniques exhibited a higher mobility of greater than 5 cm^2 V^{-1} s^{-1} [56–58]. Furthermore, decyl-substituted DNTT (C_{10}-DNTT) [59] as a π-extended derivative also formed the herringbone packing and showed an excellent mobility of up to 11 cm^2 V^{-1} s^{-1} in solution-crystallized single-crystalline thin films [60], whereas the solubility of C_{10}-DNTT is poor due to the highly π-extension of the π-core (Figure 3.8). Very recently, Hanna et al. developed asymmetrically substituted BTBT molecules, based on the molecular design for liquid crystals, exhibiting high carrier mobility by solution process as well as high thermal tolerability in devices [61].

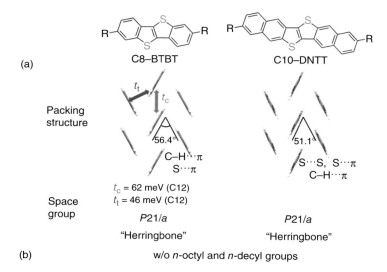

Figure 3.8 (a) Chemical structures of C_8–BTBT and C_{10}–DNTT and (b) packing structures of C_8–BTBT and C_{10}–DNTT along with the absolute transfer integral values (t_c and t_t) in the direction of column and transverse and their space group in single-crystal structures. n-Octyl and n-decyl groups are omitted for clarify.

3.6 Molecular Technology of Conceptually-new Bent-shaped π-Conjugated Cores for p-Type Organic Semiconductors

3.6.1 Bent-Shaped Heteroacenes

As described in the previous section, to achieve effective charge transfer between neighboring molecules in organic semiconductors, it is likely to design the molecules with π-extended electronic cores such as pentacene and DNTT. Introducing triisopropylsilylethynyl groups or long alkyl chains provided TIPS-Pen, C_n-BTBT, and C_n-DNTT with solubility to result in solution-processed crystalline thin films with excellent mobilities of greater than 10 cm² V⁻¹ s⁻¹ [56, 58, 62–65]. Introduction of these substituents, however, gives instability of crystal phases because the molecular design of π-cores with alkyl chains is similar to that for liquid-crystal molecules. The stability of the thin films is relatively poor against temperature, and crystal phases of TIPS-Pen, C_n-BTBT, and C_n-DNTT transit to liquid-crystal phases at around 110–130 °C [55, 66, 67], resulting in the transistor performance degradation at a certain temperature or higher. These properties of the major organic semiconducting materials indicate that it is still difficult to achieve simultaneously sufficient solubility, thermal durability, and high-mobility charge transport with the conventional molecular design strategy of π-extended cores for organic semiconductors.

Recently, our group proposed a conceptually new molecular design of π-extended V-shaped conjugated cores, dinaphtho[2,3-b:2′,3′-d]thiophene (DNT–V) containing a thiophene ring at the central position. DNT–V exhibits

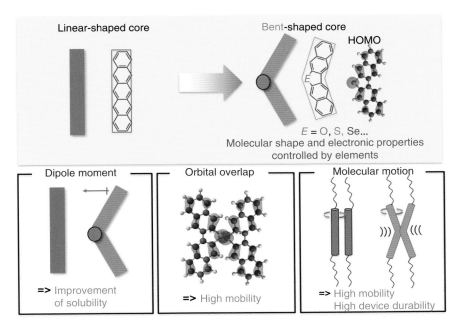

Figure 3.9 Molecular technology of bent-shaped π-electron cores.

inherent solubility because of intramolecular dipole moments and a large orbital coefficient of HOMO on the prominent sulfur atom, possessing potential for high mobility in the desired packing structure. Furthermore, the V-shaped π-core was expected to reduce molecular fluctuation to avoid fatally decreasing intermolecular transfer integrals because a bent molecular structure has larger steric hindrance than do straight π-cores such as pentacene, BTBT, and DNTT that are easily imagined to rotate (Figure 3.9).

A facile and versatile synthetic method for DNT–V and alkyl-substituted derivatives was developed, in which vary the position and length of the alkyl chains attached to the core by four steps from commercially available starting materials without any expensive metal catalysts, indicating their advantageous capability of large-scale production (Figure 3.10).

All derivatives have high chemical and thermal stability. Among them, notably, a series of C_n–DNT–VW exhibits sufficient solubility in nonhalogenated solvents such as toluene and anisole at room temperature and very high phase-transition temperatures of up to 202 °C. From the single-crystal structural analysis of DNT–V derivatives, it is found that the unique "bent structures" induced by intermolecular interactions in the crystal and the resultant herringbone packing structure (Figure 3.11). Then, from the calculations of transfer integrals based on the actual single-crystal structures, it was unveiled that the DNT–V derivatives realize effective two-dimensional orbital overlaps in the herringbone packing (Figure 3.12). Their single-crystalline films can be fabricated by a newly developed solution-crystallization method to properly evaluate the carrier mobility of DNT–VWs in the single-crystalline thin films

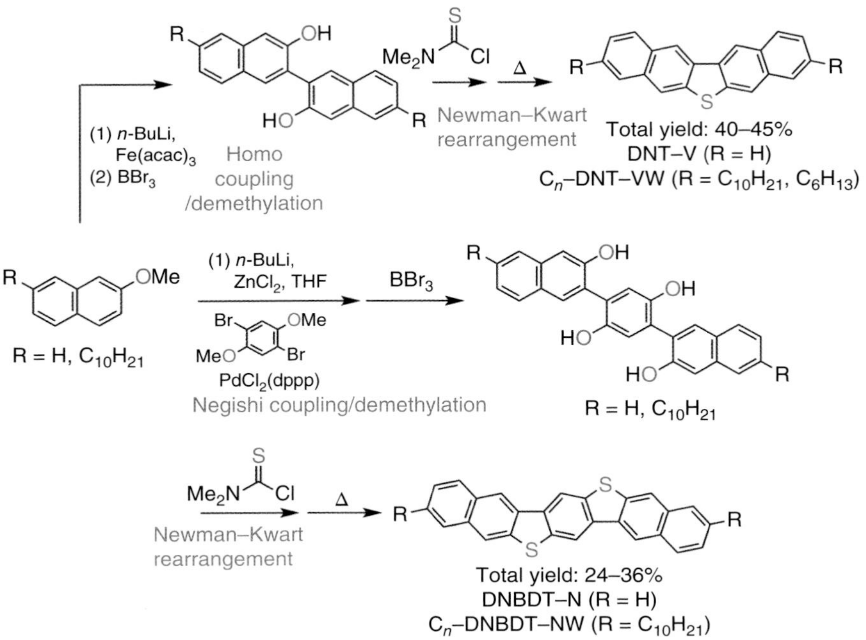

Figure 3.10 Reaction scheme of DNT–V, C$_n$–DNT–VW, DNBDT–N, and C$_n$–DNBDT–NW.

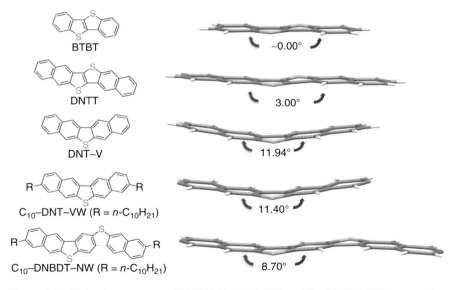

Figure 3.11 Molecular structures of DNT–V, C$_{10}$–DNT–VW, and C$_{10}$–DNBDT–NW compared with those of BTBT and DNTT in single crystals.

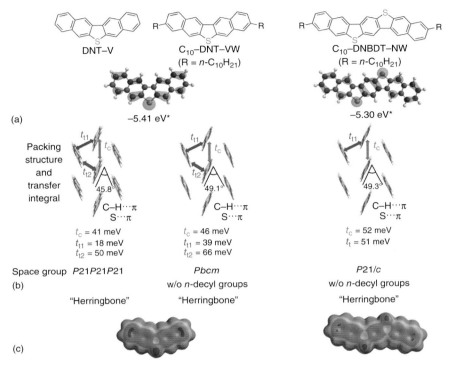

Figure 3.12 (a) Chemical structures of DNT–V, C$_{10}$–DNT–VW, and C$_{10}$–DNBDT–NW, and HOMO of DNT–V and DNBDT–N (R = H). (b) Packing structures of DNT–V, C$_{10}$–DNT–VW, and C$_{10}$–DNBDT–NW along with the absolute transfer integral values (t_c, t_t, t_{t1}, and t_{t2}) in the direction of column and transverse (transfer integrals were calculated at PBEPBE/6-31Gd level) and their space group in single-crystal structures. n-Decyl groups are omitted for clarify. (c) Electrostatic potential maps of DNT–V and DNBDT–N in the property range from −100 to +150 kJ mol^{-1} (calculated by SPARTAN 16 program).

without any influences from extrinsic and random grain boundaries [29, 68, 69]. Among the synthesized compounds, hexyl-substituted DNT–V (C$_6$–DNT–VW) shows a mobility of up to 9.5 cm^2 V^{-1} s^{-1}. Notably, the single-crystal structural analysis and their transport characteristics of single-crystalline films clearly demonstrated the correlation between the mobility and the transfer integrals as a result of good overlapping of the molecular packing geometry (Figure 3.11). Moreover, its transistors are thermally resistant at a temperature of up to 150 °C. This result is outstanding among reported high-mobility organic semiconductors and is desirable for further practical device-fabrication processes [7].

Although the highly potential sulfur-bridged V-shaped derivatives (DNT–V and C$_n$–DNT–VW) have been developed, the following drawbacks of the DNT–V core still remain: (i) a large ionization potential (5.73 eV) of DNT–V because of deficient π-conjugation length, resulting in a large absolute threshold voltage in the FET and (ii) a low molecular weight (MW) of DNT–V (MW 284) and a resultant low sublimation temperature, leading to unsatisfactory thermal durability in the form of thin films. Therefore, it is needed to further optimize the main π-core. It was speculated that the molecular concept of the V-shaped molecule could be extended to the N-shaped molecular structure (N-shaped π-core), dinaphtho[2,3-d:2′,3′-d']benzo[1,2-b:4,5-b']dithiophene (DNBDT–N) and the alkyl-substituted derivatives (C_n–DNBDT–NW) (Figure 3.12). As expected, the ionization potentials of DNBDT and C_{10}–DNBDT–NW ($n = 10$) are 5.45 and 5.24 eV, respectively, whose values are apparently improved because of the π-extended core. In terms of thermal stability, the phase-transition temperature of C_{10}–DNBDT–NW from the crystal phase is evaluated to be greater than 200 °C. The single-crystal structural analysis of the N-shaped molecule, C_{10}–DNBDT–NW, clarified the same attractive intermolecular interactions as C_n–DNT–VW. Thus, the C_{10}–DNBDT–NW molecule also showed the feature of energetically favorable *bent structures* (Figure 3.12) in single crystal. Although the unsubstituted DNBDT–N formed the unfavorable overlap in 1D π-stacking, the simple introduction of long alkyl chains to DNBDT–N core can achieve a herringbone packing structure and effective intermolecular overlap by means of van der Waals interaction among long alkyl chains (Figure 3.12). Notably, solution-crystallized FETs using C_{10}–DNBDT–NW exhibited a mobility of up to 16 $cm^2 V^{-1} s^{-1}$. In addition, as a result of thermal stress tests for the solution-crystallized FETs on polycarbonate substrates, its single-crystalline film shows no obvious performance degradation even after annealing at 200 °C (Figure 3.13). Thus, these sulfur-bridged V- and N-shaped organic semiconductor materials, which are exceptional in satisfying the properties required for printed semiconductor devices, emerge as promising candidates for the printed and flexible electronics industries.

Recently, furan, an oxygen-containing congener, has been of particular interest because of its densely packed nature, which originates from a smaller element size [70–74]. Since such promising compounds were reported, furan-based π-conjugated systems have proved to be potential candidates as organic semiconductors. Our group reported oxygen-incorporated congeners, dinaphtho[2,3-b:2′,3′-d]furan derivatives (DNF–Vs). Notably, alkyl-substituted DNF–V derivatives (C_n–DNF–VW), regardless of the positions of their substituents, exhibited a high hole mobility of greater than 1 $cm^2 V^{-1} s^{-1}$ in single-crystalline thin films through edge-casting [75]. Intriguingly, DNF–Vs show blue-emissive characteristics with a high quantum efficiency ($\Phi_F = 72\%$) even in the solid state [75, 76], whereas the sulfur-bridged congener exhibits very weak photoluminescence with Φ_F of lower than 3%. Single-crystal analysis proved that all derivatives form a typical herringbone packing and good overlaps of the charge-transporting π-cores. These results indicate that an oxygen-bridged V-shaped π-core is less susceptible to molecular displacement in an aggregated structure than its sulfur-bridge counterparts, mainly due to its smaller element size. Furthermore, Okamoto et al. described an oxygen-bridged N-shaped π-electron core, dinaphtho[2,3-d:2′,3′-d']benzo[1,2-b:4,5-b']difuran (DNBDF), as a new entity of organic semiconducting materials. Interestingly, by the introduction of flexible alkyl chains at appropriate positions, DNBDF π-cores exhibit solution processability, a highly stabilized crystal phase, high mobility, and blue luminescence in a solid. Throughout this study, C_{10}–DNBDF–NV exhibits a highly stabilized crystal phase up to 295 °C, blue emission with a moderately

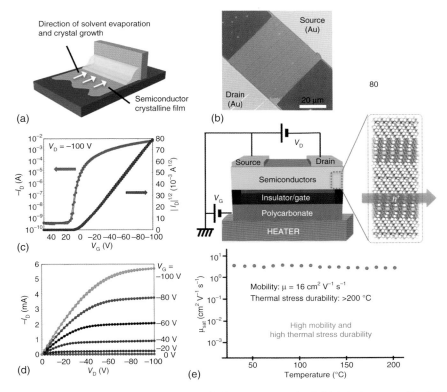

Figure 3.13 (a) Solution-process image of edge-casting method. (b) Solution-crystallized films of C_{10}–DNBDT–NW together with source and drain electrodes. (c and d) Transfer and output characteristics of the representative C_{10}–DNBDT–NW solution-crystallized transistor. (e) Thermal stress durability test of C_{10}–DNBDT–NW solution-crystallized transistor along with its molecular alignment and the device structure for the test.

high quantum yield of 42%, even with flexible long alkyl chains in the solid state, and high carrier mobility of up to 1.8 $cm^2\,V^{-1}\,s^{-1}$ in a solution-crystallized FET [77]. Thus, DNF–Vs and DNBDF–Ns with both high-efficient blue emission and high mobility are candidates for advanced light-emitting transistors rather than simple FETs.

3.7 Molecular Technology for n-Type Organic Semiconductors

n-Type electron transporting organic semiconductors, needed for the function of complementary organic circuits, OPVs, and p–n junction diodes, exhibit challenges such as low mobility, instability in air, poor solubility for efficient film-casting, and large barriers to electron injection with air-stable electrodes such as gold. n-Type semiconducting compounds are rarer than p-type ones due to the nature of conventional π-conjugated systems of the high-lying lowest unoccupied molecular orbital (LUMO) levels. Therefore, the general molecular design

72 | *3 Molecular Technology for Organic Semiconductors*

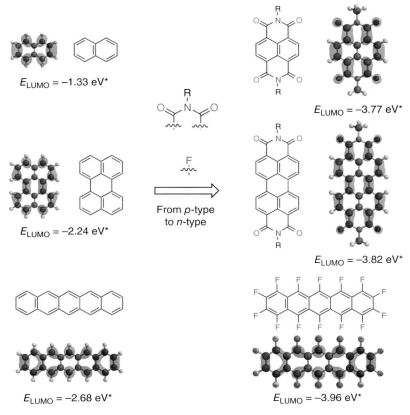

*Calculated at B3LYP/6-31Gd$^+$ level

Figure 3.14 Molecular design of n-type semiconductors: chemical structures and LUMO.

of n-type semiconductor is that acene, arylene, thiophene, and phenylene-based π-conjugated cores are functionalized with electron-withdrawing substituents such as perfluoroalkyl, cyano, fluoro, perfluoroaryl, carbonyl, and diimide groups to delocalize electrons and yield stability and low-lying LUMO [78, 79]. As representatively shown in Figure 3.14, naphthalene diimide (NDI), perylene diimide (PDI), and perfluoropentacene (F_{14}-Pen) functionalizing diimide and fluoro groups effectively have low-lying LUMO levels of −3.77, −3.82, and −3.96 eV, calculated at B3LYP/6-31Gd$^+$ level, respectively. In this section of n-type semiconductors, well-investigated NDI and PDI-based n-type semiconductors and new promising n-type benzobis(thiadiazole)-based semiconductors are described based on the molecular design (Figure 3.15).

3.7.1 Naphthalene Diimide and Perylene Diimide

In terms of device durability under ambient condition, de Leeuw et al. first reported ambient stability in n-type semiconductors in 1997 [80]. As degradation was attributed to the reaction of charge-carrying electrons with oxygen and

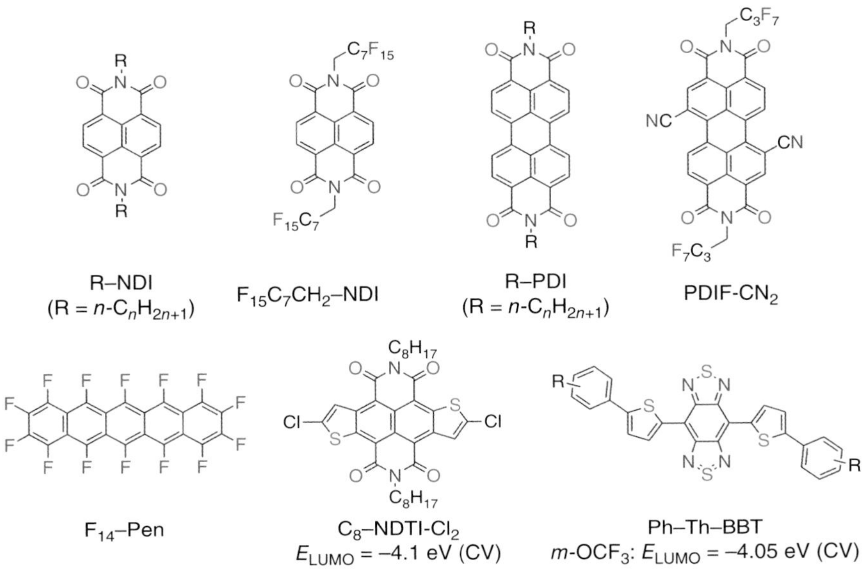

Figure 3.15 Chemical structures of representative n-type semiconductors.

water, several strategies such as lowering LUMO levels of −4.0 eV or introducing physical oxygen/water barriers have been pursued to overcome these issues. Arylene with electron-withdrawing diimide groups affords promising architectures. As one early example of arylene diimide, n-fluoroalkyl-substituted NDIs ($F_{15}C_7CH_2$–NDI) exhibited air-stable electron mobilities of up to 0.1 cm^2 V^{-1} s^{-1} due to the presence of densely packed fluoroalkyl groups, whereas the LUMO level of NDI is higher than −4.0 eV. In sharp contrast, n-alkyl-substituted NDI-based OFETs only operate under inert condition. This result clearly indicates that n-fluoroalkyl groups act as a barrier to oxygen/water penetration [81, 82]. However, depending on the oxygen pressure and film morphology, some of these n-fluoroalkyl-substituted semiconductor-based OFETs degrade over periods of days [83].

Facchetti, Marks, and coworkers demonstrated that functionalized arylene diimide cores with electron-withdrawing substituents such as halogens and cyano groups result in highly electron-deficient n-type semiconductors having low-lying LUMO levels and the packing structure favorable for charge transport [83]. Furthermore, very recently, 2,6-dichloronaphthalene diimide was sublimed in air to give a new polymorph with the two-dimensional brick-work packing (β-phase), whereas the herringbone packing motif (α-phase) is grown from organic solvents in the previous report. Then, their single-crystalline FETs based on α- and β-phase crystals exhibit electron mobilities of up to 8.6 and up to 3.5 cm^2 V^{-1} s^{-1}, respectively [84].

Regarding π-extended PDI derivatives, Horowitz et al. first reported the evidence for electron transport in PDI [85] and then Frisbie et al. reported structural and electrical transport properties of a family of π-stacking soluble organic semiconductors, alkyl-substituted perylene diimides (C_n–PDI) [86]. OFETs based on C_n–PDI-deposited films showed electron mobilities of up to 1.7 cm^2 V^{-1} s^{-1} in vacuum. The strategy of intrinsically air-stable transistors against oxygen and water is realized to stabilize charge-carrying electrons in low-lying LUMO levels. Notably, PDI functionalized with cyano groups (PDIF–CN$_2$) in the bay region exhibits highly air-stable electron mobility of 0.64 cm^2 V^{-1} s^{-1} [87]. The packing structure of PDIF-CN$_2$ forms the brick-work packing exhibiting effective orbital overlaps that take advantage of the feature of the LUMO configuration (Figure 3.16). PDIF–CN$_2$-based transistors can be fabricated from vapor-deposited and solution-processed films [87]. Furthermore, its single-crystal FETs were fabricated by lamination of the semiconductor crystal on Si–SiO$_2$/PMMA-Au gate-dielectric-contact substrates. PDIF–CN$_2$ exhibits electron mobilities of up to 6 and 3 cm^2 V^{-1} s^{-1} in vacuum and air, respectively [88].

Yamashita et al. developed novel n-type materials based on a benzobis(thiazole) (BBT) units having low LUMO level (−4.04 eV, determined by cyclic voltammetry) showing high mobility (0.77 cm^2 V^{-1} s^{-1}) and good air stability [89]. Very recently, Tokito et al. further addressed the development of soluble n-type organic semiconductors based on BBT by introducing a novel molecular design of the simple substitution of trifluoromethyl or trifluoromethoxy groups at the *meta* positions of phenyl terminal units to support sufficient solubility, creating suitable LUMO energy levels and high crystallinity. These newly designed BBT-based molecules showed electron mobilities as high as 0.61 cm^2 V^{-1} s^{-1} in solution-processed OFETs [90].

Summary and Future Outlook for Molecular Technology of Organic Semiconductors
In this chapter, the author has focused on representative p- and n-type organic semiconductors mainly for the application of OFETs. The acene- and heteroacene-based molecules for p-type semiconductors and the acenes and heteroacenes exhibiting strong electron-withdrawing groups such as diimide and fluoro groups and so on for n-type semiconductors are described. Their electronic structures such as the configuration of HOMOs and LUMOs, their energy levels, and electrostatic potential maps can be facilely visualized and understood by theoretical calculations. The packing structures of acenes and heteroacenes form either one-dimensional π-stack or two-dimensional herringbone-type packing, and the latter case is favorable for charge transport. As discussed several organic semiconducting molecules in the aggregated form, the degree of their orbital overlaps depends on the configuration of HOMO or LUMO as well as their packings. Before the transistor evaluation, the orbital overlaps in the aggregated form could be quantitatively understood by calculating transfer integrals between the neighboring molecules, facilitating the screening of highly potential organic semiconductors in this stage. Such a recent research via *molecular technology* obtained by understanding the relationship between the aggregated structure and carrier transport is crucial for accelerating development of next-generation superior organic semiconductors. As one example, very recently, outstanding results on p-type semiconducting molecules is that conceptually new bent-shaped π-cores show a mobility of more than 10 cm^2 V^{-1} s^{-1}

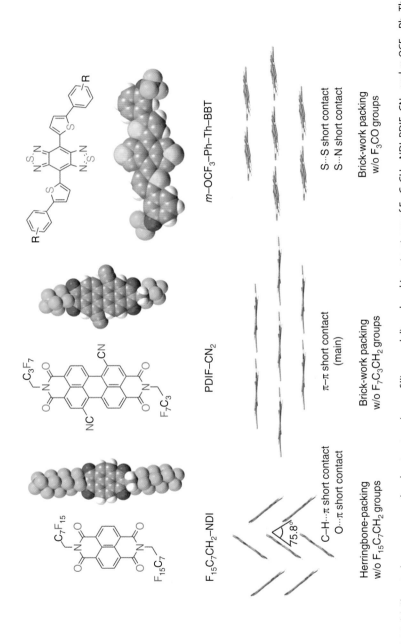

Figure 3.16 (a) Chemical structures, molecular structure (space filling model), and packing structures of $F_{15}C_7CH_2$–NDI, PDIF–CN_2, and m-OCF_3–Ph–Th–BBT. $F_{15}C_7CH_2$, $F_7C_3CH_2$, and m-OCF_3 groups for each molecule are omitted for clarity, respectively.

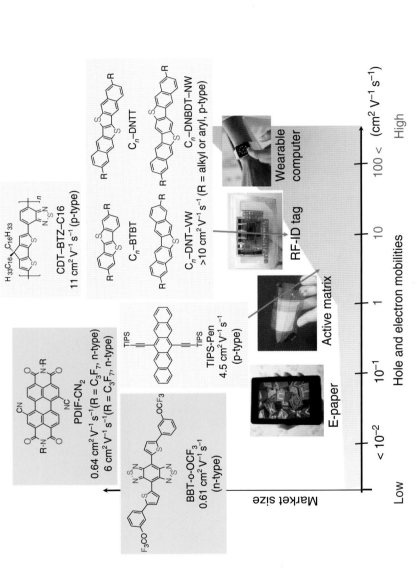

Figure 3.17 Hole and electron mobilities vs market size. Representative solution-processable organic semiconductors along with high performance and high chemical stability.

in solution-crystallized thin-film FETs along with highly crystalized phase of more than 200 °C. This result is ideal for further practical device-fabrication processes as well as for next-generation organic semiconductors. On the other hand, the mobility of n-type semiconductors is still one order lower than that of p-type semiconductors, whereas several types of n-type semiconductors have been recently reported. Therefore, toward high-electron-mobility organic semiconductors, conceptually new molecules and aggregated forms based on *molecular technology* should be emergently proposed and investigated for realizing future printable and flexible electronics applications (Figure 3.17).

References

1 Bao, Z. and Locklin, J. (2007). *Organic Field-Effect Transistors*, 1ste. Florida: CRC Press.
2 Müllen, K. and Scherf, U. (2006). *Organic Light-Emitting Devices: Synthesis, Properties and Applications*. Wiley-VCH.
3 Sun, S.-S. and Sariciftci, N.S. (2005). *Organic Photovoltaics: Mechanism, Materials, and Devices*. CRC Press.
4 Anthony, J.E. (2006). *Chem. Rev.* 106: 5028.
5 Coropceanu, V., Cornil, J., da Silva Filho, D.A. et al. (2007). *Chem. Rev.* 107: 926.
6 Takimiya, K., Shinamura, S., Osaka, I., and Miyazaki, E. (2011). *Adv. Mater.* 23: 4347.
7 Kuribara, K., Wang, H., Uchiyama, N. et al. (2012). *Nat. Commun.* 3: 723.
8 Troisi, A., Cheung, D.L., and Andrienko, D. (2009). *Phys. Rev. Lett.* 102: 116602.
9 Uemura, T., Yamagishi, M., Soeda, J. et al. (2012). *Phys. Rev. B* 85: 035313.
10 Shinar, J. (2004). *Organic Light-Emitting Devices: A Survey*. Springer.
11 Lin, P. and Yan, F. (2012). *Adv. Mater.* 24: 34.
12 Capelli, S.C., Albinati, A., Mason, S.A., and Willis, B.T.M. (2006). *J. Phys. Chem. A* 110: 11695.
13 Lusi, M., Vitorica-Yrezabal, I.J., and Zaworotko, M.J. (2015). *Cryst. Growth Des.* 15: 4098.
14 Holmes, D., Kumaraswamy, S., Matzger, A.J., and Vollhardt, K.P.C. (1999). *Chem. Eur. J.* 5: 3399.
15 Mattheus, C.C., Dros, A.B., Baas, J. et al. (2001). *Acta Crystallogr. C* 57: 939.
16 Nishio, M., Hirota, M., and Umezawa, Y. (1998). *The CH/π Interaction: Evidence, Nature, and Consequences*, vol. 1998. New York: Wiley-VCH.
17 Cox, E.G. and Smith, J.A.S. (1954). *Nature* 173: 75.
18 Nayak, S.K., Sathishkumar, R., and Row, T.N.G. (2010). *CrystEngComm* 12: 3112.
19 Tsuzuki, S., Honda, K., Uchimaru, T. et al. (2002). *J. Am. Chem. Soc.* 124: 104.
20 Tsuzuki, S., Honda, K., Uchimaru, T., and Mikami, M. (2004). *J. Chem. Phys.* 120: 647.
21 Warta, W., Stehle, R., and Karl, N. (1985). *Appl. Phys. A: Solids Surf.* 36: 163.
22 Klauk, H., Halik, M., Zschieschang, U. et al. (2002). *J. Appl. Phys.* 92: 5259.

23 Gundlach, D.J., Nichols, J.A., Zhou, L., and Jackson, T.N. (2002). *Appl. Phys. Lett.* 80: 2925.
24 Kloc, C., Simpkins, P.G., Siegrist, T., and Laudise, R.A. (1997). *J. Cryst. Growth* 182: 416.
25 Laudise, R.A., Kloc, C., Simpkins, P.G., and Siegrist, T. (1998). *J. Cryst. Growth* 187: 449.
26 Goldmann, C., Haas, S., Krellner, C. et al. (2004). *J. Appl. Phys.* 96: 2080.
27 Jurchescu, O.D., Baas, J., and Palstra, T.T.M. (2004). *Appl. Phys. Lett.* 84: 3061.
28 Jurchescu, O.D., Popinciuc, M., van Wees, B.J., and Palstra, T.T.M. (2007). *Adv. Mater.* 19: 688.
29 Takeya, J., Yamagishi, M., Tominari, Y. et al. (2007). *Appl. Phys. Lett.* 90: 102120.
30 Maliakal, A., Raghavachari, K., Katz, H. et al. (2004). *Chem. Mater.* 16: 4980.
31 Laquindanum, J.G., Katz, H.E., and Lovinger, A.J. (1998). *J. Am. Chem. Soc.* 120: 664.
32 Tang, M.L., Okamoto, T., and Bao, Z. (2006). *J. Am. Chem. Soc.* 128: 16002.
33 Yamada, K., Okamoto, T., Kudoh, K. et al. (2007). *Appl. Phys. Lett.* 90: 072102.
34 Xiao, K., Liu, Y., Qi, T. et al. (2005). *J. Am. Chem. Soc.* 127: 13281.
35 Vyas, V.S., Gutzler, R., Nuss, J. et al. (2014). *CrystEngComm* 16: 7389.
36 Takimiya, K., Ebata, H., Sakamoto, K. et al. (2006). *J. Am. Chem. Soc.* 128: 12604.
37 Yamamoto, T. and Takimiya, K. (2007). *J. Am. Chem. Soc.* 129: 2224.
38 Niimi, K., Shinamura, S., Osaka, I. et al. (2011). *J. Am. Chem. Soc.* 133: 8732.
39 ADF2008.01;SCM *Theoretical Chemistry.* Amsterdam, The Netherlands: Vrije Universiteit http://www.scm.com (last accessed 23 December 2017).
40 Senthilkumar, K., Grozema, F.C., Bickelhaupt, F.M., and Siebbeles, L.D.A. (2003). *J. Chem. Phys.* 119: 9809.
41 Prins, P., Senthilkumar, K., Grozema, F.C. et al. (2005). *J. Phys. Chem. B* 109: 18267.
42 Gundlach, D.J., Lin, Y.Y., Jackson, T.N. et al. (1997). *IEEE Electron Device Lett.* 18: 87.
43 Lin, Y.Y., Gundlach, D.J., Nelson, S.F., and Jackson, T.N. (1997). *IEEE Electron Device Lett.* 18: 606.
44 Klauk, H., Halik, M., Zschieschang, U. et al. (2003). *Appl. Phys. Lett.* 82: 4175.
45 Kelley, T.W., Boardman, L.D., Dunbar, T.D. et al. (2003). *J. Phys. Chem. B* 107: 5877.
46 Masatoshi, K. and Yasuhiko, A. (2008). *J. Phys. Condens. Matter* 20: 184011.
47 Anthony, J.E., Brooks, J.S., Eaton, D.L., and Parkin, S.R. (2001). *J. Am. Chem. Soc.* 123: 9482.
48 Payne, M.M., Parkin, S.R., Anthony, J.E. et al. (2005). *J. Am. Chem. Soc.* 127: 4986.
49 Dickey, K.C., Anthony, J.E., and Loo, Y.L. (2006). *Adv. Mater.* 18: 1721.
50 Park, S.K., Jackson, T.N., Anthony, J.E., and Mourey, D.A. (2007). *Appl. Phys. Lett.* 91: 063514.
51 Diao, Y., Tee, B.C.K., Giri, G. et al. (2013). *Nat. Mater.* 12: 665.
52 Giri, G., Park, S., Vosgueritchian, M. et al. (2014). *Adv. Mater.* 26: 487.

References

53 Lee, J., Bruzek, M.J., Thompson, N.J. et al. (2013). *Adv. Mater.* 25: 1445.
54 Purushothaman, B., Parkin, S.R., and Anthony, J.E. (2010). *Org. Lett.* 12: 2060.
55 Ebata, H., Izawa, T., Miyazaki, E. et al. (2007). *J. Am. Chem. Soc.* 129: 15732.
56 Uemura, T., Hirose, Y., Uno, M. et al. (2009). *Appl. Phys. Express* 2: 111501.
57 Yuan, Y., Giri, G., Ayzner, A.L. et al. (2014). *Nat. Commun.* 5: 3005.
58 Minemawari, H., Yamada, T., Matsui, H. et al. (2011). *Nature* 475: 364.
59 Kang, M.J., Doi, I., Mori, H. et al. (2011). *Adv. Mater.* 23: 1222.
60 Nakayama, K., Hirose, Y., Soeda, J. et al. (2011). *Adv. Mater.* 23: 1626.
61 Iino, H., Usui, T., and Hanna, J.-i. (2015). *Nat. Commun.* 6: 6828.
62 Minari, T., Kano, M., Miyadera, T. et al. (2008). *Appl. Phys. Lett.* 92: 173301.
63 Kano, M., Minari, T., and Tsukagoshi, K. (2010). *Appl. Phys. Express* 3: 051601.
64 Giri, G., Verploegen, E., Mannsfeld, S.C.B. et al. (2011). *Nature* 480: 504.
65 Liu, C., Minari, T., Lu, X. et al. (2011). *Adv. Mater.* 23: 523.
66 Kuwabara, H., Ikeda, M., and Takimiya, K. (2010). Field effect transistors provided with condensed heterocyclic aromatic compounds, WO 2010/098372 A1.
67 Chen, J., Anthony, J., and Martin, D.C. (2006). *J. Phys. Chem. B* 110: 16397.
68 Sundar, V.C., Zaumseil, J., Podzorov, V. et al. (2004). *Science* 303: 1644.
69 Reese, C., Chung, W.-J., Ling, M.-m. et al. (2006). *Appl. Phys. Lett.* 89: 202108.
70 Mitsui, C., Soeda, J., Miwa, K. et al. (2012). *J. Am. Chem. Soc.* 134: 5448.
71 Niimi, K., Mori, H., Miyazaki, E. et al. (2012). *Chem. Commun.* 48: 5892.
72 Tsuji, H., Shoyama, K., and Nakamura, E. (2012). *Chem. Lett.* 41: 957.
73 Watanabe, M., Su, W.-T., Chang, Y.J. et al. (2013). *Chem. Asian J.* 8: 60.
74 Gidron, O., Dadvand, A., Sheynin, Y. et al. (2011). *Chem. Commun.* 47: 1976.
75 Nakahara, K., Mitsui, C., Okamoto, T. et al. (2014). *Chem. Commun.* 50: 5342.
76 Mitsui, C., Kubo, W., Tanaka, Y. et al. (2017). *Chem. Lett.* 46: 338.
77 Mitsui, C., Tanaka, Y., Tanaka, S. et al. (2016). *RSC Adv.* 6: 28966.
78 Naraso, H., Nishida, J.-I., Kumaki, D. et al. (2006). *J. Am. Chem. Soc.* 128: 9598.
79 Newman, C.R., Frisbie, C.D., da Silva Filho, D.A. et al. (2004). *Chem. Mater.* 16: 4436.
80 de Leeuw, D.M., Simenon, M.M.J., Brown, A.R., and Einerhand, R.E.F. (1997). *Synth. Met.* 87: 53.
81 Katz, H.E., Lovinger, A.J., Johnson, J. et al. (2000). *Nature* 404: 478.
82 Bao, Z., Lovinger, A.J., and Brown, J. (1998). *J. Am. Chem. Soc.* 120: 207.
83 Jones, B.A., Facchetti, A., Wasielewski, M.R., and Marks, T.J. (2007). *J. Am. Chem. Soc.* 129: 15259.
84 He, T., Stolte, M., Burschka, C. et al. (2015). *Nat. Commun.* 6: 5954.
85 Horowitz, G., Kouki, F., Spearman, P. et al. (1996). *Adv. Mater.* 8: 242.
86 Chesterfield, R.J., McKeen, J.C., Newman, C.R. et al. (2004). *J. Phys. Chem. B* 108: 19281.
87 Jones, B.A., Ahrens, M.J., Yoon, M.-H. et al. (2004). *Angew. Chem. Int. Ed.* 43: 6363.
88 Molinari, A.S., Alves, H., Chen, Z. et al. (2009). *J. Am. Chem. Soc.* 131: 2462.
89 Kono, T., Kumaki, D., Nishida, J.-I. et al. (2010). *Chem. Commun.* 46: 3265.
90 Mamada, M., Shima, H., Yoneda, Y. et al. (2015). *Chem. Mater.* 27: 141.

4

Design of Multiproton-Responsive Metal Complexes as Molecular Technology for Transformation of Small Molecules

Shigeki Kuwata

Tokyo Institute of Technology, School of Materials and Chemical Technology, Department of Chemical Science and Engineering, 2-12-1-E4-1 Ookayama, Meguro-ku, Tokyo, 152-8552, Japan

4.1 Introduction

The highly efficient and selective catalysis by metalloenzymes in nature relies on the integrated functional groups in their active site. The metal and surrounding peptide residues and other prosthetic groups cooperate in substrate binding, activation, and transformation through multiple weak, noncovalent interactions including hydrogen bonds [1]. Such sophisticated biological catalysis has motivated considerable efforts of synthetic chemists to design new polydentate ligands that allow to locate proton-responsive sites at appropriate positions around the metal center [2, 3]. These ligands are directed not only to emulate the structures and functions of the metalloenzymes but to develop novel man-made catalysts. This chapter aims to discuss the future prospects of multiproton-responsive complexes thus designed as a molecular technology for chemical transformation of inert molecules, emphasizing contributions from our research group. First, some representative biological systems, where the participation of the functional groups around the metal center is evident or suggested in their catalysis, are portrayed. The following sections are dedicated to the synthesis and reactivities of bioinspired complexes with two or more proton-responsive sites fixed around the metal by rigid chelate scaffolds such as pincer-type and tripodal ligands. Chemical transformations with this class of compounds are also described.

4.2 Cooperation of Metal and Functional Groups in Metalloenzymes

The active site of metalloenzymes generally contains versatile proton-responsive units ligating to the metal or residing in the proximity of the metal without direct bonding (i.e. in the second coordination sphere). In most cases, these functional groups are essential components in the catalysis. This section briefly describes the catalytic roles of the proton-responsive functional groups in some representative metalloenzymes.

Molecular Technology: Energy Innovation, Volume 1, First Edition.
Edited by Hisashi Yamamoto and Takashi Kato.
© 2018 Wiley-VCH Verlag GmbH & Co. KGaA. Published 2018 by Wiley-VCH Verlag GmbH & Co. KGaA.

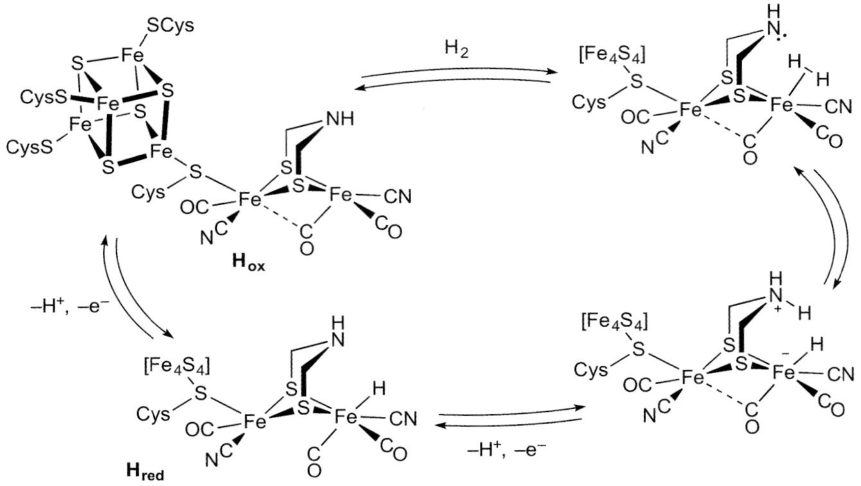

Scheme 4.1 Catalytic cycle proposed for [FeFe] hydrogenases.

4.2.1 [FeFe] Hydrogenase

Hydrogenases are metalloenzymes that transform protons into molecular hydrogen reversibly at remarkable reaction rates for using hydrogen as a source of low-potential electrons [4, 5]. Hydrogenases are classified by the metal composition in their active site; [FeFe] hydrogenases feature an azadithiolato-bridged FeFe core with a dangling Fe_4S_4 cubane-type unit (Scheme 4.1). In the catalytic cycle, the Fe atom distal to the cubane unit in H_{ox} first undergoes coordination of dihydrogen. Subsequent heterolytic cleavage of the H—H bond would be assisted by the Brønsted basic amine moiety, which is placed appropriately in the second coordination sphere thanks to the chelate structure. Finally, protons and electrons are released to regenerate H_{ox}. As each step including the proton shuttling with the pendant amine is reversible, both H_2 production and H_2 oxidation are catalyzed by the [FeFe] hydrogenases, which are often more active in H_2 production.

4.2.2 Peroxidase

Peroxidases are heme-containing enzymes that catalyze oxidation of various substrates with hydrogen peroxide, a by-product from aerobic metabolism [6–8]. A currently accepted mechanism is depicted in Scheme 4.2. The uncoordinated, distal histidine operates as a Brønsted acid-base catalyst to promote proton migration in the coordinated hydrogen peroxide (**1** to **2**). Following heterolytic cleavage of the O—O bond leads to the formation of a high-valent oxo intermediate known as compound I, which then oxidizes the substrate. This step can be recognized as a proton-coupled electron transfer (PCET) from the histidine and heme to the hydrogen peroxide substrate. The arginine residue in the second coordination sphere additionally facilitates the coordination of

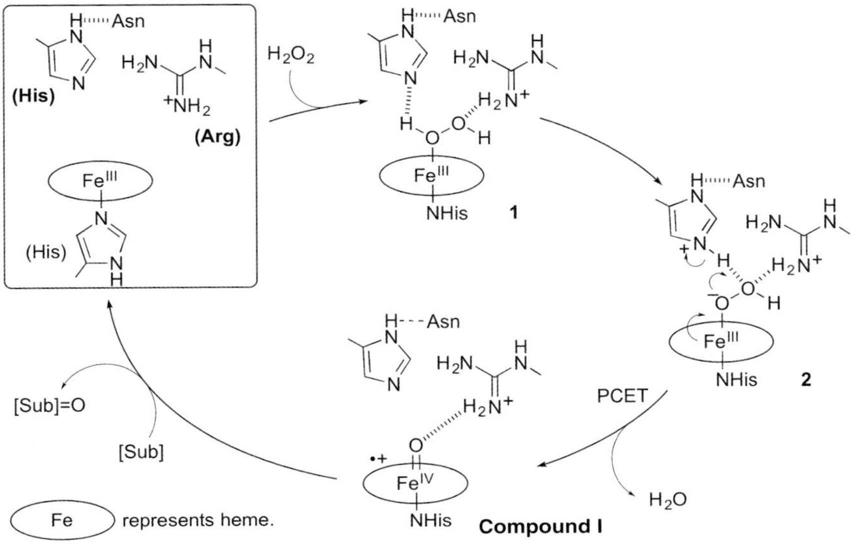

Scheme 4.2 Mechanism of peroxidases.

hydrogen peroxide and stabilizes the intermediate **2** and compound I through hydrogen bonds. The mechanism nicely illustrates the participation of plural protic groups in the chemical transformation without their direct coordination.

4.2.3 Nitrogenase

While industrial production of ammonia has been performed by the Haber-Bosch process at high temperature and pressure, nitrogenase enzymes in nature fix atmospheric nitrogen at room temperature according to Scheme 4.3a, in a catalytic manner [9, 10]. The molybdenum-dependent nitrogenases contain an Fe_7MoS_9C cluster unit known as the FeMo cofactor (FeMo-co) in their active site (Scheme 4.3b). Although the mechanism of the reduction of N_2 to ammonia on the FeMo-co is not yet fully understood, initial binding of N_2 to Fe sites is strongly implicated. A leading mechanism recently proposed is illustrated in Scheme 4.3c [11]. Consecutive fourfold e^-/H^+ transfer to the FeMo-co first takes place to give the E_4 intermediate having two bridging hydrides and two protic hydrosulfido ligands. Subsequent reductive elimination of H_2 would generate an electron-rich iron center, which binds dinitrogen through π-back donation. The π-acidity of the N_2 ligand may be further enhanced by the neighboring hydrosulfido protons, and at the same time, the π-back donation increases the Brønsted basicity of the N_2 ligand. Ultimately, PCET from the FeMo-co to chemically inert dinitrogen occurs to afford the diazene intermediate $E_4(N_2H_2)$. This species would further undergo multiple e^-/H^+ transfer to produce ammonia as the final product. This mechanism implies the significance of the metal-ligand cooperation to avoid high barriers on the reaction coordinate, which are often encountered in challenging chemical transformations such as dinitrogen reduction.

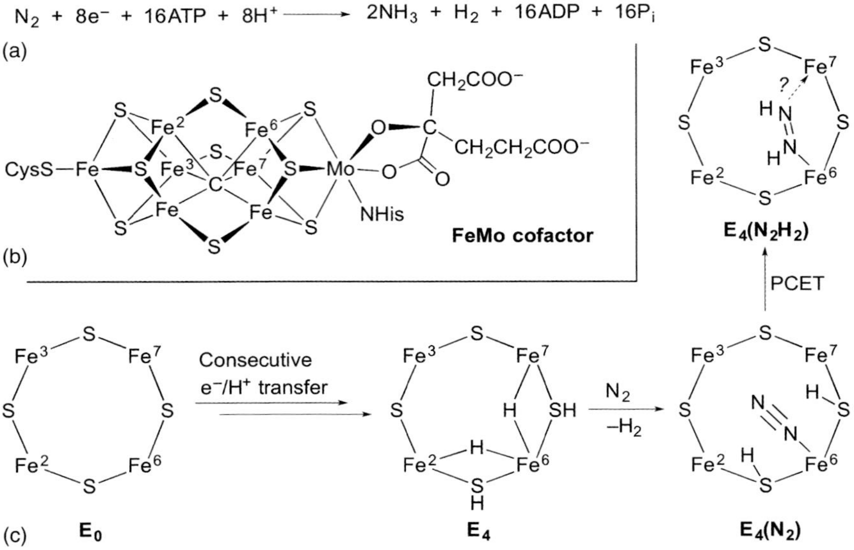

$$N_2 + 8e^- + 16ATP + 8H^+ \longrightarrow 2NH_3 + H_2 + 16ADP + 16P_i$$

Scheme 4.3 (a) Enzymatic nitrogen fixation. (b) Structure of FeMo-co in molybdenum-dependent nitrogenases. (c) Proposed mechanism for reduction of N_2 on FeMo-co. E_n denotes the FeMo-co, where n gives the number of electrons transferred.

4.3 Proton-Responsive Metal Complexes with Two Appended Protic Groups

Multidentate ligands furnished with two or more proton-responsive pendant groups effectively introduce multifunctionality around the metal, which is found in the active site of metalloenzymes. The numbers, acidity, and relative orientation of the appended functional groups built in the ligand should greatly affect the properties of the crafted molecular architecture. This section focuses on bidentate and tridentate ligands furnished with two protic pendant groups. Protic pyrazole [12, 13] and 2-hydroxypyridine [14, 15] are mainly discussed as the appended functional group because these units with some resemblance to biomolecules can locate a reasonably acidic hydrogen atom in the second coordination sphere. In addition, their delocalized π-system would allow a facile and bidirectional proton transfer that exerts a significant impact on the metal-centered reactivity through smooth switching between a dative L-type and monoanionic X-type modes (Scheme 4.4). The metal-ligand cooperating reactivities of the proton-responsive complexes are also described.

4.3.1 Pincer-Type Bis(azole) Complexes

The meridional, tridentate pincer-type ligands have been used as templates for predictable construction of rigid coordination environments with notable stability [16]. 2,6-Bis(pyrazol-3-yl)pyridines **3**, which furnish two proton-responsive pendant arms, are thus attractive modules to construct a metal-ligand cooperating platform [17–19].

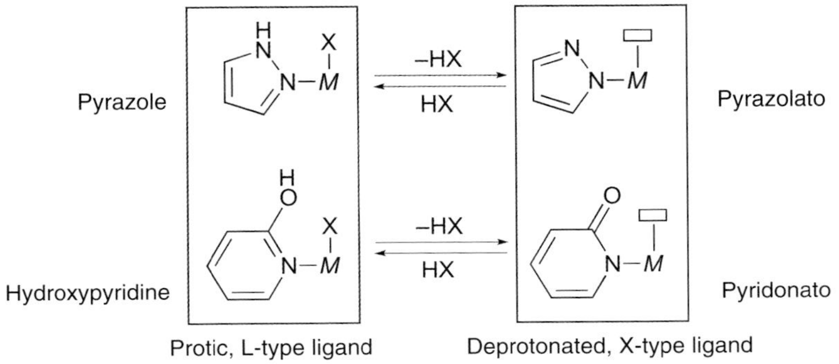

Scheme 4.4 Reversible deprotonation of pyrazole and hydroxypyridine ligands.

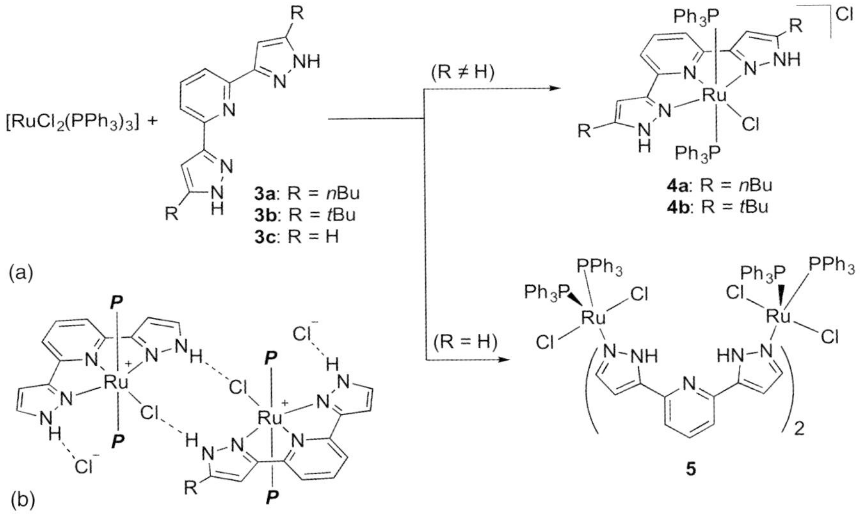

Scheme 4.5 (a) Synthesis of ruthenium complexes **4** and **5** bearing protic NNN pincer-type ligands **3**. (b) Hydrogen bond interactions in protic pincer-type complexes **4**. $P = PPh_3$.

Thiel [20] and we [21] independently demonstrated that the bis(pyrazolyl) pyridines **3a** and **3b** reacted with $[RuCl_2(PPh_3)_3]$ to give the pincer-type complexes **4** with two protic NH groups at the position β to the metal (Scheme 4.5a). Crystallographic study revealed the presence of multiple hydrogen bonds depicted in Scheme 4.5b, suggesting the Brønsted acidic nature of the NH groups in the second coordination sphere. Interestingly, the analogous reaction of the protic pincer-type ligand **3c** without substituents at the 5-positions of the pyrazole rings affords the dinuclear complex **5**, where two molecules of **3c** bridge the two metal atoms with an unexpected tautomeric form. The selective formation of **5** may be rationalized by the reduced steric hindrance at the pyrazole nitrogen atoms distal to the pyridine as well as the increased numbers of intramolecular NH···Cl and NH···pyridine hydrogen bonds.

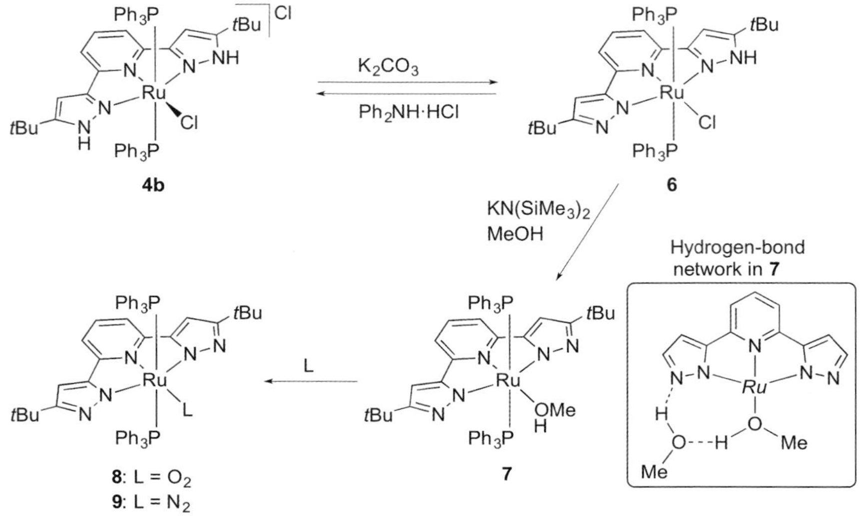

Scheme 4.6 Reversible deprotonation and subsequent transformations of protic pincer-type complex **4b**. $Ru = Ru(PPh_3)_2$.

Complexes **4** and their derivatives catalyze hydrogenation [20, 22] and transfer hydrogenation [20, 23, 24] of acetophenone. Direct participation of the NH groups in the catalysis, however, remains debatable because an *N*-allylated homologue of **4a** also exhibits catalytic activity toward transfer hydrogenation [25]. Nevertheless, some stoichiometric reactions have clearly substantiated the proton-responsive nature of **4**. Thus, reversible deprotonation of the cationic complex **4b** has been achieved by a weak base to give the uncharged pyrazolato-pyrazole complex **6** (Scheme 4.6) [21]. Even twofold deprotonation of **4b** in methanol takes place to afford the bis(pyrazolato) methanol complex **7**. X-ray analysis of **7** has disclosed that hydrogen bonds including a co-crystalized methanol molecule in the second coordination sphere stabilize the methanol coordination, as illustrated in Scheme 4.6. Still the methanol ligand is so labile that it readily replaced by O_2 and even N_2 to yield the side-on peroxo complex **8** and end-on dinitrogen complex **9**, respectively. Coordination of a weak π-accepting ligand N_2 suggests the increased electron donation of the proton-responsive ligand upon deprotonation, although the high N—N stretching frequency of 2160 cm^{-1} of **9** implies that the π-back donation is insufficient for further reductive transformation of N_2.

The stoichiometric deprotonation reactions described above show that the pincer-type bis(pyrazolyl)pyridine ruthenium complexes **4** provide two protons outward. Given the increased electron density of the deprotonated complex, the proton transfer to the substrate in the second coordination sphere may well be coupled with electron transfer from the metal center to the substrate, as in the PCET process proposed in the nitrogenase catalysis (vide supra).

We have recently demonstrated that the iron complex **10a** bearing the bis(pyrazolyl)pyridine ligand **3b** catalyzes the N—N bond cleavage of a

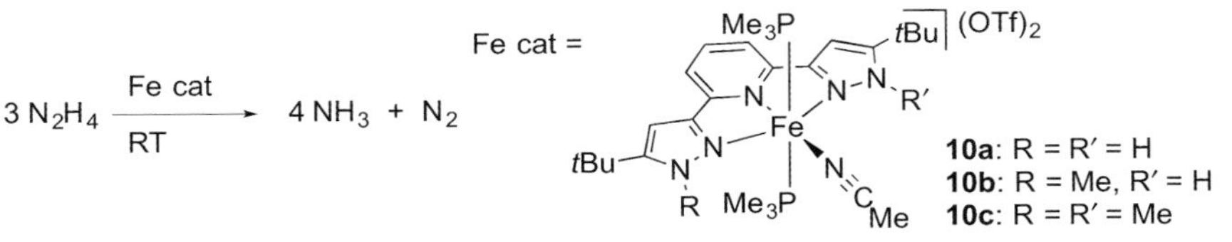

4.3 Proton-Responsive Metal Complexes with Two Appended Protic Groups

$$3\ N_2H_4 \xrightarrow[\text{RT}]{\text{Fe cat}} 4\ NH_3 + N_2$$

Fe cat = [structure] (OTf)$_2$

10a: R = R′ = H
10b: R = Me, R′ = H
10c: R = R′ = Me

Scheme 4.7 Catalytic disproportionation of hydrazine with protic pincer-type iron complex **10a**.

nitrogenase substrate, hydrazine (Scheme 4.7) [26]. The reactions with the mono- and dimethylated homologues **10b** and **10c** produce less than stoichiometric amounts of the product, implying the significance of the NH groups provided by the multiproton-responsive pincer-type ligand **3b**. The plausible mechanism for the catalytic disproportionation is shown in Scheme 4.8. The pyrazole NH group in the second coordination sphere facilitates heterolytic N—N bond cleavage of the coordinated hydrazine in **11** through a hydrogen bond to afford the Fe(IV) amido species **12** and ammonia. This process can be compared with the O—O cleavage through PCET in peroxidases (vide supra). The second pyrazole NH group in the protic pincer ligand operates as an acid-base catalyst to promote substitution of the amido ligand in **12** by the hydrazine to give the hydrazido(1−) species **13**. Subsequent PCET from the hydrazido ligand to the pyrazolato Fe(IV) unit would afford the Fe(II) diazene complex **14**. The phenyldiazene complex **14b** has been actually isolated when phenylhydrazine (R = Ph, R′ = H) was used as the substrate.

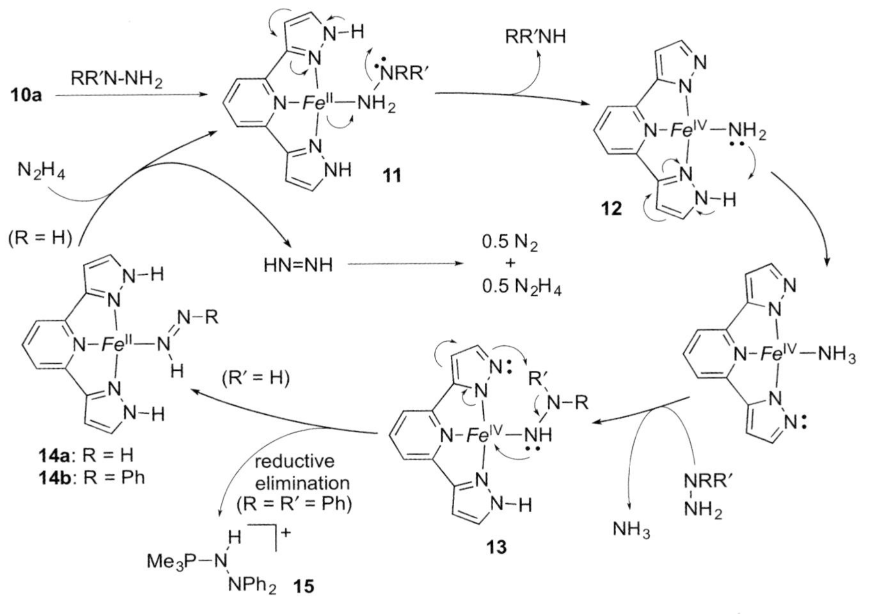

Scheme 4.8 Proposed mechanism for catalytic disproportionation of hydrazine with protic pincer-type complex **10a**. $Fe = Fe(PMe_3)_2{}^{2+}$. The *tert*-butyl groups in the pincer ligand have been omitted in the mechanistic scheme.

The crystal structure of **14b** revealed the phenyldiazene ligand benefits from stabilization by a hydrogen-bonding network with the two pyrazole NH units and counteranion. On the other hand, the reaction of 1,1-diphenylhydrazine (R = R′ = Ph) with **10a**, wherein the distal hydrogen for the PCET is unavailable, yields the hydrazinophosphonium salt **15** as a result of reductive elimination from **13**. Finally, substitution of the diazene ligand in **14** by hydrazine completes the catalytic cycle, and the free diazene immediately undergoes bimolecular disproportionation into dinitrogen and hydrazine. This biorelevant catalysis by the multiproton-responsive complex **10a** provides insights into the catalytic role of the two well-positioned NH groups in the second coordination sphere, including participation in bidirectional PCET.

Thanks to the modularity of the pincer-type ligand [16, 27], unsymmetrical pincer-type complexes bearing two different protic pendants have been developed and their proton-responsive character has been evaluated. The protic pincer-type complexes **17**, which contain a protic pyrazole and isoelectronic *N*-heterocyclic carbene (protic NHC [12, 13, 28, 29]) arms, are synthesized by chelation-assisted formal tautomerization of the imidazole proligands **16** (Scheme 4.9) [30, 31]. Complexes **17** undergo reversible deprotonation upon

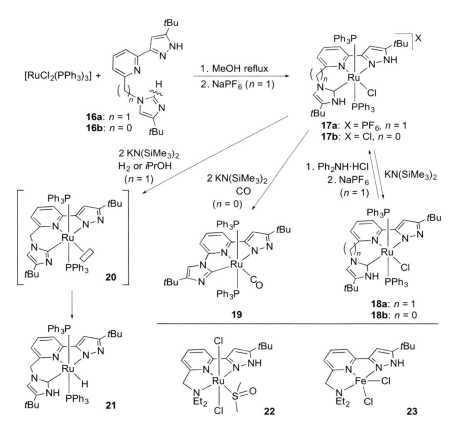

Scheme 4.9 Synthesis and reactivities of unsymmetric, multiproton-responsive pincer-type complexes.

treatment with an equimolar amount of a base, giving the pyrazolato-NHC complexes **18**. The selective loss of the pyrazole proton in **17**, which has been confirmed unambiguously by the ^{1}H NMR spectra with the aid of ^{15}N-labeling and X-ray analysis, indicates that the protic pyrazole is more Brønsted acidic than the protic NHC pendant.

Twofold deprotonation of the five-membered chelate complex **17b** under carbon monoxide yields the pyrazolato-imidazolyl carbonyl complex **19** (Scheme 4.9). Comparison of the CO stretching frequencies between **19** and the related bis(pyrazolato) complexes derived from **4b** shows that the imidazolyl unit is more electron-donating than the pyrazolato group. On the other hand, the six-membered chelate complex **17a** reacted with a base in the presence of H_2 or 2-propanol to afford the hydrido complex **21** most likely through metal-ligand cooperating hydrogenation or hydrogen transfer to the coordinatively unsaturated pyrazolato-imidazolyl complex **20** [32, 33].

We have also synthesized the unsymmetrical protic pincer-type ruthenium and iron complexes **22** and **23** featuring a tertiary amine pendant [22]. The amino group does not appear to operate as a hemilabile, proton-responsive site because its dissociation in the ruthenium complex **22** is not observed even at an elevated temperature in the ^{1}H NMR criteria. Complex **22** catalyzes hydrogenation of acetophenone, although the catalytic activity is lower than that of the bis(pyrazole) complex **4b**.

Pyrazole-assisted cyclometalation is also applicable for the synthesis of protic NCN pincer-type complexes, which feature the electron-donating central aryl group. The reactions of 1,3-bis(pyrazolyl)benzenes **24** with [RuCl(OAc)(PPh$_3$)$_3$] result in selective C—H metalation at the 2-position of the central benzene ring to afford the NCN pincer-type ruthenium complexes **25** (Scheme 4.10) [24]. The iridium complex **26** is obtained in a similar manner. The stronger trans influence of the monoanionic NCN pincer-type ligand is evidenced by the longer M-Cl distance in comparison with those in the NNN pincer complex **27**. The enhanced electron donation of the pincer ligand also enables facile conversion of the ruthenium complexes **25** to the carbonyl complexes **28**. In addition, comparison of the CO stretching frequencies with the NNN analogues indicates the increased electron density in the NCN pincer-type complexes.

4.3.2 Bis(2-hydroxypyridine) Chelate Complexes

A 2-hydroxypyridine moiety is known to exist in the active site of [Fe] hydrogenases, which catalyze only H_2 heterolysis without redox, and to be involved in H_2 activation as a cooperating unit as in the azadithiolato ligand in [FeFe] hydrogenases (vide supra) [4, 5, 14]. In parallel with the recognition, a number of complexes containing 2-hydroxypyridine derivatives have been developed as metal-ligand bifunctional catalysts [14, 15].

As a (2-hydroxypyridine)-based multiproton-responsive complex, Papish and coworkers reported the half-sandwich-type ruthenium complex **29** bearing a 6,6′-dihydroxy-2,2′-bipyridine (6-DHPB) ligand (Scheme 4.11) [34]. The two OH groups are free from coordination but engaged in hydrogen bonds with the counteranion and solvated methanol molecule in the crystal. Complex **29** catalyzes transfer hydrogenation of acetophenones with 2-propanol. Notably,

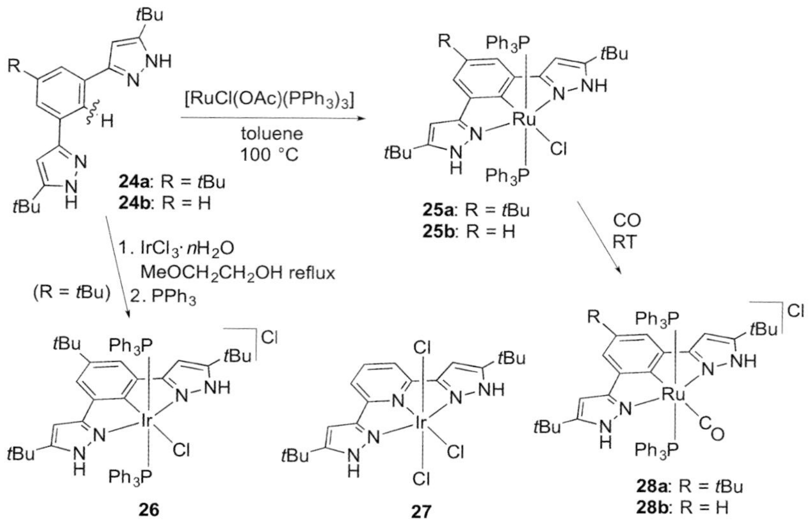

Scheme 4.10 Synthesis of protic NCN pincer-type complexes.

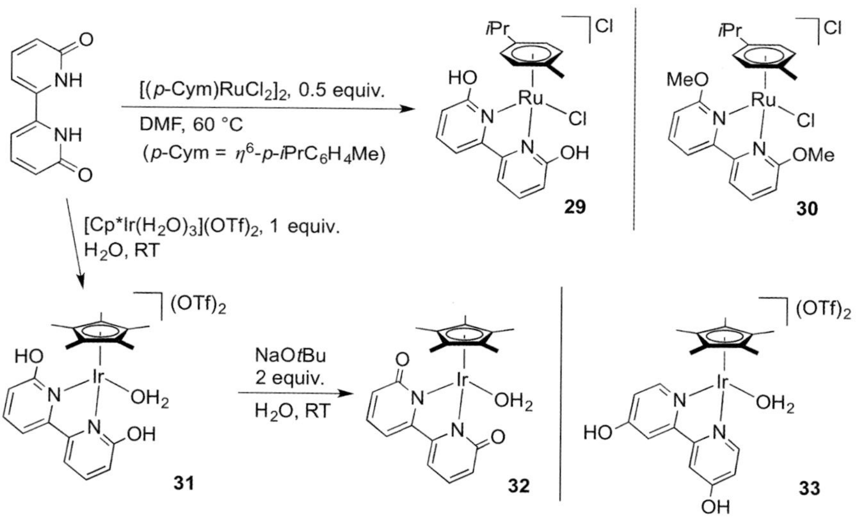

Scheme 4.11 Synthesis of N,N-chelate bis(hydroxypyridine) complexes **29** and **31**.

the protic complex **29** exhibits a catalytic activity superior to those of the aprotic dimethoxy derivative **30** and unsubstituted bipyridine complex in transfer hydrogenation with sodium formate in MeOH–H_2O (1 : 9), implicating a significant role of the proton-responsive sites in the catalysis.

Independently, Fujita and Yamaguchi published the isoelectronic iridium complex **31** [35] as a logical extension of their intensive studies on the catalysis of

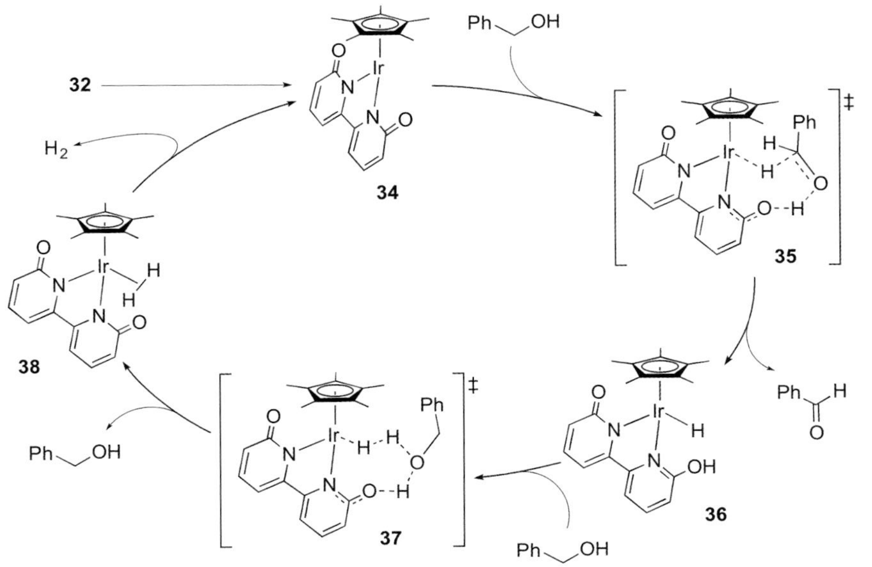

Scheme 4.12 Proposed mechanism for catalytic dehydrogenative oxidation of benzyl alcohol with bis(pyridonato) complex **32**.

mono(hydroxypyridine) iridium complexes (Scheme 4.11). They demonstrated that **31** and the doubly deprotonated complex **32** [36] catalyze dehydrogenative oxidation of primary and secondary alcohols to the corresponding carbonyl compounds without base additives. The proposed mechanism for the acceptorless alcohol dehydrogenation by the much superior catalyst **32** (TON of up to 275 000) is shown in Scheme 4.12 [37]. After dissociation of the aqua ligand in **32**, the resultant coordinatively unsaturated intermediate **34** undergoes a hydride transfer from the substrate alcohol coupled with a proton transfer to the pyridonato oxygen in the second coordination sphere (**35**), which gives the hydrido-hydroxypyridine complex **36**. Owing to the assistance of an external alcohol to the proton shift (**37**), the dihydrogen complex **38** would be formed. Finally, liberation of hydrogen gas regenerates **34**. In line with the mechanism, the 4,4′-dihydroxy-2,2′-bipyridine (4-DHPB) complex **33** without proton-responsive sites in the second coordination sphere exhibits lower catalytic activity [35]. Fujita, Yamaguchi [38, 39], and others [40, 41] have further applied this class of multiproton-responsive complexes for various dehydrogenative transformations of alcohols and amines.

Since their pioneering work on hydrogen storage using formic acid and formate by 4-DHPB complexes [42], Himeda and coworkers have extensively investigated the catalysis of hydroxypyridine chelate complexes [43, 44]. Hull, Himeda, and Fujita disclosed that the tetrahydroxybipyrimidine-bridged dinuclear iridium complex **39** containing both 4- and 6-DHPB motifs catalyzes dehydrogenation of formic acid under acidic conditions (Scheme 4.13) [45]. Meanwhile, the reaction in the reverse direction, hydrogenation of CO_2 to formate, is catalyzed by the deprotonated form **40** under basic conditions.

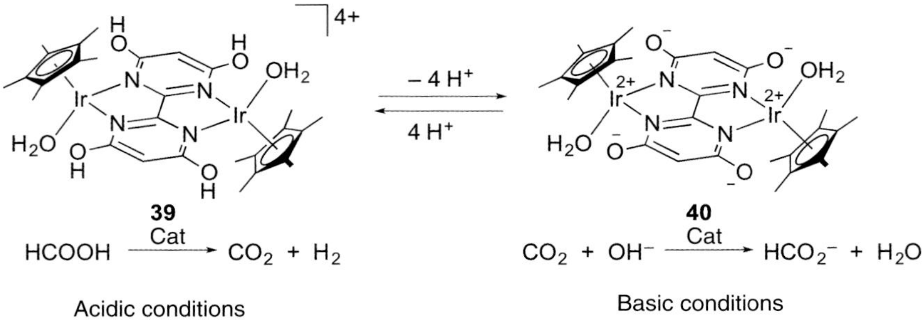

Scheme 4.13 Dehydrogenation of formic acid and hydrogenation of CO_2 catalyzed by tetrahydroxybipyrimidine complexes **39** and **40**.

Following these findings, Himeda and Fujita investigated a series of 6-DHPB complexes capable of hydrogenation of CO_2 and dehydrogenation of formic acid, as mononuclear prototypes of the excellent catalysts **39** and **40**. Schemes 4.14 illustrates the proposed mechanism for the catalytic dehydrogenation of formic acid with [Cp*Ir(6-dhpb)]$^{2+}$ (the cationic part of **31**) [46]. The mechanism quite resembles that for the dehydrogenative oxidation of alcohols shown in Scheme 4.12. The coordinatively unsaturated intermediate **34** generated by deprotonation undergoes hydrogen transfer from formic acid with the aid of the pyridonato oxygen atom in the second coordination sphere, which affords the

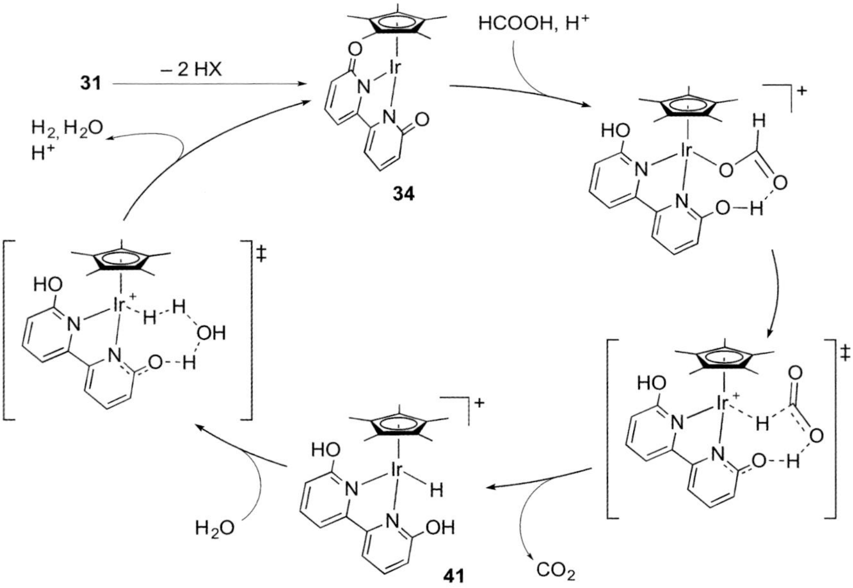

Scheme 4.14 Proposed mechanism for catalytic dehydrogenation of formic acid with bis(hydroxypyridine) complex **31**.

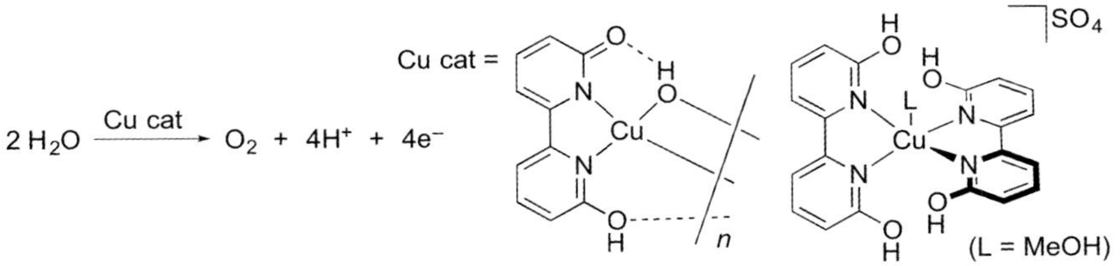

Scheme 4.15 Electrochemical water oxidation catalyzed by bis(hydroxypyridine)copper complexes.

hydrido complex **41** along with CO_2. Subsequent water-assisted proton relay leads to the formation of H_2 and **34**. The reverse reaction, CO_2 hydrogenation catalyzed by the bis(pyridonato) complex **32**, would take place in a similar manner [46]. Quite recently, the role of alkali metal in the CO_2 hydrogenation as well as the absence of significant outer-sphere interaction in the formic acid dehydrogenation under acidic conditions has been proposed for these catalyst systems [47].

Recently, Lin [48] and Papish [49] reported that the 6-DHPB copper complexes catalyze electrochemical water oxidation (Scheme 4.15). Some iridium complexes related to **31** are also known to catalyze this reaction [50–52].

Szymczak recently synthesized the multiproton-responsive ruthenium complex **42** bearing two hydroxypyridine moieties on a pincer-type platform (Scheme 4.16) [53]. The two OH groups form intramolecular hydrogen bonds with the chlorido ligand. Complex **42** catalyzes transfer hydrogenation of ketones with 2-propanol. On the basis of experiments using the carbonyl derivative **43** with higher stability during the catalysis, a mechanism shown in Scheme 4.17 has been proposed [54]. The base- and cation-dependence of the catalysis as well as NMR detection of a hydrido ligand trans to the phosphine supports initial formation of **44**. Subsequent hydrido transfer to the ketone substrate would be assisted by electrostatic interaction between the carbonyl oxygen atom and the dangling alkaline metal in the second coordination sphere of the ruthenium atom (**45**). After exchange of alcohol on the alkaline metal to release the reduction product, hydride transfer to the ruthenium atom regenerate **44**. H_2 heterolysis on a related bis(hydroxypyridine) pincer-type ruthenium complex has also been reported recently [55].

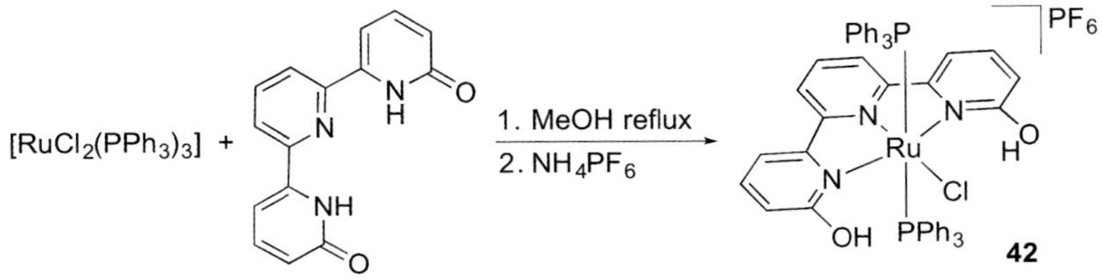

Scheme 4.16 Synthesis of ruthenium complex **42** bearing hydroxypyridine-based protic pincer-type ligand.

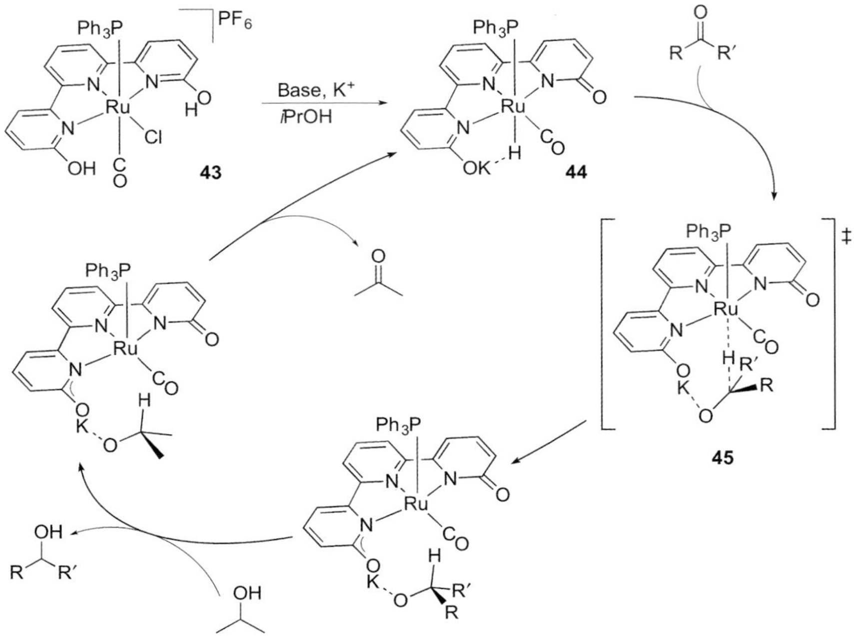

Scheme 4.17 Proposed mechanism for catalytic transfer hydrogenation of ketone with protic pincer-type complex **43**.

4.4 Proton-Responsive Metal Complexes with Three Appended Protic Groups on Tripodal Scaffolds

Efficient accumulation of more than two proton-responsive units to mimic the complicated biological coordination sphere would require new ligand topology other than the bidentate chelate and meridional pincer-type skeleton. For this purpose, tertiary amine-based tripodal ligands furnished with three proton-responsive arms have been used widely owing to their accessibility and structural diversity.

We have recently developed a tris(pyrazol-3-ylmethyl)amine as a new entry of the multiproton-responsive ligands (Scheme 4.18) [56]. The chlorido-bridged dinuclear ruthenium complex **46** serves as a useful synthetic precursor. When **46** is heated with triphenylphosphine in toluene, the mononuclear phosphine complex **47** is obtained. Interestingly, the reaction in methanol affords the isomer **48**, in which the incoming phosphine ligand resides in the position cis to the amine nitrogen atom in the tripodal ligand [57]. The notable dependence of the product on the reaction solvent may be explained by the hydrogen bond network including the protic solvent, which renders the chlorido ligand in the pyrazole rings inert toward ligand substitution. The Brønsted acidity of the proton-responsive ligand has been substantiated by, for example, reversible deprotonation of **47**, which gives the pyrazolato-bis(pyrazole) complex **49** [56].

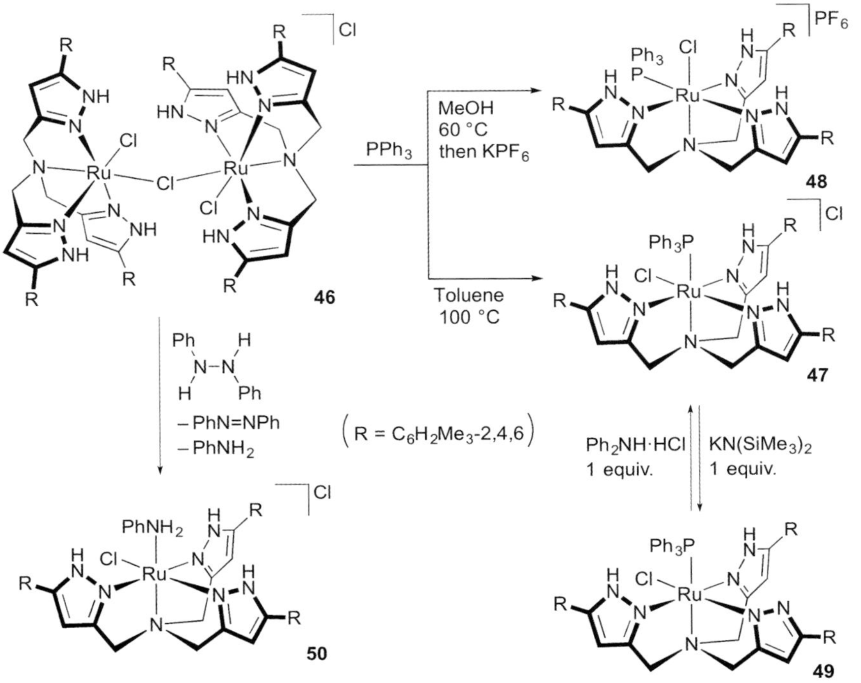

Scheme 4.18 Reactions of tris(pyrazolylmethyl)amine ruthenium complex **46**.

When an excess amount of 1,2-diphenylhydrazine (DPH) is added to **46**, catalytic N—N bond cleavage of DPH takes place to afford azobenzene and aniline, along with the aniline complex **50**. The aprotic tris(pyridylmethyl)amine (TPA) complex [Ru(tpa)(MeCN)$_2$]$^{2+}$ shows no catalytic activity, indicating the significance of the proton-responsive sites in the second coordination sphere.

Szymczak reported the copper complexes **51** and **52** bearing a hydroxypyridine-based protic tripodal ligand (Scheme 4.19) [58, 59]. The crystallographic analysis and ^{19}F NMR spectroscopy revealed that the Cu(I)-F interaction in **52a** is very weak, whereas the three strong hydrogen bonds with the multiproton-responsive ligand capture the fluoride anion [59]. The fluoride ion in the second coordination sphere of **52a** is successfully replaced upon treatment with a silyl nitrite to give the Cu(II) aqua complex **55** along with nitric oxide [60]. Importantly, both of the aprotic [CuF(PPh$_3$)$_3$] and electron-deficient Cu(II) complex **51a** do not yield sufficient amounts of nitric oxide under otherwise identical reaction conditions. These control experiments suggest that the reaction involves 2H$^+$/e$^-$ transfer from the multiproton-responsive copper complex to the nitrite anion through intermediates like **53** and **54**. The result would also provide some insight into the mechanism of copper nitrite reductases that catalyze the one-electron reduction of nitrite to nitric oxide.

In their seminal work on the coordination chemistry of multiproton-responsive tripodal ligands [61], Borovik demonstrated that the unsymmetric

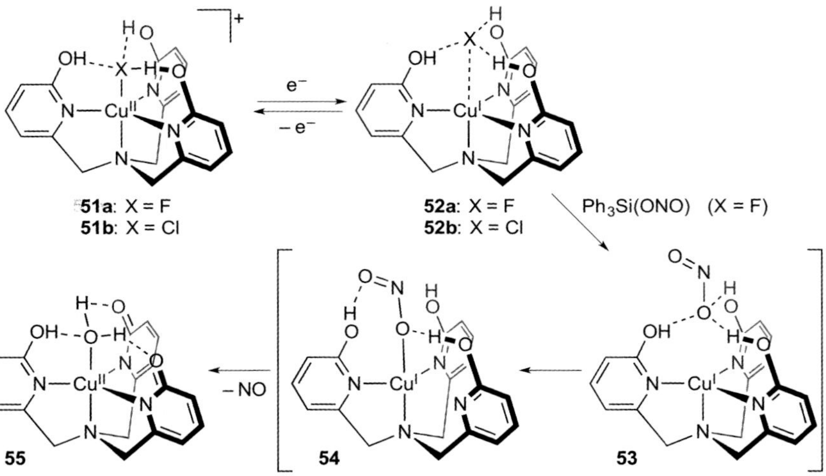

Scheme 4.19 Reactions of tris(hydroxylpyridylmethyl)amine copper complexes **52**.

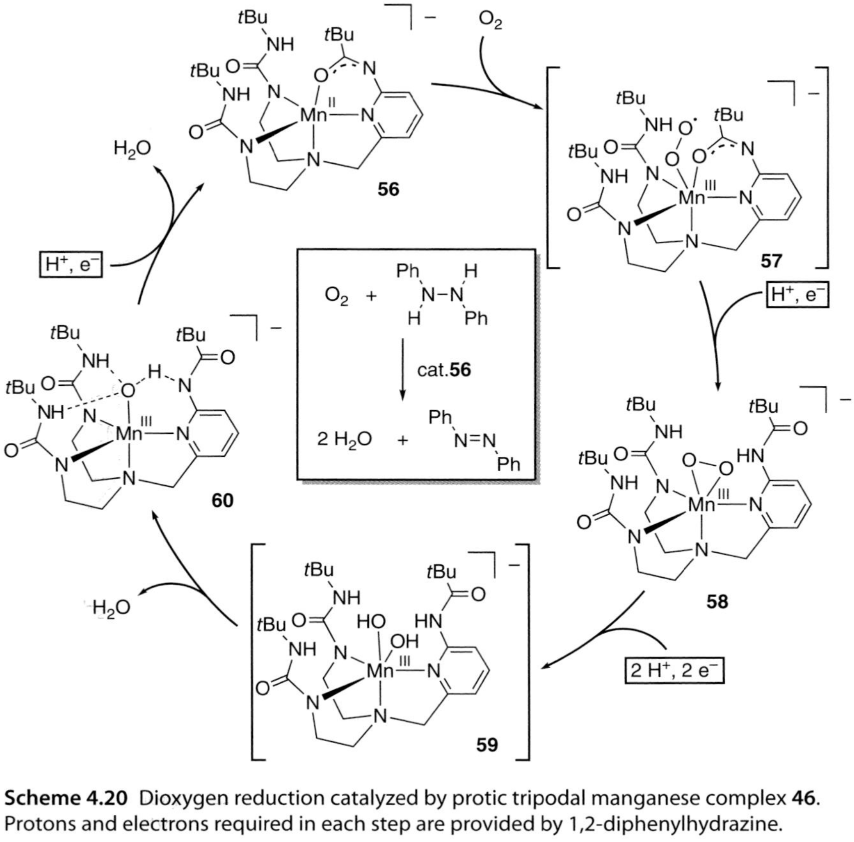

Scheme 4.20 Dioxygen reduction catalyzed by protic tripodal manganese complex **46**. Protons and electrons required in each step are provided by 1,2-diphenylhydrazine.

tris(carboxamide)manganese complex **56** catalyzes the O_2 reduction with DPH (Scheme 4.20) [62]. The isolable Mn(II) complex **56** first reacts with O_2 to give the unstable superoxo Mn(III) adduct **57**. Subsequent PCET from DPH leads to the formation of azobenzene and the peroxo Mn(III) complex **58**, in which the carboxamide arms are all protonated. Complex **58** further undergoes PCET-induced, homolytic O—O bond cleavage to afford bis(hydroxo) species **59**. Proton shift from one of the carboxamide arm to the hydroxo ligand would release water to give the hydroxo-carboxamide or oxo-carboxamidato Mn(III) complex **60**, which has been isolated and crystallographically characterized. Finally, PCET from DPH again takes place to regenerate the Mn(II) carboxamidato complex **56**. The multiproton-responsive coordination sphere crafted by the protic tripodal ligand is thus involved in the catalysis through stabilization of intermediates and promotion of proton-electron transfer.

Fout recently developed a C_3-symmetric tripodal amine ligand featuring three pyrrole-imine arms as depicted in Scheme 4.21. The tautomerism of the pendant arm to the (amino)azafulvene form has some similarity with the character of pyrazole and hydroxypyridine, and hence, the multiproton-responsive reactivities would be expected for this tripodal ligand. The reaction of $[Fe(OTf)_2(MeCN)_2]$ with the ligand precursor leads to the formation of the triflato complex **61**, in which the arms tautomerize to the (amino)azafulvene form and the NH groups rotate away from the apical triflato ligand [63]. Treatment of **61** with an equimolar amount of $(n\text{Bu}_4\text{N})\text{NO}_2$ yields the Fe(III) oxo complex **62**, wherein the NH groups in the arms all face to the terminal

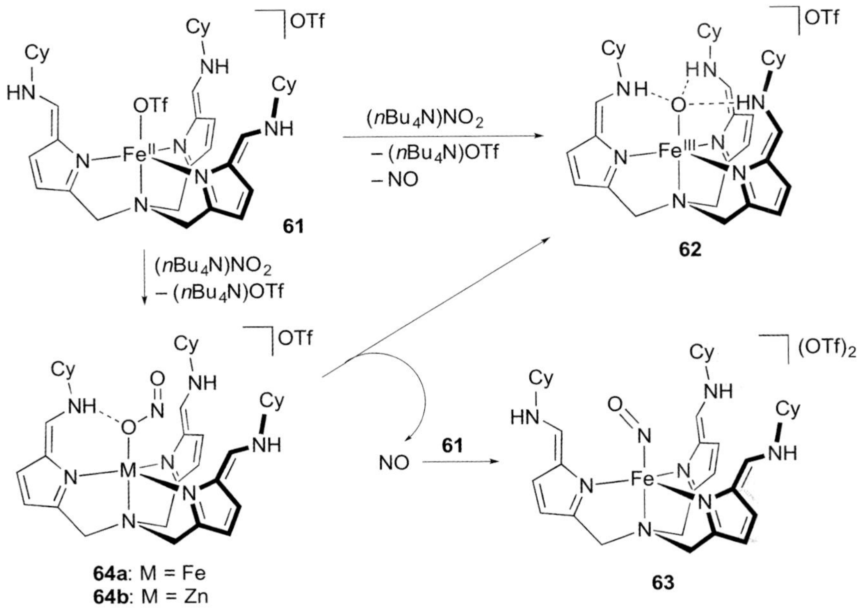

Scheme 4.21 Reactions of iron complex **61** bearing protic tripodal ligand.

oxo ligand and are engaged in intramolecular hydrogen bonding [64]. Loss of nitric oxide under mild conditions has been confirmed by the reaction with only half amount of the nitrite salt, which resulted in the formation of **62** and an equimolar amount of the nitrosyl complex **63**. Considering isolation of the (nitrito)zinc complex **64b**, the release of nitric oxide most likely occurs through the nitrito intermediate **64a** involving an intramolecular hydrogen bond in the second coordination sphere. As an elegant extension of this chemistry, quite recently Fout disclosed the catalytic reduction of nitrate and perchlorate ions with DPH [65].

4.5 Summary and Outlook

The significance of highly organized multifunctionality in the active sites of metalloenzymes is now well recognized. Multiproton-responsive ligands provide a powerful strategy to construct bioinspired coordination spaces featuring directed and multiple noncovalent interactions in a well-defined manner. In addition to structural replication of the biological systems, some intriguing reactivities toward ubiquitous and biologically relevant molecules such as H_2, CO_2, and O_2, and hydrazines have been elicited from multiproton-responsive complexes. Coupled proton/electron transfer appears to be a characteristic of these complexes. Further development of multiproton-responsive ligands will open new avenues for crafting molecular architectures for challenging chemical transformations under mild reaction conditions. These studies should include quantitative evaluation of the thermodynamic parameters, such as pK_a, of this class of ligands.

Acknowledgments

This work was supported by the PRESTO program on "Molecular Technology and Creation of New Functions" from JST (Grant Number JPMJPR14K6). I wish to thank all my coworkers who have contributed to the results presented here.

References

1 (a) Bertini, I., Gray, H.B., Stiefel, E.I., and Valentine, J.S. (2007). *Biological Inorganic Chemistry. Structure and Reactivity*. Sausalito: University Science Books. (b) Crichton, R.R. (2012). *Biological Inorganic Chemistry. A New Introduction to Molecular Structure and Function*, 2ee. Oxford: Elsevier.
2 Crabtree, R.H. (2011). Multifunctional ligands in transition metal catalysis. *New J. Chem.* 35 (1): 18–23.
3 Khusnutdinova, J.R. and Milstein, D. (2015). Metal-ligand cooperation. *Angew. Chem. Int. Ed.* 54 (42): 12236–12273.

4 Schilter, D., Camara, J.M., Huynh, M.T. et al. (2016). Hydrogenase enzymes and their synthetic models: the role of metal hydrides. *Chem. Rev.* 116 (15): 8693–8749.
5 Lubitz, W., Ogata, H., Rüdiger, O., and Reijerse, E. (2014). Hydrogenases. *Chem. Rev.* 114 (8): 4081–4148.
6 Erman, J.E. and Vitello, L.B. (2002). Yeast cytochrome c peroxidase: mechanistic studies via protein engineering. *Biochim. Biophys. Acta* 1597 (2): 193–220.
7 Hiner, A.N.P., Raven, E.L., Thorneley, R.N.F. et al. (2002). Mechanisms of compound I formation in heme peroxidases. *J. Inorg. Biochem.* 91 (1): 27–34.
8 Poulos, T.L. (2014). Heme enzyme structure and function. *Chem. Rev.* 114 (7): 3919–3962.
9 Hu, Y. and Ribbe, M.W. (2016). Nitrogenases - a tale of carbon atom(s). *Angew. Chem. Int. Ed.* 55 (29): 8216–8226.
10 Hoffman, B.M., Lukoyanov, D., Yang, Z.-Y. et al. (2014). Mechanism of nitrogen fixation by nitrogenase: the next stage. *Chem. Rev.* 114 (8): 4041–4062.
11 Hoffman, B.M., Lukoyanov, D., Dean, D.R., and Seefeldt, L.C. (2013). Nitrogenase: a draft mechanism. *Acc. Chem. Res.* 46 (2): 587–595.
12 Kuwata, S. and Ikariya, T. (2011). β-Protic pyrazole and N-heterocyclic carbene complexes: synthesis, properties, and metal-ligand cooperative bifunctional catalysis. *Chem. Eur. J.* 17 (13): 3542–3556.
13 Kuwata, S. and Ikariya, T. (2014). Metal-ligand bifunctional reactivity and catalysis of protic N-heterocyclic carbene and pyrazole complexes featuring β-NH units. *Chem. Commun.* 50 (92): 14290–14300.
14 Moore, C.M., Dahl, E.W., and Szymczak, N.K. (2015). Beyond H_2: exploiting 2-hydroxypyridine as a design element from [Fe]-hydrogenase for energy-relevant catalysis. *Curr. Opin. Chem. Biol.* 25: 9–17.
15 Wang, W.-H., Muckerman, J.T., Fujita, E., and Himeda, Y. (2013). Hydroxy-substituted pyridine-like N-heterocycles: versatile ligands in organometallic catalysis. *New J. Chem.* 37 (7): 1860–1866.
16 (a) Szabó, K.J. and Wendt, O.F. ed. (2014). *Pincer and Pincer-Type Complexes: Applications in Organic Synthesis and Catalysis*. Weinheim: Wiley-VCH; (b) van Koten, G. and Milstein, D. ed. (2013). *Organometallic Pincer Chemistry*. Heidelberg: Springer.
17 Cook, L.J.K., Mohammed, R., Sherborne, G. et al. (2015). Spin state behavior of iron(II)/dipyrazolylpyridine complexes. New insights from crystallographic and solution measurements. *Coord. Chem. Rev.* 289–290: 2–12.
18 Halcrow, M.A. (2014). Recent advances in the synthesis and applications of 2,6-dipyrazolylpyridine derivatives and their complexes. *New J. Chem.* 38 (5): 1868–1882.
19 Craig, G.A., Roubeau, O., and Aromi, G. (2014). Spin state switching in 2,6-bis(pyrazol-3-yl)pyridine (3-bpp) based Fe(II) complexes. *Coord. Chem. Rev.* 269: 13–31.
20 Jozak, T., Zabel, D., Schubert, A., and Thiel, W.R. (2010). Ruthenium complexes bearing N—H acidic pyrazole ligands. *Eur. J. Inorg. Chem.* (32): 5135–5145.

21 Yoshinari, A., Tazawa, A., Kuwata, S., and Ikariya, T. (2012). Synthesis, structures, and reactivities of pincer-type ruthenium complexes bearing two proton-responsive pyrazole arms. *Chem. Asian. J.* 7 (6): 1417–1425.

22 Toda, T., Kuwata, S., and Ikariya, T. (2015). Synthesis and structures of ruthenium and iron complexes bearing an unsymmetrical pincer-type ligand with protic pyrazole and tertiary aminoalkyl arms. *Z. Anorg. Allg. Chem.* 641 (12–13): 2135–2139.

23 Roberts, T.D. and Halcrow, M.A. (2016). Supramolecular assembly and transfer hydrogenation catalysis with ruthenium(II) complexes of 2,6-di(1H-pyrazol-3-yl)pyridine derivatives. *Polyhedron* 103: 79–86.

24 Toda, T., Saitoh, K., Yoshinari, A. et al. (2017). Synthesis and structures of NCN pincer-type ruthenium and iridium complexes bearing protic pyrazole arms. *Organometallics* 36 (6): 1188–1195.

25 Ghoochany, L.T., Farsadpour, S., Sun, Y., and Thiel, W.R. (2011). New N,N,N-donors resulting in highly active ruthenium catalysts for transfer hydrogenation at room temperature. *Eur. J. Inorg. Chem.* (23): 3431–3437.

26 Umehara, K., Kuwata, S., and Ikariya, T. (2013). N—N bond cleavage of hydrazines with a multiproton-responsive pincer-type iron complex. *J. Am. Chem. Soc.* 135 (18): 6754–6757.

27 Younus, H.A., Ahmad, N., Su, W., and Verpoort, F. (2014). Ruthenium pincer complexes: ligand design and complex synthesis. *Coord. Chem. Rev.* 276: 112–152.

28 Jahnke, M.C. and Hahn, F.E. (2015). Complexes with protic (NH,NH and NH,NR) N-heterocyclic carbene ligands. *Coord. Chem. Rev.* 293–294: 95–115.

29 Jahnke, M.C. and Hahn, F.E. (2015). Complexes bearing protic N-heterocyclic carbenes: synthesis and applications. *Chem. Lett.* 44 (3): 226–237.

30 Toda, T., Kuwata, S., and Ikariya, T. (2014). Unsymmetrical pincer-type ruthenium complex containing β-protic pyrazole and N-heterocyclic carbene arms: comparison of Brønsted acidity of NH groups in second coordination sphere. *Chem. Eur. J.* 20 (31): 9539–9542.

31 Toda, T., Yoshinari, A., Ikariya, T., and Kuwata, S. (2016). Protic N-heterocyclic carbene versus pyrazole: rigorous comparison of proton- and electron-donating abilities in a pincer-type framework. *Chem. Eur. J.* 22 (46): 16675–16683.

32 Miranda-Soto, V., Grotjahn, D.B., DiPasquale, A.G., and Rheingold, A.L. (2008). Imidazol-2-yl complexes of Cp*Ir as bifunctional ambident reactants. *J. Am. Chem. Soc.* 130 (40): 13200–13201.

33 Miranda-Soto, V., Grotjahn, D.B., Cooksy, A.L. et al. (2011). A labile and catalytically active imidazol-2-yl fragment system. *Angew. Chem. Int. Ed.* 50 (3): 631–635.

34 Nieto, I., Livings, M.S., Sacci, J.B. III, et al. (2011). Transfer hydrogenation in water via a ruthenium catalyst with OH groups near the metal center on a bipy scaffold. *Organometallics* 30 (23): 6339–6342.

35 Kawahara, R., Fujita, K., and Yamaguchi, R. (2012). Dehydrogenative oxidation of alcohols in aqueous media using water-soluble and reusable Cp*Ir catalysts bearing a functional bipyridine ligand. *J. Am. Chem. Soc.* 134 (8): 3643–3646.

36 Kawahara, R., Fujita, K., and Yamaguchi, R. (2012). Cooperative catalysis by iridium complexes with a bipyridonate ligand: versatile dehydrogenative oxidation of alcohols and reversible dehydrogenation–hydrogenation between 2-propanol and acetone. *Angew. Chem. Int. Ed.* 51 (51): 12790–12794.

37 Zeng, G., Sakaki, S., Fujita, K. et al. (2014). Efficient catalyst for acceptorless alcohol dehydrogenation: interplay of theoretical and experimental studies. *ACS Catal.* 4 (3): 1010–1020.

38 Fujita, K., Kawahara, R., Aikawa, T., and Yamaguchi, R. (2015). Hydrogen production from a methanol-water solution catalyzed by an anionic iridium complex bearing a functional bipyridonate ligand under weakly basic conditions. *Angew. Chem. Int. Ed.* 54 (31): 9057–9060.

39 Fujita, K., Tanaka, Y., Kobayashi, M., and Yamaguchi, R. (2014). Homogeneous perdehydrogenation and perhydrogenation of fused bicyclic N-heterocycles catalyzed by iridium complexes bearing a functional bipyridonate ligand. *J. Am. Chem. Soc.* 136 (13): 4829–4832.

40 Wang, R., Fan, H., Zhao, W., and Li, F. (2016). Acceptorless dehydrogenative cyclization of o-aminobenzyl alcohols with ketones to quinolines in water catalyzed by water-soluble metal-ligand bifunctional catalyst [Cp*(6,6′-[OH]$_2$bpy)(H$_2$O)][OTf]$_2$. *Org. Lett.* 18 (15): 3558–3561.

41 Roy, B.C., Chakrabarti, K., Shee, S. et al. (2016). Bifunctional Ru(II)-complex-catalysed tandem C—C bond formation: efficient and atom economical strategy for the utilization of alcohols as alkylating agents. *Chem. Eur. J.* 22 (50): 18147–18155.

42 Himeda, Y., Onozawa-Komatsuzaki, N., Sugihara, H. et al. (2004). Half-sandwich complexes with 4,7-dihydroxy-1,10-phenanthroline: water-soluble, highly efficient catalysts for hydrogenation of bicarbonate attributable to the generation of an oxyanion on the catalyst ligand. *Organometallics* 23 (7): 1480–1483.

43 Onishi, N., Xu, S., Manaka, Y. et al. (2015). CO_2 hydrogenation catalyzed by iridium complexes with a proton-responsive ligand. *Inorg. Chem.* 54 (11): 5114–5123.

44 Wang, W.-H., Himeda, Y., Muckerman, J.T. et al. (2015). CO_2 hydrogenation to formate and methanol as an alternative to photo- and electrochemical CO_2 reduction. *Chem. Rev.* 115 (23): 12936–12973.

45 Hull, J.F., Himeda, Y., Wang, W.-H. et al. (2012). Reversible hydrogen storage using CO_2 and a proton-switchable iridium catalyst in aqueous media under mild temperatures and pressures. *Nat. Chem.* 4 (5): 383–388.

46 Ertem, M.Z., Himeda, Y., Fujita, E., and Muckerman, J.T. (2016). Interconversion of formic acid and carbon dioxide by proton-responsive, half-sandwich Cp*Ir^{III} complexes: a computational mechanistic investigation. *ACS Catal.* 6 (2): 600–609.

47 Siek, S., Burks, D.B., Gerlach, D.L. et al. (2017). Iridium and ruthenium complexes of N-heterocyclic carbene- and pyridinol-derived chelates as catalysts for aqueous carbon dioxide hydrogenation and formic acid dehydrogenation: the role of the alkali metal. *Organometallics* 36 (6): 1091–1106.

48 Zhang, T., Wang, C., Liu, S. et al. (2014). A biomimetic copper water oxidation catalyst with low overpotential. *J. Am. Chem. Soc.* 136 (1): 273–281.

49 Gerlach, D.L., Bhagan, S., Cruce, A.A. et al. (2014). Studies of the pathways open to copper water oxidation catalysts containing proximal hydroxy groups during basic electrocatalysis. *Inorg. Chem.* 53 (24): 12689–12698.

50 Lewandowska-Andralojc, A., Polyansky, D.E., Wang, C.-H. et al. (2014). Efficient water oxidation with organometallic iridium complexes as precatalysts. *Phys. Chem. Chem. Phys.* 16 (24): 11976–11987.

51 DePasquale, J., Nieto, I., Reuther, L.E. et al. (2013). Iridium dihydroxybipyridine complexes show that ligand deprotonation dramatically speeds rates of catalytic water oxidation. *Inorg. Chem.* 52 (16): 9175–9183.

52 Zhang, T., deKrafft, K.E., Wang, J.-L. et al. (2014). The effects of electron-donating substituents on [Ir(bpy)Cp*Cl]$^+$: water oxidation versus ligand oxidative modifications. *Eur. J. Inorg. Chem.* 2014 (4): 698–707.

53 Moore, C.M. and Szymczak, N.K. (2013). 6,6′-Dihydroxy terpyridine: a proton-responsive bifunctional ligand and its application in catalytic transfer hydrogenation of ketones. *Chem. Commun.* 49 (4): 400–402.

54 Moore, C.M., Bark, B., and Szymczak, N.K. (2016). Simple ligand modifications with pendent OH groups dramatically impact the activity and selectivity of ruthenium catalysts for transfer hydrogenation: the importance of alkali metals. *ACS Catal.* 6 (3): 1981–1990.

55 Geri, J.B. and Szymczak, N.K. (2015). A proton-switchable bifunctional ruthenium complex that catalyzes nitrile hydroboration. *J. Am. Chem. Soc.* 137 (40): 12808–12814.

56 Yamagishi, H., Nabeya, S., Ikariya, T., and Kuwata, S. (2015). Protic ruthenium tris(pyrazol-3-ylmethyl)amine complexes featuring a hydrogen-bonding network in the second coordination sphere. *Inorg. Chem.* 54 (24): 11584–11586.

57 Yamagishi, H., Konuma, H., and Kuwata, S. (2017). Stereoselective synthesis of chlorido-phosphine ruthenium complexes bearing a pyrazole-based protic tripodal amine ligand. *Polyhedron* 125: 173–178.

58 Moore, C.M., Quist, D.A., Kampf, J.W., and Szymczak, N.K. (2014). A 3-fold-symmetric ligand based on 2-hydroxypyridine: regulation of ligand binding by hydrogen bonding. *Inorg. Chem.* 53 (7): 3278–3280.

59 Moore, C.M. and Szymczak, N.K. (2015). Redox-induced fluoride ligand dissociation stabilized by intramolecular hydrogen bonding. *Chem. Commun.* 51 (25): 5490–5492.

60 Moore, C.M. and Szymczak, N.K. (2015). Nitrite reduction by copper through ligand-mediated proton and electron transfer. *Chem. Sci.* 6: 3373–3377.

61 Cook, S.A. and Borovik, A.S. (2015). Molecular designs for controlling the local environments around metal ions. *Acc. Chem. Res.* 48 (8): 2407–2414.

62 Shook, R.L., Peterson, S.M., Greaves, J. et al. (2011). Catalytic reduction of dioxygen to water with a monomeric manganese complex at room temperature. *J. Am. Chem. Soc.* 133 (15): 5810–5817.

63 Matson, E.M., Bertke, J.A., and Fout, A.R. (2014). Isolation of iron(II) aqua and hydroxyl complexes featuring a tripodal H-bond donor and acceptor ligand. *Inorg. Chem.* 53 (9): 4450–4458.

64 Matson, E.M., Park, Y.J., and Fout, A.R. (2014). Facile nitrite reduction in a non-heme iron system: formation of an iron(III)-oxo. *J. Am. Chem. Soc.* 136 (50): 17398–17401.

65 Ford, C.L., Park, Y.J., Matson, E.M. et al. (2016). A bioinspired iron catalyst for nitrate and perchlorate reduction. *Science (New York, N.Y.)* 354 (6313): 741–743.

5

Photo-Control of Molecular Alignment for Photonic and Mechanical Applications

Miho Aizawa[1], Christopher J. Barrett[1,2], and Atsushi Shishido[1,3],**

[1] Tokyo Institute of Technology, Laboratory for Chemistry and Life Science, R1-12, 4259 Nagatsuta, Midori-ku, Yokohama 226-8503, Japan
[2] McGill University, Department of Chemistry, 801 Sherbrooke Street West, Montreal, QC H3A 0B8, Canada
[3] PRESTO, JST, 4-1-8 Honcho, Kawaguchi, Saitama 332-0012, Japan

5.1 Introduction

The development of stimuli-responsive functional materials is among the key goals of modern materials science. The structure and properties of such switchable materials can be designed to be controlled by various stimuli, among which light is frequently the most powerful trigger. Light is a gentle energy source that can target materials remotely, with extremely high spatial and temporal resolution, easily and cheaply. Light control over molecular alignment in particular has in recent years attracted particular interest, due to potential applications as reconfigurable photonic elements and optical-to-mechanical energy conversion. In this chapter, we highlight some recent examples and emerging trends in this exciting field of research, focusing on liquid crystals (LCs), liquid-crystalline polymers, and photo-chromic organic materials, which we hope will help stimulate more interest toward the development of light-responsive materials and their successful application to a wide variety of current and future high-tech applications in optics, photonics, and energy harvesting and conversion.

Materials based on LCs have emerged as an especially versatile host material for optical effects; often described as a "fourth state of matter," LCs are now ubiquitous in our everyday lives, as they lie at the heart of practically all of our present visual display devices. Lying in between isotropic liquids and crystalline solids, LCs combine both liquid-like high mobility and the orientational order of solids achievable. They can also "communicate" with each other via molecular cooperative motions, and sympathetic orientation of LC molecules can be readily controlled externally, which enables their use as "macroscopic molecular switches" and actuators with precise control over refractive index and anisotropy. LC materials thus possess immense technological potential because of this inducible and reversible anisotropy and cooperative motion [1–4]. By incorporation of cleverly designed photo-responsive molecules into the LCs, this molecular

* corresponding author

Molecular Technology: Energy Innovation, Volume 1, First Edition.
Edited by Hisashi Yamamoto and Takashi Kato.
© 2018 Wiley-VCH Verlag GmbH & Co. KGaA. Published 2018 by Wiley-VCH Verlag GmbH & Co. KGaA.

alignment can be precisely controlled over large areas and enables the fabrication of specialty light-responsive materials for photonic applications such as optical switching and signal processing, lasing, and actuation [5–8]. Light-controlled functional switching processes in particular have been attracting great attention from both Materials Chemistry and Optical Physics research scientists. Specific advantages of photo-induced functional systems for application in devices include facile non-contact influence, superior spatio-temporal resolution, and multifunctional operability. Effectively, it is these new functionalities and advantages of photo-responsive molecular materials that have enabled many of the devices we now use daily [9–12].

One of the most elegant and effective of these light-control schemes devised is based on reversible photo-switching through incorporation of photo-chromic units doped or functionalized into a host LC material system. Azobenzene derivatives, for example, through their reversible *trans-cis* geometric isomerization [13], can be used to construct molecular-level switchable systems [14] to control self-assembly and aggregation of various supramolecular systems [15] or the structure and function of biomolecules [16, 17] and to design stimuli-responsive macroscopic actuators, micro-motors, and photo-deformable materials. In other examples, similar simple absorbing dyes can be used to bring about enhanced response of the material system to optical fields [18, 19]. This is especially effective in the important field of photorefractive systems, where photo-alignment has been developed effectively to become a major component of NLO photo-functional materials research [20].

Reversible photo-induced molecular alignment enabled by azobenzene dates back to early holographic recording experiments by the groups of Todorov [21] and Wendorff [22] in the 1980s. Ichimura et al. were then the first to propose a "command surface" concept to control the orientation of bulk LC molecules by means of an adjacent photo-responsive surface [23, 24]. Tazuke et al. then demonstrated photo-chemical phase transitions achievable with light, that is, that a photo-isomerization process can disrupt the LC phase and turn the material isotropic [25]. In the early 2000s, cross-linked LC systems emerged as a new mechanically improved class of materials for photo-alignment control. Finkelmann et al. demonstrated the photo-contraction of LC elastomers [26], and the Ikeda group demonstrated a large photo-induced bending of cross-linked LC films [27, 28]. Key to those macroscopic LC actuators is an amplification of the molecular-level switching motion through cooperative effects that take place in LCs. Interestingly, such photo-induced cooperative effects were also reported in photo-chromic crystals by the Irie group [29]. Compared with elastomers, crystals can exhibit faster photo-mechanical actuation because of their higher Young's modulus, enabling them as an emerging new class of "harder" but flexible photo-mechanical materials.

The goal of this chapter is to review and highlight, through selected recent examples, some significant leading examples of photo-control that has achieved over molecular alignment and motion in the design of macroscopic photo-switchable materials and light-driven actuators, and some emerging trends evolving in new directions of mechanical applications. Several lengthy comprehensive reviews on both photo-responsive LCs and photo-mechanical materials have appeared during the past few years [6, 30–36], which we recommend for more detailed research reading, and along with this chapter, we hope will serve to demonstrate that this field has both a long history, yet is also still timely and exciting for new directions of application. We start by outlining some recent advances in photo-chemical photo-alignment control using dyes, then present some new trends in photo-physico-chemical alignment control in dye-free systems, and then conclude with some key reports of application of these systems for actuation and devices, finally summarizing some remaining challenges and future perspectives.

5.2 Photo-Chemical Alignment

It has been more than 30 years since early research toward light-induced molecular alignment control in photo-responsive azobenzene-containing polymer films started receiving wide attention and the first practical demonstrations for various photonic applications. It was 1984 when Todorov et al. first observed that birefringence could be induced reversibly by linearly polarized light and thus demonstrated polarization-holographic gratings in amorphous poly(vinyl alcohol) films doped with methyl orange dye [21], and in 1987 Eich and Wendorff et al. extended these studies to liquid-crystalline polymers that enabled reversible holographic gratings inscribed with higher diffraction efficiency [22]. The first demonstration of photo-alignment control using nondye LC materials via surface photochemistry was reported by Ichimura et al. in 1988 [23], where they observed that the *trans-cis* isomerization of azobenzene molecules anchored on a 2D substrate could induce a homeotropic-to-planar reorientation of 3D bulk LC molecules adjacent. They termed this clever functional surface a "command surface," evoking the image of many transparent "soldier" LC molecules in bulk following the orientational orders of the azo "commanders" from the surface (Figure 5.1). Shortly after this first report, Gibbons et al. [38], Dyadyusha et al. [39, 40], and Schadt et al. [41] nearly simultaneously demonstrated similar photo-chemical alignment success using an azo-dye-doped polyimide and a photo-cross-linkable polymer surface, respectively. Inspired by these pioneering early studies, synthesis of new photo-aligning materials and development of new photo-alignment technologies has become one of the hottest recent topics of applied LC science. Following these pioneering studies, many other photo-chromic and photo-cross-linkable materials were reported shortly thereafter, as reviewed well in 2000 and then again in 2002 [42, 43].

A wide range of applications were proposed and demonstrated by many groups in the two decades following these first reports, for example, optical storage, holography, optical switching, and display technologies, in response to evolving technology needs [44–47]. More recently, photo-alignment methods were introduced into the industrial production of LC displays as a facile alternative to alignment via rubbing techniques [48], and today, these advanced surface photo-alignment techniques for controlling the orientation of LCs are now standard in industry as strong competition for conventional rubbing-based alignment layers in display technologies [49]. Compared with physical techniques,

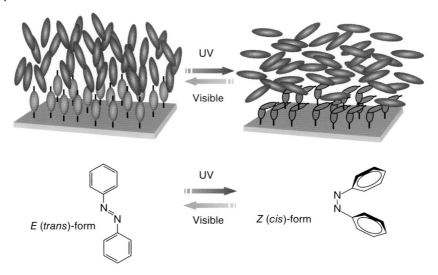

Figure 5.1 Schematic illustration of reversible photo-alignment by a "command surface." *Source*: From Seki 2014 [37]. Reproduced with permission of Nature Publishing Group.

photo-alignment layers enable remote and reversible control over anisotropic molecular arrangement, which in turn allows patterning and better control over optical properties, for example, light propagation through LCs [20, 37, 50–52]. From the standpoint of fundamental research, photo-alignment control in polymeric LC films is still of great interest in next-generation photonic applications [53]. Toward both these applied and more fundamental research efforts, we now highlight recent examples of current state-of-the-art photo-alignment control of LC polymers and then some new photo-alignment materials that undergo other photo-chemical reactions.

Generally speaking, the photo-alignment control of LC materials has been based on either photo-induction of anisotropy in thin alignment layers, as so-called "command surfaces" [24] or by doping small amounts of photo-responsive units into the bulk of the LC material [48]. Either way, an azo or other dye needs to be incorporated into the device to transduce the incident light into molecular orientation. A new strategy, employing a molecular command system at the free surface as opposed to the substrate, was recently proposed by the Seki group [54], which can offer some distinct advantages. Here, they blended a small amount of an azobenzene-containing block copolymer (azo-BCP) into a non-photo-responsive LC polymer. The LC polymer segregates to the LC-air interface and spontaneously assumes a homeotropic alignment on clean quartz slides and then acts as a "free-surface command layer" that allows for in-plane photo-alignment control of the LC polymer with linearly polarized light. With this film system, in-plane photo-patterning can thus be achieved with linearly polarized light. The free-surface command layers can then be removed after the photo-alignment process, effectively providing a dye-free system after this removal step, which can then be reapplied onto secondary material surfaces. Using a standard inkjet printing process, for example, onto various material

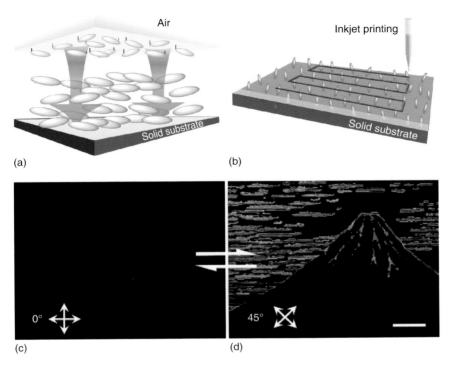

Figure 5.2 (a) Schematic illustration of a photo-alignment system using a photo-responsive layer at the free surface. (b) Illustration of the inkjet printing procedure of a free-surface command layer on a homeotropically aligned, non-photo-responsive LC polymer. (c and d) Examples of birefringence patterning by this inkjet alignment method. Scale bar, 200 μm. *Source*: From Fukuhara et al. 2014 [54]. https://www.nature.com/articles/ncomms4320. Licensed under CC BY 4.0.

surfaces, they demonstrated that photo-aligned fine patterning and arbitrary "designer" images of aligned mesogens in the polymer films are possible and facile, with high spatial resolution, as demonstrated in Figure 5.2. This clever and general strategy is applicable to a wide variety of material systems and importantly requires no pretreatment or modification of the substrate surface. Building on this first report, Nakai et al. then further demonstrated a reversible out-of-plane photo-switching effect from a free-surface command skin layer [55], using a homeotropic-planer reversible alignment of a cyanobiphenyl side chain polymer film, controlled by the photo-isomerization of an azobenzene command polymer placed at the free surface.

A further recent development in photo-alignment control that should be highlighted is the use of azobenzene-containing ferroelectric LC polymers as switchable second-order nonlinear optical (NLO) materials. Here, photo-induced changes in molecular alignment can provide several advantageous secondary functions, and we next introduce the use of azobenzene-containing ferroelectric LC polymers as switchable second-order NLO materials [56]. Such NLO effects can be difficult to achieve in bulk because second-order NLO materials must necessarily possess a non-centrosymmetric molecular alignment of the molecular constituents. Organic materials have further advantages for use as NLO materials because unlike their traditional inorganic crystal counterparts, the NLO response of organic molecules can be reversibly switched by photo-chemical or electro-chemical stimuli. Toward this goal, alignment methods based on azobenzene-triggered macroscopic order-disorder molecular phase changes in cross-linked ferroelectric LC polymers were proposed recently that provide both high-contrast and reversible switching in cross-linked ferroelectric LC polymers, as shown in Figure 5.2. To address the significant problem of the large optical extinction coefficient of azobenzene moieties, two-photon excitation was applied for efficient switching through the bulk of the material [56], which also offers a significant improvement on the reversibility and repeatability of the switching and also affords high switching contrast that can be achieved via two-photon methods.

New materials have also been developed to accommodate specification requirements demanded from various specialized photonic applications. Photo-chemically reactive materials with requirements of extremely low color, for example, or extremely high thermal stability, can be crucial for display technologies with greater color fidelity needs or for use in more extreme environments [48]. Another thrust of photo-chemical alignment research is in developing new photo-responsive materials beyond traditional azobenzene dyes. Schadt et al., for example, have reported photo-alignment of LCs with polyvinylcinnamate films with linearly polarized light [41, 57], and the Kawatsuki group has also systematically investigated the photo-alignment behavior of photo-cross-linkable LC polymers containing cinnamate and cinnamic acid (CA) derivatives [58]. Recently, they reported a new photo-reactive liquid-crystalline polymer composed of 4-methoxy-N-benzylideneaniline (NBA) side groups that showed photo-induced molecular reorientation for holographic applications [59], where a large birefringence of 0.11 was obtained in a perfectly colorless system. Moreover, they explored this photo-induced reorientation of composite films consisting of a photo-inactive polymethacrylate with benzoic acid (BA) side groups and photo-responsive monomeric materials such as CA or NBA derivatives [60, 61]. H-bonding between BA and CA/NBA side groups was encouraged and played an important role in achieving a sufficient amplification of the photo-alignment. In other similar reports, the same group demonstrated a facile fabrication of a photo-alignable polymer film achieved by coating two non-photo-reactive materials employing the free surface [62, 63]. Here, they produced photo-alignable NBA side groups in selective areas by coating phenylamine derivatives onto a polymethacrylate film with phenylaldehyde side groups, where neither coating materials have photo-reactivity. Most recently, the same group further achieved a control of homeotropic/homogeneous alignment by means of top-coating aromatic molecules combined with irradiation with linearly polarized light [64]. Via this process, they demonstrated a precise control of in-plane and out-of-plane orientations using cheap and facile inkjet printing of aromatic molecules.

Stumpe and coworkers have recently demonstrated other novel types of colorless photo-active molecules for the induction of anisotropy, with impressive dichroism values reported of up to 0.2 [65]. Here, "photo-rotors" containing a photo-sensitive ethane unit flanked by donor and acceptor substituents were used, elongated by cyclohexane ring systems for forming rod-like LCs. This photo-rotor molecule has practically no absorption in the visible region; therefore, high-quality colorless anisotropic optical films can be fabricated by this process, with high induced dichroism. Finally, Kosa et al. reported various naphthopyran-containing LCs that can undergo order-increasing photo-induced phase transitions with large photo-induced dichroism, with potential implications in ophthalmic applications [66]. These naphthopyran-based compound exhibited photo-induced conformational change of the photo-switchable molecule from a closed form to an open form, which leads to order-increasing phase transitions. By doping the naphthopyran molecules into a LC host, it was observed that the phase transition temperature of the host materials was shifted to lower temperature with the closed form shape of the naphthopyran. These photo-induced shifts of phase transition induced by the photo-responsive molecular conformational changes then leads to the order-increasing phase transitions. Isotropic-to-nematic and nematic-to-smectic phase transitions have both been demonstrated by UV light irradiation as shown in Figure 5.3. Furthermore, the naphthopyran molecules of this system demonstrated unprecedented

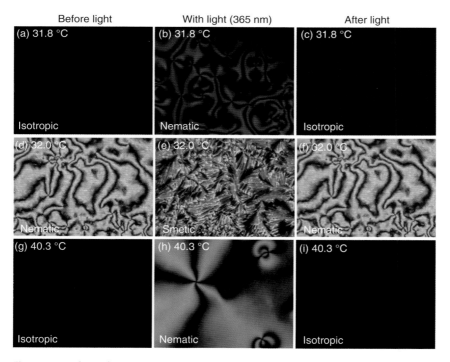

Figure 5.3 Polarized optical micrographs of phototropic phase transitions. Illustrations of the photo-induced phase transitions from isotropic to nematic (a–c), nematic to smectic (d–f), and isotropic to nematic (g–i). *Source*: From Kosa et al. 2012 [66]. Reproduced with permission of Nature Publishing Group.

changes in the order parameter of the dye, with potential applications in ophthalmic devices such as photo-chromic and polarized variable transmission sunglasses.

5.3 Photo-Physical Alignment

Research toward photo-physical systems, that is, materials in which the photo-alignment control is achieved by using non-photo-chromic groups, has also enjoyed strong recent progress. Compared with photo-chemical systems, non-photo-isomerizable dyes provide several unique advantages: First, the optical molecular reorientation only takes place when the incident light intensity is above a well-defined threshold, which allows writing and reading using the same wavelength without disturbing the molecular alignment. Second, photo-physical processes use an order-to-order process, which is more simple and facile, and also allows more precise control over LC alignment. Third, a non-photo-isomerizable system is more stable in many cases compared with azobenzene-containing systems because the conformational changes taking place in the photo-stationary state of azobenzene-doped LCs may give rise to reorientation instabilities and fluctuations. Photo-physical alignment methods have been developed primarily with NLO systems. The NLO class describes the behavior of light propagation through nonlinear media, meaning that optical output effects can depend strongly nonlinearly on the optical input and can be completely absent before a threshold is reached [67, 68]. As a result of the mass introduction of laser technology in the 1960s, various new and advanced applications enabled by NLO processes have been developed, such as frequency conversion [69, 70], multi-photon absorption [71, 72], self-phase modulation [73, 74], and self-focusing [75, 76]. However, the generation of such optical nonlinearities typically is challenging for soft organic materials because of the requirement of very high light intensities and costly for the high-power laser sources required. Several groups have thus worked with crystals [77, 78], high T_g polymers [79, 80], and robust LCs [81] that exhibit high optical nonlinearity to induce NLO effects with inexpensive and low-power lasers. In particular, some LC systems received great attention as NLO devices due to the strong LC order amplifying an optical nonlinearity [82]. So photo-physical reorientation has thus been extensively studied in dye-doped LCs because of such strong light-intensity-dependent refractive index changes that can be induced due to high LC molecular alignment. This "orientational optical nonlinearity" gives rise to interesting NLO phenomena such as self-phase modulation of light beams and the potential to generate optical solitons [68, 73, 81]. Compared with photo-chemical systems, non-photo-isomerizable systems may provide some unique characteristics, allowing for molecular reorientation only above a clear threshold intensity and reduced reorientation instabilities and fluctuations in the photo-stationary state. Several material systems with high optical nonlinearity have been reported [83–86].

Around 1980, three research groups almost simultaneously developed nematic liquid crystals (NLCs) that showed remarkably high optical nonlinearity, of up to nine orders of magnitude higher than that what had previously observed for usual materials [87–89]. This phenomenon was caused by an increase of the photo-induced molecular director orientation of LCs and the resultant huge change in the refractive index [87]. Marrucci offered a physical mechanism for this photo-induced NLCs molecular reorientation [82], by suggesting that when a homeotropic LC is vertically irradiated with linearly polarized light, the NLC molecular orientation directs along the polarization direction, which leads to the strong homogeneous orientation. At the same time, rotation of the molecular director prevented by surface anchoring and bulk elasticity is opposed. The final molecular director was thus determined based on the balance between these opposing torques, and the rotation of the molecular director leads to a rotation of birefringence axes, thus to a net high refractive index change. Following these first discoveries of remarkably high optical nonlinearity of NLCs, Janossy et al. reported that a large optical nonlinearity improvement in NLCs was possible by doping a small amount of anthraquinone dye [90]. The enhanced optical nonlinearity in such dye-doped NLC was two orders magnitude larger than that of non-doped NLC, as the molecular polarizability of photo-excited dyes enhanced the torque to rotate NLC. Thanks to this dye-induced optical nonlinearity enhancement, the input power-level threshold necessary to induce the reorientation was decreased to safer (and cheaper) levels for NLO effects [18].

Various dyes were then developed to further enhance the nonlinearity of LCs, where it was discovered that the chemical structure of dyes strongly affected the molecular reorientation efficiency [18, 84]. In 2000, Zhang et al. showed that oligothiophene (TR5) could also work successfully as photo-functional dye for the enhancement of optical nonlinearity of LCs [19]. Small amounts of TR5 were doped into both polar LCs (5CB) and nonpolar LCs (MBBA), and the threshold light intensity for photo-induced molecular reorientation was dramatically decreased in both systems. This research provides a novel way to control the LC orientation with light of conveniently low intensity. Moreover, because TR5 is a molecule with high fluorescence, this system can be expected to be useful for other interesting photonic applications requiring emission. In 2004, Lucchetti et al. reported a strong refractive index change caused by molecular reorientation that was increased with low light intensities in the range of $nW\,cm^{-2}$ [85]. They employed azobenzene-doped LC glass cells with several types of surface treatment and investigated their optical nonlinearity. In this report, the highest value of nonlinear refractive index reported so far in LC materials was achieved with untreated cells, suggesting that nonlinear response is enhanced by weaker anchoring conditions. These record-setting optical nonlinearities were determined via a holographic grating setup, which restricted the practical use of the optical nonlinearity due to its requirement for a two-beam interference. In contrast, a self-focusing effect for evaluating the threshold intensity is triggered more simply by single-beam irradiation, so can be expected to have a wider variety of applications [81]. In these systems, optical properties were evaluated using the diffraction ring pattern formed by self-phase modulation. The number of rings indicated a refractive index modulation at different light intensities, and the threshold intensity was defined as that reached when the first diffraction ring formed [89].

More recently, a conceptually new approach to enhance the nonlinearity of doped LCs, based on polymer stabilization, was proposed. Here, Aihara et al. found that TR5-doped LCs stabilized by photo-polymerization reduced the threshold intensity [91]. They combined oligothiophene molecules as low-molecular-weight absorbing dopants with polymer-stabilized LCs (PSLC) and investigated optical nonlinearity using the self-focusing effect. In this system, they employed homeotropic-aligned cells and observed that the threshold intensity became six times lower when compared with conventional homeotropic LCs. Moreover, Wang et al. investigated the NLO effects of hybrid-aligned oligothiophene-doped PSLC films [92], where again, self-diffraction rings were formed by photo-induced molecular reorientation at the lowest light intensity of 400 mW cm^{-2}, which indicates that hybrid-aligned PSLC decreases the threshold intensity by a factor of 8.5 compared with conventional homeotropic-aligned LCs. The decreasing threshold intensity was explained by a decrease in total surface anchoring. They also demonstrated that the self-focusing effect due to molecular reorientation could be induced with a simple 1-mW-battery-operated laser pointer as shown in Figure 5.4. Most recently, Usui et al. investigated a different approach for reducing light intensity for reorientation by modifying a substrate surface that controls initial molecular orientation in polymer-stabilized nematic LCs doped with oligothiophene [93]. This report showed that the threshold intensity for inducing molecular reorientation of PSLCs was greatly reduced by carefully controlling the surface treatment of glass substrates. The optical nonlinearity owing to molecular reorientation was quantitatively evaluated here by the measurement of the self-diffraction rings arising from self-focusing and self-phase modulation effects. The threshold intensity decreased as surface anchoring was weakened, as for the cells treated with a 0.003 wt% silane coupler solution, the threshold intensity was reduced by 30% compared with the highest silane coupler concentration. Further decreases in the threshold intensity of NLO in dye-doped LCs or PSLCs will lead to the development of photonic materials and devices, such as a novel material that shields only high-intensity light, of great interest to wearable optics, sunshields, and smart windows.

In addition to enhancing the optical nonlinearity (and decreasing the light intensity at which NLO self-phase modulation takes place), in such systems, it is also possible to lock in the molecular alignment through photo-polymerization, enabling the inscription of fixed photonic elements. Here, Yaegashi et al. created micro-lens arrays by combining photo-induced reorientation of dye-doped LC-LC mixtures and simultaneous photo-polymerization [94]. These fabricated microlens arrays have a polarization-selectivity and arrays with various lattice patterns were obtained by controlling the polarization directions of the incident beam. Conventional photo-alignment control has traditionally been triggered by photo-chemical or photo-cross-linkable molecules, and their further development depends on new fundamental developments in molecular design. Additionally, new strategies in material design, including free-surface optimization in azobenzene systems, and polymer-stabilization-enhanced photo-physical processes, have the potential for more flexibility in optimization of material performance and may well offer new routes for next-generation photo-alignment control.

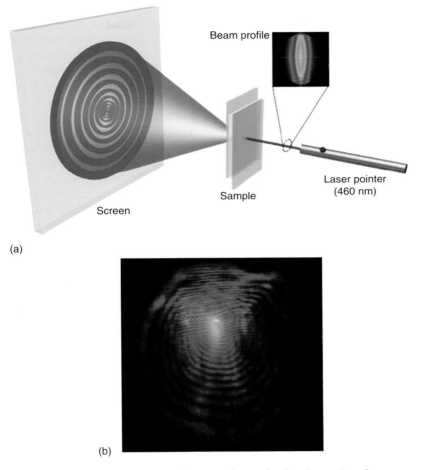

Figure 5.4 (a) Schematic diagram of the optical setup for the observation of laser-pointer-driven nonlinear optical effects in a hybrid-aligned cell. (b) Photograph of a ring pattern generated from a hybrid-aligned cell with a common 1 mW handheld laser pointer. *Source*: From Wang et al. 2015 [92]. https://www.nature.com/articles/srep09890. Licensed under CC BY 4.0.

5.4 Photo-Physico-Chemical Alignment

As described in the previous section, photo-induced molecular alignment methods using both photo-chemical and photo-physical systems have been extensively studied by many research groups, who have demonstrated various successful applications. These methods, however, still have some difficult challenges to overcome, such as a requirement of complicated (thus tedious, costly) procedures for fabricating molecular alignment layers commercially and subsequently aligning the LCs in a high-throughput mass production facility. In addition, specific photo-reactive compounds are needed for induction of the molecular alignment, which can often introduce unwanted mechanical or optical properties. The required irradiation with linearly polarized light is also not always desired or optimal. Thus, to address these concerns, there is great interest in developing novel photo-control methods of molecular alignment method by using photo-physico-chemical systems, that is, using light to induce an orientation, yet in completely dye-free systems that can be optimized instead for optical and mechanical properties, and amenable to cheap and facile high-throughput existing fabrication facilities [95].

The first of these new general dye-free photo-physico-chemical methods was proposed in 2016 [95]. In this new method, molecular alignment was induced by masked photo-polymerization of a monomer and cross-linker in a completely dye-free system, where masked photo-polymerization brings about molecular transport toward irradiated or unirradiated regions, depending on gradients in the chemical potential (Figure 5.5). This alignment process is common to a wide range of existing photo-polymerizable materials and does not require alignment layers or polarized light. This new paradigm builds upon some

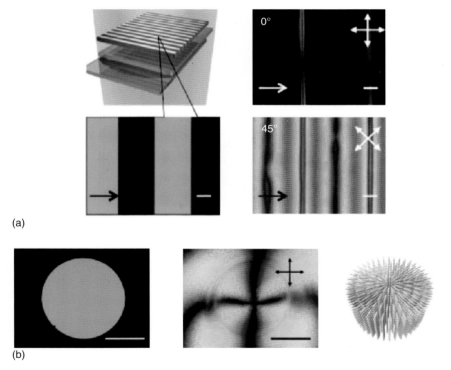

Figure 5.5 (a) Unidirectional molecular alignment behavior of the polymer film photo-polymerized with a line-space photomask. Illustration of the photo-polymerization process and a micrograph of the line-space photomask (left). Polarized optical micrographs of the polymerized film (right). (b) Two-dimensional (2D) molecular alignment by photo-polymerization with a pinhole photomask. Micrographs of the pinhole (left) and 2D aligned film (center). Illustration of the alignment direction of the fabricated polymer film (right). Source: From Hisano et al. 2016 [95]. http://iopscience.iop.org/article/10.7567/APEX.9.072601/meta. Licensed under CC BY 4.0.

previous research that was conducted to find such an integration of molecules is achieved by polymers using photo-polymerization systems. Historically, starting in 1976, Tomlinson et al. first fabricated volume-phase holograms and grating devices with photo-polymer systems [96]. The two-way diffusion induced by polymerization of the more reactive monomers at the exclusion of the less reactive species from the irradiated regions led to a gradient of the chemical composition and enabled the inscription of the gratings. The Bunning group also demonstrated holographic Bragg grating materials inscribable in a polymer dispersed LC system in 1993 [97]. Gratings were formed here by the changes in the chemical potential of the system induced by photo-polymerization. The LC separates as a distinct phase in submicrometer droplets in the regions of polymerization. The fabricated grating is thus a superposition of a polymer density grating and a LC droplet grating. Following this, Broer et al. investigated similar photo-polymerization-induced-diffusion systems. In 1995, they produced stable optical filters by cross-linking cholesteric molecules by photo-polymerization [98]. They balanced monomer diffusion by controlling the photo-polymerization rate for the formation of a pitch gradient and obtained wide-band reflective polarizers. More recently, they fabricated various polymeric relief microstructures by photo-polymerization with photomasks [99]. In this paper, they investigated the influence of various conditions such as structure period, energy dose, development temperature, film thickness, and photo-polymer blend composition. The same group then rationalized the phenomenology of this system by a diffusion-polymerization model [100]. According to this model, the polymerization-induced monomer-concentration gradient together with diffusivity differences, cross-linking properties, interaction between the different components, and the surface free energy together determine the migration of monomer and therefore the final relief structure. Despite these previous reports, however, still no one has reported the direct induction of molecular alignment through diffusion, with no alignment layer.

Molecular alignment induced by masked photo-polymerization of a monomer and cross-linker in a completely dye-free system could now be considered a leading technique for future applications, where masked photo-polymerization brings about molecular transport toward irradiated or unirradiated regions, depending on gradients in the chemical potential [95]. This is the first report that masked polymerization with nonpolarized light enables the precise control of molecular alignment without using a conventional alignment layer. In this process, they used a mixture of an optically anisotropic acrylate monomer and an isotropic dimethacrylate cross-linker. The molecular alignment direction depends on the shape of the photomask because the alignment direction is controlled by the vector direction normal to the boundary between the irradiated and unirradiated regions and the resultant molecular diffusion. As shown in Figure 5.5, uniform 1D or 2D molecular alignment was achieved by this photo-physico-chemical alignment technique. Moreover, they investigated the effect of photo-polymerization temperature, and it was revealed that this alignment is realized not by the cooperative effect of molecules or elongation of the film surface arising from embossing but by the shear stress arising from molecular diffusion. This novel alignment method revealed that photo-polymerization-induced molecular alignment can be a unique and improved alternative approach for precisely aligning molecules with various complex patterns.

5.5 Application as Photo-Actuators

In addition to all of the applications as orientable materials, one interesting side-use of many of these light-responsive systems, especially those that rely on reversible shape changes, is the mechanical forces and stresses that can result reversibly from irradiation. If cleverly applied, these molecular forces can be amplified into macroscopic shape-shifting, actuating "artificial muscles," or "molecular machines." Actuators are systems in which energy is converted from any input stimulus into useful mechanical motion, and light-activated photo-reversible systems in particular can often be easily designed and synthesized, and many are able to undergo large deformation upon relatively low input stimulus [101, 102]. Photo-mechanical actuation, where light energy is converted into mechanical shape changes of the material, is particularly promising for devices, due to the possibility for precisely defined, noncontact actions triggered by low-cost light sources or even by sunlight. Various light-induced shape changes have been achieved in shape-memory polymers [103], carbon-nanotube-containing composites and bilayers [104, 105], and cross-linked polymers and elastomers incorporating photo-chromic molecules [106–108]. Now, we focus on photo-chromic actuators, in the same materials as those of the photo-chemical molecular alignment techniques that have developed orientation applications. One example of interesting photo-mechanical behavior is a bending of twisted-nematic elastomers to act as light actuators.

Efficient control over the molecular alignment is a key for achieving and optimizing large-scale photo-mechanical effects, yet can be different when comparing the photo-mechanical effect in amorphous and liquid-crystalline polymer systems. In amorphous azobenzene-containing polymers, due to the lack of cooperative molecular motions, photo-induced dimensional changes only in the range of 1% can be achieved, and in general, they photo-expand with light. In LC polymers and elastomers, however, reversible uniaxial photo-contraction may reach 15–20% [26, 109, 110], due to coupling between photo-modification of molecular alignment and large mechanical deformation provided by the cross-linked polymer network, and in general, these systems photo-contract. The light-induced forces generated both systems, which are brought about by surface strains in three-dimensional deformed films [111], are often large enough to do significant work against external load, and are being able to fuel, for example, plastic motors [112], robotic-arm movements [113], and catapult motions [114]. An interesting new development in the design of photo-actuators is the expansion from linear contraction and in-plane bending to out-of-plane twisting and helical motions. An inspiration for increasing the complexity of the photo-induced motions perhaps derives from nature: Various biological "engines" are built upon twisting or helical motions [115, 116], and several animals combine bending, twisting, and sweeping motions to generate power for efficient flight [117, 118]. White and coworkers were the first to emulate and achieve such combined in-plane oscillation and out-of-plane twisting in artificial azo-containing LC polymer networks. Oscillation (or bending) coupled with out-of-plane twisting occurs at intermediate angles, due to the combined strain and shear gradients caused by nonuniform light absorption through the thickness of the cantilever [119].

Twisted-nematic elastomers have proven particularly interesting in terms of their photo-mechanical behavior, as demonstrated by Harris et al. that the chirality associated with the twisted molecular alignment may produce a coiling motion of elastomeric cantilevers [120]. Broer and coworkers also reported an azobenzene-doped photo-polymerized film with twisted networks that showed uniaxial bending or helical coiling deformation modes after UV irradiation [120]. The initial shape of twisted-nematic cantilevers is sensitive to their dimensions: Depending on the aspect ratio, they can adopt either helicoidal or spiral shapes [121, 122]. The light-triggered twisting/coiling motions were further extended by Wie et al. who reported photo-mechanical responses of both twisted-nematic and hybrid-aligned cantilevers [123]. A further step toward biomimicry was recently taken by Katsonis et al., who fabricated spring-like photo-actuators and demonstrated various types of complex motions such as helix winding, unwinding, and inversion, all photo-induced as shape changes of azobenzene-containing LC polymer films (Figure 5.6) [124]. Here, using spring-like polymeric films, molecular movement was converted and amplified into controlled and reversible twisting motions, that by careful sample preparation techniques these LC polymer photo-actuators comprised regions that exhibit different dynamic behavior, where these light-induced conformation changes were determined and controlled by their alignment directions. Such motion mimics the movement of plant tendrils found, for example, in wild cucumbers. Moreover, they reported photo-switching behavior and thermal stability of photo-activated molecular deformation systems on a larger macroscopic scale [125]. By employing fluorinated azobenzenes as basic switching elements to slow the thermal reconversion from cis back to trans, the fluorinated switches activated by visible light retained their photo-chemical shape for more than 8 days.

Newer interesting developments of azobenzene photo-mechanical systems derive from the use of either complex LC order or LC elastomer microstructures. Complex-ordered and patterned LC polymer networks have been most notably studied by Broer and coworkers [126, 127], and engineering of complex molecular ordered and patterned LC polymer networks and elastomers into proof-of-principle real devices has been studied by both Broer and White groups [128]. Here, they used LC cells containing cinnamate-based photo-alignment layers that were irradiated through a photomask with a wedge-shaped opening while slowly rotating the cells, using a programming of LCN and LCE materials to localize their mechanical response to generate surface features or local shape changes. Using these cells for photo-polymerization, various complex-ordered free-standing films with continuous change of the LC alignment direction were

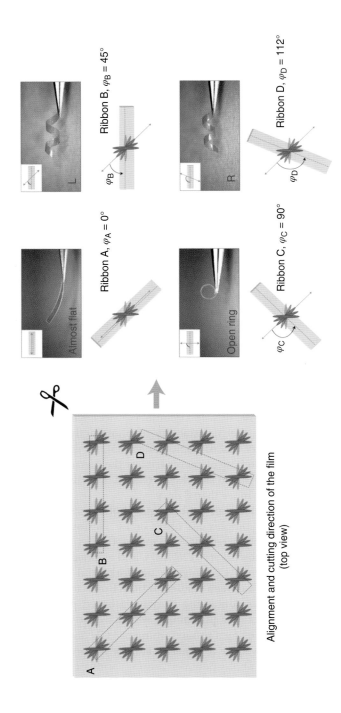

Figure 5.6 Photo-actuation modes of azobenzene-containing cross-linked LC polymer ribbons, exhibiting a complex twisting photo-mechanical response. *Source:* From Iamsaard et al. 2014 [124]. Reproduced with permission of Nature Publishing Group.

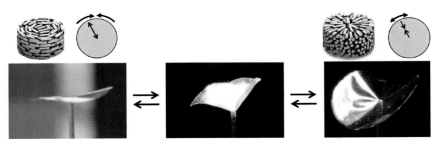

Figure 5.7 Actuation behavior of cross-linked LC polymer films with the azimuthal and radial alignment upon heating with an IR lamp. The arrows along the radius and the azimuth indicate the direction of deformation. *Source*: From de Haan et al. 2012 [126]. Reproduced with permission of John Wiley and Sons.

achieved. And most recently, Broer et al. reported a heat-driven mechanical effect of azimuthal and radial aligned actuator films with an IR lamp (Figure 5.7) [126]. By using patterned molecular alignment, interesting photo-thermal actuators exhibiting, for example, checkerboard patterns upon stimulus, were fabricated, where the deformation directions of the actuator films were observed to be different, depending on the alignment patterns [127].

Topological defects of similar liquid-crystalline polymer networks were also investigated and reported by White and coworkers [129]. They then demonstrated several types of surface topographies fabricated by exposing the azobenzene-functionalized LCN films in a controlled manner, applying these patterned alignment techniques to liquid-crystalline elastomers [130, 131], where the polymer films with voxelated circular director profiles showed a thermomechanical response. Upon heating, conical actuation was observed in the polymer film, and on cooling, the deformation recovers, yielding the initial flat film reversibly (Figure 5.8). Most recently, they reported a photo-responsive and reversible shape change in elastic films prepared with azo-LCE [132]. Here, photo-induced reversible shape morphing between 2D and 3D shapes was demonstrated. This complex alignment method can be addressed both remotely and selectively, which has great advantages for a variety of applications. Finally and most recently, Yu et al. reported a clever new strategy to manipulate fluid slugs by light-driven tubular microactuators fabricated from photo-responsive liquid-crystalline polymers [133]. Photo-responsive asymmetric deformation of the actuator induces capillary forces for liquid propulsion as shown in Figure 5.9. They fabricated several shapes of micro-actuators such as straight, serpentine, helical, and "Y"-shaped from a mechanically robust linear LC polymer. Moreover, they also created a light-driven micro-swimming "robot" with a gripper based on this new functional material to realize more complex movements like swimming, grabbing, carrying, and transport [134]. These materials represent remotely light-driven effects achieving complex driving and controlling of various micro-robotic motions and structures, and therefore, these materials are expected to hold great potential for applications as micro-machines and light-powered robotics.

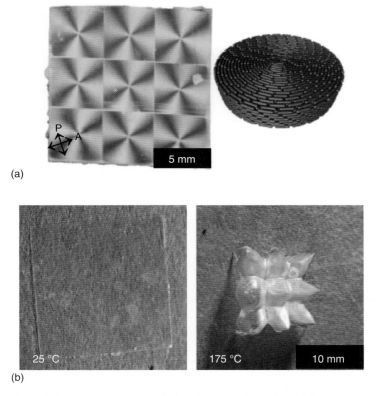

(a)

(b)

Figure 5.8 (a) Photograph of LCE film with nine +1 topological defects between crossed polarizers and illustration of the alignment direction around the defect. (b) Actuation behavior of the LCE film. Nine cones arise from the LCE film upon heating, and the film becomes reversibly flat upon cooling. *Source*: From Ware et al. 2015 [130]. Reproduced with permission of The American Association for the Advancement of Science.

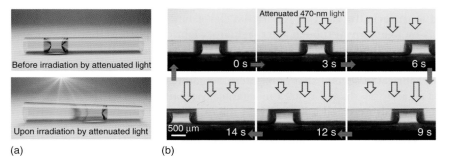

(a) (b)

Figure 5.9 (a) Illustrations showing a light-driven motion of liquid "slugs" inside a tubular microactuator driven by photo-deformation. (b) Photographs of the reversible motion of a silicone oil slug inside a microtube actuated by 470 nm light. *Source*: From Lv et al. 2016 [133]. Reproduced with permission of Nature Publishing Group.

5.6 Conclusions and Perspectives

Precise and reversible control over molecular alignment and shape has emerged in recent years as a crucial key requirement for a wide array of optics applications such as reconfigurable photonic elements and optical-to-mechanical energy conversion. Soft materials such as LCs and liquid-crystalline polymers have also emerged as some of the most promising and exciting classes of host materials to achieve such photo-switching effectively and efficiently, where molecular properties can be separately tailored and tuned for each specific application or device. As a stimulus for reversible control over these properties, light can be considered as superior, for a direct transfer of photon energy into mechanical motion with no moving parts. Light is also an ideal triggering mechanism as it can be localized in time and space to allow for remote activation of a system, is cheap and easy to employ, and visible light is also an inherently gentle non-damaging stimulus. Thus, for alignment control, actuation, and mechanical motion, photo-functional soft materials are of greatest recent interest. Azobenzene is the emerging leader among the small class of photo-reversible molecules, and soft azobenzene-containing materials show great promise for next-generation photonic and photo-mechanical devices. In this chapter, we have introduced some key current research areas of photo-driven molecular alignment methods and highlighted some of their recent applications using photo-chemical, photo-physical, and photo-physico-chemical systems. Photo-chemical and photo-physical alignment processes especially benefit from well-established theoretical understanding. The latest class, photo-physico-chemical alignment methods, where alignment shear stress arises from molecular diffusion, are also now just being approached theoretically to help rationalize and thus optimize this new type of molecular alignment system. Twisting, aligning, and bending materials with light is an exciting effect in particular that can offer important and significant advantages to many applied fields and warrants much further study and application. Continued efforts toward the development of molecular alignment control with light can open new possibilities and opportunities for new future applications of functional soft materials for next-generation effects and devices.

References

1 Fleischmann, E.K. and Zentel, R. (2013). *Angew. Chem. Int. Ed.* 52: 8810.
2 Bisoyi, H.K. and Kumar, S. (2011). *Chem. Soc. Rev.* 40: 306.
3 Goodby, J.W. (2011). *Liq. Cryst.* 38: 1363.
4 Kato, T., Mizoshita, N., and Kishimoto, K. (2006). *Angew. Chem. Int. Ed.* 45: 38.
5 De Sio, L., Tabiryan, N., Bunning, T. et al. (2013). *Prog. Optics* 58: 1.
6 Yu, H. and Ikeda, T. (2011). *Adv. Mater.* 23: 2149.
7 White, T.J., McConney, M.E., and Bunning, T.J. (2010). *J. Mater. Chem.* 20: 9832.
8 Coles, H. and Morris, S. (2010). *Nat. Photon.* 4: 676.

9 Priimagi, A., Barrett, C.J., and Shishido, A. (2014). *J. Mater. Chem. C* 2: 7155.
10 Seki, T. (2016). *J. Mater. Chem. C* 4: 7895.
11 Xiao, K., Kong, X.-Y., Zhang, Z. et al. (2016). *J. Photochem. Photobiol. C* 26: 31.
12 Bisoyi, H.K. and Li, Q. (2014). *Acc. Chem. Res.* 47: 3184.
13 Dhammika Bandara, M.H. and Burdette, S.C. (2012). *Chem. Soc. Rev.* 41: 1809.
14 Russew, M.M. and Hecht, S. (2010). *Adv. Mater.* 22: 3348.
15 Yagai, S. and Kitamura, A. (2008). *Chem. Soc. Rev.* 37: 1520.
16 Beharry, A.A. and Woolley, G.A. (2011). *Chem. Soc. Rev.* 40: 4422.
17 Goulet-Hanssens, A. and Barrett, C.J. (2013). *J. Polym. Sci. Part A* 51: 3058.
18 Janossy, I. and Lloyd, A.D. (1991). *Mol. Cryst. Liq. Cryst.* 203: 77.
19 Zhang, H., Shiino, S., Shishido, A. et al. (2000). *Adv. Mater.* 12: 1336.
20 Yaroshchuk, O. and Reznikov, Y. (2012). *J. Mater. Chem.* 22: 286.
21 Todorov, T., Nikolova, L., and Tomova, N. (1984). *Appl. Opt.* 23: 4309.
22 Eich, M., Wendorff, J.H., Reck, B., and Ringsdorf, H. (1987). *Makromol. Chem. Rapid Commun.* 8: 59.
23 Ichimura, K., Suzuki, K., Seki, T. et al. (1988). *Langmuir* 4: 1214.
24 Ichimura, K. (2000). *Chem. Rev.* 100: 1847.
25 Tazuke, S., Kurihara, S., and Ikeda, T. (1987). *Chem. Lett.* 16: 911.
26 Finkelmann, H., Nishikawa, E., Pereira, G.G., and Warner, M. (2001). *Phys. Rev. Lett.* 87 (015501).
27 Ikeda, T., Nakano, M., Yu, Y. et al. (2003). *Adv. Mater.* 15: 201.
28 Yu, Y., Nakano, M., and Ikeda, T. (2003). *Nature* 425: 145.
29 Kobatake, S., Takami, S., Muto, H. et al. (2007). *Nature* 446: 778.
30 Yu, H. (2014). *Prog. Polym. Sci.* 39: 781.
31 Seki, T. (2013). *Macromol. Rapid. Commun.* 35: 271.
32 Mahimwalla, Z., Yager, K.G., Mamiya, J. et al. (2012). *Polym. Bull.* 69: 967.
33 Irie, M., Fukaminato, T., Matsuda, K., and Kobatake, S. (2014). *Chem. Rev.* 114: 12174.
34 Abendroth, J.M., Bushuyev, O.S., Weiss, P.S., and Barrett, C.J. (2015). *ACS Nano* 9: 7746.
35 Naumov, P., Chizhik, S., Panda, M.K. et al. (2015). *Chem. Rev.* 115: 12440.
36 Nagano, S. (2016). *Chem. Rec.* 16: 378.
37 Seki, T. (2014). *Polym. J.* 46: 751.
38 Gibbons, W.M., Shannon, P.J., Sun, S.-T., and Swetlin, B.J. (1991). *Nature* 351: 49.
39 Dyadyusha, A., Kozinkov, V., Marusii, T. et al. (1991). *Ukr. Fiz. Zh.* 36: 1059.
40 Dyadyusha, A.G., Marusii, T.Y., Reshetnyak, V.Y. et al. (1992). *JETP Lett.* 56: 17.
41 Schadt, M., Schmitt, K., Kozinkov, V., and Chigrinov, V. (1992). *Jpn. J. Appl. Phys.* 31: 2155.
42 Natansohn, A. and Rochon, P. (2002). *Chem. Rev.* 102: 4139.
43 Delaire, J.A. and Nakatani, K. (2000). *Chem. Rev.* 100: 1817.
44 Natansohn, A. and Rochon, P. (1999). *Adv. Mater.* 11: 1387.
45 Ikeda, T. (2003). *J. Mater. Chem.* 13: 2037.

46 Shibaev, V., Bobrovsky, A., and Boiko, N. (2003). *Prog. Polym. Sci.* 28: 729.
47 Shishido, A. (2010). *Polym. J.* 42: 525.
48 Chigrinov, V.G., Kozenkov, V.M., and Kwok, H.S. (2008). *Photoalignment of Liquid-Crystalline Materials: Physics and Applications*, Wiley SID Series in Display Technology. Wiley.
49 Miyachi, K., Kobayashi, K., Yamada, Y., and Mizushima, S. (2010). *SID Sym. Dig. Tech. Papers* 41: 579.
50 Seki, T., Nagano, S., and Hara, M. (2013). *Polymer* 54: 6053.
51 Wei, B.Y., Hu, W., Ming, Y. et al. (2014). *Adv. Mater.* 26: 1590.
52 Tabiryan, N.V., Nersisyan, S.R., Steeves, D.M., and Kimball, B.R. (2010). *Opt. Photonics News* 21: 40.
53 Yu, H. (2014). *J. Mater. Chem. C* 2: 3047.
54 Fukuhara, K., Nagano, S., Hara, M., and Seki, T. (2014). *Nat. Commun.* 5: 3320.
55 Nakai, T., Tanaka, D., Hara, M. et al. (2016). *Langmuir* 32: 909.
56 Priimagi, A., Ogawa, K., Virkki, M. et al. (2012). *Adv. Mater.* 24: 6410.
57 Schadt, M., Seiberle, H., and Schuster, A. (1996). *Nature* 381: 212.
58 Kawatsuki, N. (2011). *Chem. Lett.* 40: 548.
59 Kawatsuki, N., Matsushita, H., Kondo, M. et al. (2013). *APL Materials* 1: 022103.
60 Minami, S., Kondo, M., and Kawatsuki, N. (2016). *Polym. J.* 48: 267.
61 Fujii, R., Kondo, M., and Kawatsuki, N. (2016). *Chem. Lett.* 45: 673.
62 Kawatsuki, N., Miyake, K., and Kondo, M. (2015). *ACS Macro Lett.* 4: 764.
63 Kawatsuki, N., Miyake, K., Ikoma, H. et al. (2015). *Polymer* 77: 239.
64 Miyake, K., Ikoma, H., Okada, M. et al. (2016). *ACS Macro Lett.* 5: 761.
65 Rosenhauer, R., Kempe, C., Sapich, B. et al. (2012). *Adv. Mater.* 24: 6520.
66 Kosa, T., Sukhomlinova, L., Su, L. et al. (2012). *Nature* 485: 347.
67 Shen, Y.R. (1984). *The Principle of Nonlinear Optics*. New York: John Wiley & Sons.
68 Tabiryan, N.V., Sukhov, A.V., and Zel'dovich, B.Y. (1986). *Mol. Cryst. Liq. Cryst.* 136: 1.
69 Dalton, L.R., Sullivan, P.A., and Bale, D.H. (2010). *Chem. Rev.* 110: 25.
70 Kumar, R.A. (2013). *J. Chem.* 2013: 154862.
71 Pawlicki, M., Collins, H.A., Denning, R.G., and Anderson, H.L. (2009). *Angew. Chem. Int. Edit.* 48: 3244.
72 Park, S.H., Yang, D.Y., and Lee, K.S. (2009). *Laser Photo. Rev.* 3: 1.
73 Peccianti, M. and Assanto, G. (2012). *Phys. Rep.* 516: 147.
74 Dudley, J.M., Genty, G., and Coen, S. (2006). *Rev. Mod. Phys.* 78: 1135.
75 McLaughlin, D.W., Muraki, D.J., and Shelley, M. (1996). *J. Phys. D* 97: 471.
76 Wan, W., Dylov, D.V., Barsi, C., and Fleischer, J.W. (2010). *Opt. Lett.* 35: 2819.
77 Sasaki, T., Mori, Y., Yoshimura, M. et al. (2002). *Mater. Sci. Eng. R* 30: 1.
78 Vijayan, N., Babu, R.R., Gopalakrishnan, R. et al. (2004). *J. Cryst. Growth* 262: 490.
79 Yesodha, S.K., Sadashiva Pillai, C.K., and Tsutsumi, N. (2004). *Prog. Poly. Sci.* 29: 45.
80 Cho, M.J., Choi, D.H., Sullivan, P.A. et al. (2008). *Prog. Poly. Sci.* 33: 1013.

81 Khoo, I.C. (2009). *Phys. Rep.* 471: 221.
82 Marrucci, L. (2002). *Liq. Cryst. Today* 11: 1.
83 Khoo, I.C. (1996). *IEEE J. Quantum Elect.* 32: 525.
84 Marrucci, L., Paparo, D., Maddalena, P. et al. (1997). *J. Chem. Phys.* 107: 9783.
85 Lucchetti, L., Di Fabrizio, M., Francescangeli, O., and Simoni, F. (2004). *Opt. Commun.* 233: 417.
86 Budagovsky, I.A., Zolot'ko, A.S., Ochkin, V.N. et al. (2008). *J. Exp. Theor. Phys.* 106: 172.
87 Zel'dovich, B.Y., Pilipetskii, N.F., Sukhov, A.V., and Tabiryan, N.V. (1980). *JETP Lett.* 31: 263.
88 Zolot'ko, A.S., Kitaeva, V.F., Kroo, N. et al. (1980). *JETP Lett.* 32: 158.
89 Durbin, S.D., Arakelian, S.M., and Shen, Y.R. (1981). *Phys. Rev. Lett.* 47: 1411.
90 Jánossy, I., Lloyd, A.D., and Wherrett, B.S. (1990). *Mol. Cryst. Liq. Cryst.* 179: 1.
91 Aihara, Y., Kinoshita, M., Wang, J. et al. (2013). *Adv. Opt. Mater.* 1: 787.
92 Wang, J., Aihara, Y., Kinoshita, M. et al. (2015). *Sci. Rep.* 5 (9890).
93 Usui, K., Katayama, E., Wang, J. et al. (2017). *Polym. J.* 49: 209.
94 Yaegashi, M., Kinoshita, M., Shishido, A., and Ikeda, T. (2007). *Adv. Mater.* 19: 801.
95 Hisano, K., Kurata, Y., Aizawa, M. et al. (2016). *Appl. Phys. Express* 9: 072601.
96 Tomlinson, W.J., Chandross, E.A., Weber, H.P., and Aumiller, G.D. (1976). *Appl. Opt.* 15: 534.
97 Sutherland, R.L., Natarajan, L.V., Tondiglia, V.P., and Bunning, T.J. (1993). *Chem. Mater.* 5: 1533.
98 Broer, D.J., Lub, J., and Mol, G.N. (1995). *Nature* 378: 467.
99 Sánchez, C., de Gans, B.J., Kozodaev, D. et al. (2005). *Adv. Mater.* 17: 2567.
100 Leewis, C.M., de Jong, A.M., van IJzendoorn, L.J., and Broer, D.J. (2004). *J. Appl. Phys.* 95: 4125.
101 Mirfakhrai, T., Madden, J.D.W., and Baughman, R.H. (2007). *Mater. Today* 10: 30.
102 Ohm, C., Brehmer, M., and Zentel, R. (2010). *Adv. Mater.* 22: 3366.
103 Lendlein, A., Jiang, H.Y., Junger, O., and Langer, R. (2005). *Nature* 434: 879.
104 Ahir, S.V. and Terentjev, E.M. (2005). *Nat. Mater.* 4: 491.
105 Zhang, X., Yu, Z., Wang, C. et al. (2014). *Nat. Commun.* 5: 2983.
106 Ikeda, T., Mamiya, J., and Yu, Y. (2007). *Angew. Chem. Int. Ed.* 46: 506.
107 Kroener, H., White, T.J., Tabiryan, N.V. et al. (2008). *Mater. Today* 11: 34.
108 van Oosten, C.L., Bastiaansen, C.W.M., and Broer, D.J. (2009). *Nat. Mater.* 8: 677.
109 Hogan, P.M., Tajbakhsh, A.R., and Terentjev, E.M. (2002). *Phys. Rev. E* 65 (041720).
110 Li, M.H., Keller, P., Li, B. et al. (2003). *Adv. Mater.* 15: 569.
111 Akamatsu, N., Tashiro, W., Saito, K. et al. (2014). *Sci. Rep.* 4 (5377).
112 Yamada, M., Kondo, M., Mamiya, J. et al. (2008). *Angew. Chem. Int. Ed.* 47 (4986).

113 Cheng, F., Yin, R., Zhang, Y. et al. (2010). *Soft Matter* 6: 3447.
114 Lee, K.M., Kroener, H., Vaia, R.A. et al. (2011). *Soft Matter* 7: 4318.
115 Armon, S., Efrati, E., Kupferman, R., and Sharon, E. (2011). *Science* 333: 1726.
116 Gerbode, S.J., Puzey, J.R., McCormick, A.G., and Mahadevan, L. (2012). *Science* 337: 1087.
117 Warrick, D.R., Tobalske, B.W., and Powers, D.R. (2005). *Nature* 435: 1094.
118 Hendrick, T.L., Cheng, B., and Deng, X. (2009). *Science* 324: 252.
119 Lee, K.M., Smith, M.L., Kroener, H. et al. (2011). *Adv. Funct. Mater.* 21: 2913.
120 Harris, K.D., Cuypers, R., Scheibe, P. et al. (2005). *J. Mater. Chem.* 15: 5043.
121 Sawa, Y., Ye, F., Urayama, K. et al. (2011). *Proc. Nat. Acad. Sci.* 108: 6364.
122 Lee, K.M., Bunning, T.J., and White, T.J. (2012). *Adv. Mater.* 24: 2839.
123 Wie, J.J., Lee, K.M., Smith, M.L. et al. (2013). *Soft Matter* 9: 9303.
124 Iamsaard, S., Asshoff, S.J., Matt, B. et al. (2014). *Nat. Chem.* 6: 229.
125 Iamsaard, S., Anger, E., Abhoff, S.J. et al. (2016). *Angew. Chem., Int. Ed.* 55: 9908.
126 de Haan, L.T., Sánchez-Somolinos, C., Bastiaansen, C.W.M. et al. (2012). *Angew. Chem. Int. Ed.* 51: 12469.
127 de Haan, L.T., Gimenez-Pinto, V., Konya, A. et al. (2013). *Adv. Funct. Mater.* 24: 1251.
128 White, T.J. and Broer, D.J. (2015). *Nat. Mater.* 14: 1087.
129 McConney, M.E., Martinez, A., Tondiglia, V.P. et al. (2013). *Adv. Mater.* 25: 5880.
130 Ware, T.H., McConney, M.E., Wie, J.J. et al. (2015). *Science* 347: 982.
131 Ware, T.H. and White, T.J. (2015). *Polym. Chem.* 6: 4835.
132 Ahn, S., Ware, T.H., Lee, K.M. et al. (2016). *Adv. Funct. Mater.* 26 (5819).
133 Lv, J., Liu, Y., Wei, J. et al. (2016). *Nature* 537: 179.
134 Huang, C., Lv, J., Tian, X. et al. (2015). *Sci. Rep.* 5 (17414).

6

Molecular Technology for Chirality Control: From Structure to Circular Polarization

Yoshiaki Uchida[1], Tetsuya Narushima[2], and Junpei Yuasa[3]

[1] Osaka University, Graduate School of Engineering Science, 1-3 Machikaneyama-cho, Toyonaka, Osaka 560-8531, Japan
[2] National Institutes of Natural Sciences (NINS), Institute for Molecular Science (IMS), 38 Nishigo-Naka, Myodaiji, Okazaki 444-8585, Japan
[3] Nara Institute of Science and Technology, Graduate School of Materials Science, 8916-5 Takayama-cho, Ikoma, Nara 630-0192, Japan

Chirality control is significantly important as one of the molecular technologies; molecular chirality affects both light propagation behaviors and intermolecular interactions in the molecular materials. Chiral molecules influence the polarization of light and form chiral superstructures spontaneously; e.g. chiral molecules show optical rotation, and chiral biomolecules such as DNA and chiral liquid crystalline molecules form helical superstructures. Moreover, the chiral superstructures also affect the light propagation behaviors. In addition, even achiral molecules also show optical rotation in a magnetic field, and they also form chiral superstructures like bent core liquid crystals. In contrast, most of the chiral molecules are silent in circular dichroism (CD) spectroscopy in visible-light wavelength ranges, and even chiral molecules form achiral superstructures like chiral smectic A phases. It is important to understand correspondence relationships between chirality of substances and light propagation behaviors; the controllability of molecular and superstructural chirality could enable us to create unique optical and photonic devices.

In particular, circularly polarized light is indivisibly united with asymmetry of the propagating media; asymmetric media induces circularly polarized light and vice versa. Some chiral molecules show circularly polarized luminescence (CPL), whereas circularly polarized light induces some asymmetric reactions to give optically active products. The former is important for some applications such as security inks, whereas the latter has been thought to be one of the candidates of the origin of homochirality of life. Thus, the molecular technologies on the circularly polarized light have attracted a great deal of attention in organic, biological, and materials chemistry. There are several kinds of sources of circularly polarized light: chiral molecules, materials in a magnetic field, chirality of molecular aggregates, and two-dimensional chirality of nanostructures.

Molecular Technology: Energy Innovation, Volume 1, First Edition.
Edited by Hisashi Yamamoto and Takashi Kato.
© 2018 Wiley-VCH Verlag GmbH & Co. KGaA. Published 2018 by Wiley-VCH Verlag GmbH & Co. KGaA.

In this chapter, we introduce the latest molecular technologies of the fabrication of molecules, molecular aggregations, and nanostructures, those of the detection of CPL, CD, magnetic circular polarization of luminescence (MCPL), and magnetic circular dichroism (MCD), which could be useful as circularly polarized light sources for a variety of applications, and the imaging technology that can visualize the spatial distribution of chirality.

6.1 Chiral Lanthanide(III) Complexes as Circularly Polarized Luminescence Materials

6.1.1 Circularly Polarized Luminescence (CPL)

Circularly polarized light is a combination of two perpendicular electromagnetic plane waves (vertically and horizontally polarized light) having a $\lambda/4$ phase difference in equal amplitude. The sign of phase difference ($+\lambda/4$ or $-\lambda/4$) results in the opposite handedness of circular polarization: left and right circularly polarized light (LCP and RCP), which are chiral with mirror images of one another (Figure 6.1). This chiral nature of circularly polarized light is the fundamental of the chiral optics [1–3]. At the same time, the chiral characteristics of circularly polarized light facilitate an intuitive understanding of CPL as a chiroptical phenomenon arising from chiral luminescent systems (e.g. simple organic molecules, coordination compounds, and supramolecular polymers) [4–9]. This phenomenon has attracted growing interest mainly due to the development of smart optical materials such as 3D displays, information processing based on communication of spin information, and security devices [10–13]. A different aspect of CPL is an emission analogous technique of CD, which has a wide range of applications in various fields [14–17]. Most notable use of CD spectroscopy may be investigation of the secondary structure of proteins. Conversely, CPL spectroscopy is yet to be fully exploited. However, this optical technique provides a rich source of insight for the local chiral environment and intrinsically is a much more sensitive with higher time resolution, which allows monitoring changes in rapid biological processes that occur on timescales ranging from micro- to milliseconds [18–24]. Furthermore, chiral sensing of biologically important small molecules (e.g. amino acids and sugars) by induced circularly polarized luminescence (iCPL) is a new avenue of research in this field [25–27].

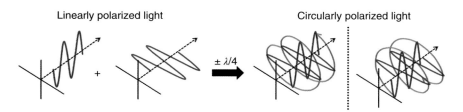

Figure 6.1 Circularly polarized light obtained from vertically and horizontally polarized light.

6.1.2 Theoretical Explanation for Large CPL Activity of Chiral Lanthanide(III) Complexes

Development of chiral luminescent materials with excellent chiroptical performance of CPL activity should be a top priority objective in CPL research. The CPL activity of luminescent compounds is usually quantified by the luminescence dissymmetry factor g_{lum} as defined in Eq. (6.1), where I_L and I_R are left and right CPL intensities, respectively. In CPL spectroscopy, ΔI $(= I_L - I_R)$ and I_{total} $(= I_L + I_R)$ are not absolute quantities because their measured values depend on the sensitivity of the spectrometer and the experimental conditions

$$g_{lum} = 2(I_L - I_R)/(I_L + I_R) \tag{6.1}$$

(such as applied voltage of the photomultiplier tube). Conversely, the ratio of the two terms, g_{lum} defined in Eq. (6.1), is an absolute value characteristic of each compound. This is the reason why the CPL activity is usually discussed in terms of g_{lum}. The absolute maximum value of g_{lum} is 2, which implies that the compound emits only LCP or RCP. We may redefine the CPL activity by rotational strength R_{ab} as defined in the Rosenfeld equation (Eq. (6.2)), where P_{ab} and M_{ab} are the electric and magnetic dipole transition moment vectors relative to

$$R_{ab} = \mathrm{Im}(P_{ab} \cdot M_{ba}) \tag{6.2}$$

the transition a → b, respectively. The rotational strength R_{ab} can be written with reference to the scalar product between P_{ab} and M_{ba} as given in Eq. (6.3), where τ_{ab} is the angle between the

$$R_{ab} = |P_{ab}||M_{ba}|\cos\tau_{ab} \tag{6.3}$$

electric and magnetic dipole transition moment vectors. On the one hand, the scalar product of P_{ab} and M_{ba} can be zero in an achiral molecule (Figure 6.2). Conversely, in the transition moments of a chiral molecule, P_{ab} and M_{ba} have components that are parallel or antiparallel ($P_{ab} \cdot M_{ba} \neq 0$). For instance, the prototypical chiral hexahelicene molecule has the almost parallel transition moments, which causes a large optical rotation [28]. The luminescence dissymmetry factor g_{lum} can be rewritten in terms of the electric and magnetic dipole transition moments as defined in Eq. (6.4). This equation simply shows that the absolute maximum value of g_{lum} ($|g_{lum}| = 2$) can be

$$g_{lum} = (4|P_{ab}||M_{ba}|\cos\tau_{ab})/(|P_{ab}|^2 + |M_{ba}|^2) \tag{6.4}$$

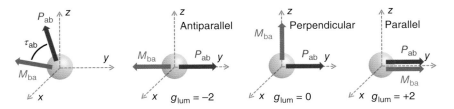

Figure 6.2 Electric and magnetic dipole transition moment vectors. Antiparallel, perpendicular, and parallel transition moments are shown.

obtained when assuming that the electric and magnetic dipole transitions have the same amplitude ($|P_{ab}| = |M_{ba}|$) in the parallel or antiparallel ($\tau_{ab} = 0°$ or $+180°$) (Figure 6.2).

The contribution of the magnetic dipole transition is much smaller than that of the electric dipole transition ($|P_{ab}| \gg |M_{ba}|$) in a spin-allowed transition (i.e. fluorescence) of a simple organic molecule. In such a case, the $|M_{ba}|^2$ term is negligible with respect to $|P_{ab}|^2$, and Eq. (6.4) can be approximated to Eq. (6.5). As can be seen from the form of Eq. (6.5), one may expect a large g_{lum}

$$g_{lum} = (4|M_{ba}|/|P_{ab}|) \cos \tau_{ab} \tag{6.5}$$

value in magnetic-allowed but electric dipole-forbidden transitions. The dominant selection rule for magnetic dipole transitions is $\Delta J = 0, \pm 1$ (excluding $J = J' = 0$) without change of parity, which is contrary to the selection rule for electric dipole transitions required a change in parity of the electron wave function [29]. Because of this requirement, chiral lanthanide(III) complexes are especially promising for chiroptical materials, for which large g_{lum} values can be obtained for the magnetically allowed but electric dipole-forbidden intraconfigurational f–f transitions [8, 9]. Actually, chiral lanthanide(III) complexes display g_{lum} values of 0.1–1, whereas simple chiral organic molecules or macromolecules typically show values of 10^{-3} to 10^{-2}.

6.1.3 Optical Activity of Chiral Lanthanide(III) Complexes

Because the intraconfigurational f–f transitions are electrically forbidden (parity-forbidden), the extinction coefficients of lanthanides are quite small, and hence, the lanthanide salts themselves are colorless. Nevertheless, most of lanthanide(III) complexes show efficient luminescence under light irradiation, mainly because of two reasons. The first one is simply due to a so-called "antenna effect" in which the organic ligands absorb light and transfer the excitation energy to the lanthanide(III) ions [30]. The second one is due to a crystal field resulting in a relaxation of the selection rules (induced electric dipole transitions), in which the ligands provide a dissymmetric environment and a static potential of odd parity around the lanthanide(III) ion. We shall not enter the further discussion for how the crystal field perturbation contributes to relax the selection rule for electric dipole transitions by mixing states having different parities (e.g. 4f and 5d orbitals) [29, 31], but only consider its chiroptical contribution to the magnetically allowed intraconfigurational f–f transitions.

The chiral ligands or induced chiral fields provide chiral environment around the lanthanide(III) ion, being the origin of the chiroptical properties (CPL) of lanthanide. The processes gaining the CPL activity has been conventionally represented with reference to static and dynamic coupling mechanisms (vide infra), but there is still no commonly accepted theory [8, 9]. Perhaps it is safe to conclude that different factors contribute to the CPL activity lanthanide, where the luminescence dissymmetry factor g_{lum} varies (ranging 0.1–1) depending on the coordination environment or the types of chiral ligands employed (Figure 6.3). In this context, a series of chiral europium(III) [Eu^{III}] camphorate complexes may be the most successful case

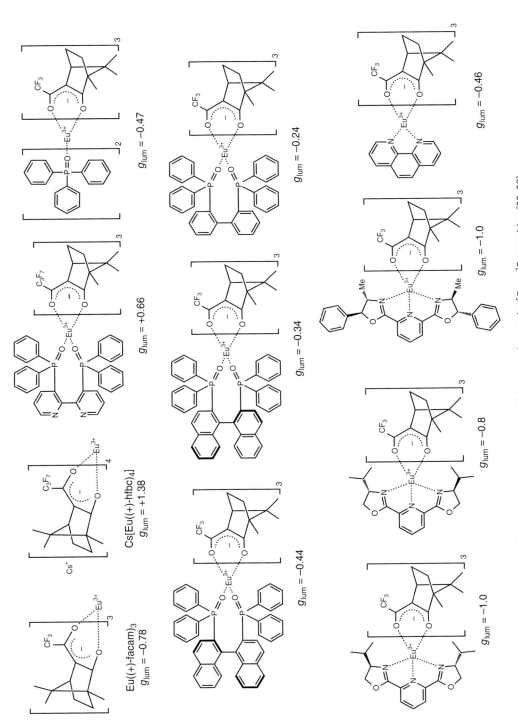

Figure 6.3 A series of chiral europium(III) camphorate complexes and their g_{lum} values at the $^5D_0 \rightarrow {}^7F_1$ transition [32–38].

of chiral lanthanide(III) compounds yielding large g_{lum} values [32–38]. A prototypical example is tris(3-trifluoro-acetyl-(+)-camphorate)europium(III), [Eu((+)-facam)$_3$], which is the commercially available NMR shift reagent. The [Eu((+)-facam)$_3$] in dry DMSO displays the g_{lum} value of −0.78 at the $^5D_0 \rightarrow ^7F_1$ emission line ($\Delta J = 1$) [32–34]. The $^5D_0 \rightarrow ^7F_1$ transition is the magnetically allowed f–f transition of Eu^{III}. Recently, Kaizaki and coworkers reported a tetrakis(3-heptafluoro-butylryl-(+)-camphorate) europium(III) complex, Cs[Eu((+)-hfbc)$_4$], with the excellent g_{lum} value of 1.38, meaning that close to 70% of europium(III) emitted LCP at the $^5D_0 \rightarrow ^7F_1$ transition of Eu^{III} ions [35]. The g_{lum} value of 1.38 is enormously prominent among camphorate Eu^{III} complexes including our own studies. In such an extraordinary CPL, the approximation formula defined in Eq. (6.5) is not a precise statement representing for g_{lum}. Hence, instead of Eq. (6.5), Eq. (6.4) can be applied to this situation. Equation (6.4) implies that the exceptional g_{lum} value of 1.38 can be obtained such that the electric and magnetic dipole transitions have close to the same amplitude ($|P_{ab}| \fallingdotseq |M_{ba}|$) with the dipole angle τ_{ab} close to 0°. Perhaps both factors (the amplitude of $|P_{ab}|$ and $|M_{ba}|$ and the dipole angle τ_{ab}) contribute to the excellent g_{lum} value of 1.38. When assuming that P_{ab} and M_{ba} are parallel ($\tau_{ab} = 0°$) similar to the case of hexahelicene derivatives (vide supra), the g_{lum} value of 1.38 will be obtained with $|P_{ab}|/|M_{ba}| = 0.4$ (or $|M_{ba}|/|P_{ab}| = 0.4$). Conversely, when we assume the same amplitude of the electric and magnetic dipole transitions ($|P_{ab}| = |M_{ba}|$), $g_{lum} = 1.38$ corresponds to the dipole angle $\tau_{ab} = 46°$.

For a pure magnetic dipole transition, $|P_{ab}||M_{ba}| = 0$ is expected; therefore, the CPL activity in the $^5D_0 \rightarrow ^7F_1$ transition of Eu^{III} requires a nonvanishing electric dipole transition amplitude in the magnetically allowed f–f transition. Perhaps there are two possible ways of interpreting that the $^5D_0 \rightarrow ^7F_1$ transition borrows electric dipole character ("J mixing" and "dynamic coupling"). When the Eu^{III} ion is located in nonspherically symmetric ligand environment, the crystal-field perturbation causes the mixing of the wave functions of sublevels with different J values (J mixing) [31]. This means that the "$^7F_1$" state has the "$^7F_2$" character to some extent; therefore, the $^5D_0 \rightarrow ^7F_1$ transition contains contributions from the $^5D_0 \rightarrow ^7F_2$, which is the electric dipole transition of Eu^{III}. Recently, the process giving rise to the electric dipole transition has been represented with reference to a dynamic coupling (ligand polarization) mechanism [8, 9, 39]. In this model, ligand-centered electric dipole transitions (such as ππ* transitions of the diketonate ligands) couple with metal-centered magnetic dipole transitions. Namely, the f–f transition causes the charge distribution of the nearby ligands, which polarizes the ligands and induces electric dipole. In such a case, the CPL activity of Eu^{III} can be generated not only by the dissymmetric environment of the first coordination sphere but also by the chiral chromophoric substrates in the second coordination sphere. Interestingly, it was suggested that Cs[Eu((+)-hfbc)$_4$] showing the excellent g_{lum} value of 1.38 has an almost perfect antiprismatic geometry (nearly achiral arrangement) with a twist angle of −41.4° in solution [39]. This anomaly can hardly be interpreted only with respect to a static coupling mechanism because the static coupling model assumed that the perturbed wave functions are localized on the lanthanide ion; therefore, the CPL activity can be induced directly by the static potential of the first coordination sphere around the metal center. The dynamic coupling mechanism may rationalize the anomalous behavior of Cs[Eu((+)-hfbc)$_4$].

6.1.4 CPL of Chiral Lanthanide(III) Complexes for Frontier Applications

As we have discussed in the previous section, chiral lanthanide(III) complexes are fascinating candidates for chiroptical materials showing the excellent performance of CPL achieving the advanced chiral photonics. Recently, their exceptional CPL has started to be used in circularly polarized electroluminescence (CPEL). Bari and coworkers have developed a novel organic light-emitting diode (OLED) (OLED) with highly intrinsic, circularly polarized Eu^{III} red emission [13]. They selected Cs[Eu((+)-hfbc)$_4$] showing the excellent g_{lum} value of 1.38 as the chiral emitter. Interestingly, this extremely large g_{lum} value of 1.38 is maintained even after dispersing Cs[Eu((+)-hfbc)$_4$] in a polyvinyl carbazole (PVK) polymeric film.

On the other hand, Cs[Eu((+)-hfbc)$_4$] has a low luminescence quantum yield (several percent as powder or in solutions), which is sometimes trade-off with its excellent g_{lum} performance. Recently, we found that the luminescence quantum yield of Cs[Eu((+)-hfbc)$_4$] is increased (almost twice) by supramolecular polymerization of the complex [40]. The supramolecular polymerization generates the one-dimensional helical chain, leading to suppression of nonradiative processes in Cs[Eu((+)-hfbc)$_4$]. Interestingly, our own findings suggested that the supramolecular polymerization also results in the increase of g_{lum}, where the g_{lum} value of 1.38 was updated to 1.41.

6.2 Magnetic Circular Dichroism and Magnetic Circularly Polarized Luminescence

In addition to the abovementioned natural CD and CPL, externally applied field can break the symmetry of materials, which causes the difference between the propagation behaviors of circularly polarized lights in the materials. For example, in a magnetic field, the absorption and emission of materials are circularly polarized: magnetic CD (MCD) [41] and CPL (MCPL) [42]. As the magnetic field induces the asymmetry of the materials, in principle, any materials should show MCD [43]. In a similar way, any materials showing emission should show MCPL [44, 45]. Thus, circularly polarized light is obtained using MCD and MCPL. Furthermore, as the magnetic field is easily controllable, MCD and MCPL enable us to switch circularly polarized light. In fact, as optical circulators and isolators, magneto-optical materials are incorporated in devices [46, 47].

As components of optical devices, the intensities of MCD and MCPL have to be strong enough to control the polarization of propagating light. The intensities of MCD and MCPL depend on the magnetization and the absorption and emission of materials, respectively. Thus, ferromagnetic materials, almost all of which are inorganic solids, are the first choice of the materials of the components because of their large magnetization inducing strong MCD and MCPL

[46, 47]. In contrast, molecular materials such as metal complexes and all-organic compounds with strong and sharp absorption and emission bands should show applicable MCD and MCPL, no matter how weak their magnetic susceptibilities are [48]. Furthermore, the molecular materials showing MCD and MCPL are suitable to use with biomaterials and to be incorporated into flexible electronic devices [49].

Here, the origins of the MCD and MCPL are explained, and the examples of molecular materials showing MCD and MCPL and molecular technology to design them and control MCD and MCPL are introduced.

6.2.1 Magnetic–Field-induced Symmetry Breaking on Light Absorption and Emission

Magnetic fields break chiral symmetry of light propagating in materials by affecting the atomic or molecular orbitals related to light absorption and emission like abovementioned molecular chirality. The theories of MCD and MCPL focus on the electronic states of the materials; the applied magnetic fields affect the ground and excited electronic states related to the absorption and emission [44, 50, 51]. The molecular theories have been powerful tools to discuss the electronic states of the molecular materials since the first MCD measurement of all-organic molecules, benzene, toluene, and so on [52]. As the theme of this section is the molecular technology for chirality control from structures to circular polarization, here the molecular theory of MCD is outlined. First, we have to start from the electronic states related to absorption. Absorption occurs when the transition from the ground state to excited states. The intensity of MCD depends on the intensity of both the electric and magnetic transition dipole moments for the absorption of light.

When the ground or excited states are splitting in a magnetic field, MCD occurs. The following equation is standard to discuss MCD spectra, which consist of three terms [48, 53].

$$\frac{\Delta A}{E} = 152.5 Bcl \left[A_1 \frac{df}{dE} + \left(B_0 + \frac{C_0}{kT} \right) f \right], \quad (6.6)$$

where ΔA is the differential absorbance of LCP (A_L) and RCP (A_R) as shown in Figure 6.4, B is the field strength, c is the concentration, l is the path length, E represents the energy coordinate for the entire spectral band, and f is a normalized band shape function. The A_1, B_0, and C_0 terms provide information about the splitting of the absorption bands for LCP and RCP, the field-induced mixing of zero-field states, and the Zeeman- splitting-based population in the ground state. As C_0 term is due to Zeeman splitting, its intensity depends on temperature. Meanwhile, as A_1 and B_0 terms are not due to paramagnetic nature, their intensities are independent of temperature.

MCPL is also divided into the same three terms as MCD, A_1, B_0, and C_0. However, the initial and final states are excited and ground state, respectively.

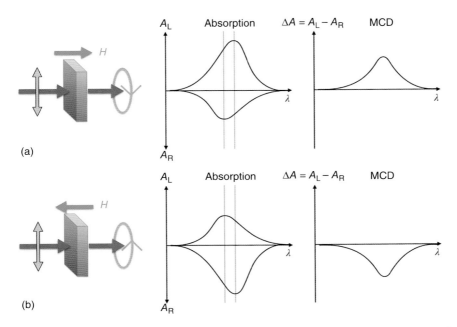

Figure 6.4 Schematics of magnetic circular dichroism spectra. (a) When a material is put in the applied magnetic field along the direction of light propagation, the absorption spectrum of LCP is different from that of RCP; both peak wavelength and intensity are different between them. (b) When the direction of the magnetic field is reversed, the absorption spectra of LCP and RCP are replaced with each other.

6.2.2 Molecular Materials Showing MCD and MCPL and Applications

Now, optical circulators and isolators have been made of inorganic materials. However, wearable devices have recently attracted attention, and organic materials and biomaterials are suitable for these devices. In this situation, MCD and MCPL of organic materials and biomaterials have received attention [49]. MCD and MCPL are quite important not only for understanding the electronic structures of materials [54] but also as working principles of optical elements for telecommunications and calculations [55]. Inorganic materials and simple small molecules showing MCD and MCPL have been mainly investigated because of the strength of their MCD and MCPL and the simplicity of the electronic structures [45, 52]. However, biocompatible organic materials with larger and more complicated functional moieties should be suitable for biodevices.

As such molecular materials, metal complexes with metal ions showing MCD and/or MCPL and large π-conjugated materials have been vigorously investigated [31, 48]. Thus far, small radical molecules like ˙OH and ˙CH have been fundamentally studied [56]. In the future, a variety of functional metal complexes and functional organic radicals showing paramagnetic C_0 term of MCD will probably be more noteworthy because C_0 term is the largest one and its sign is independent of wavelength, which is appropriate for the creation of new functions.

As the most simple examples showing new function arising from the coexisting of MCD and other properties, magneto-chiral dichroism (MChD) is well known [57–60]. When a certain material is exhibiting MCD and other properties such as CD or CPL, lights show nonreciprocal behaviors in the material; this phenomenon is called magnetochiral effects [61–63]. In particular, MChD of all-organic molecules are suitable for biodevices [64]. To use this phenomenon for optical elements, other types of CD and CPL coming from larger structures than molecules as shown in the following sections might have promise, too [65, 66].

6.3 Molecular Self-assembled Helical Structures as Source of Circularly Polarized Light

Intermolecular interactions can induce the symmetry breaking of the propagation of circularly polarized lights. This is attributed not to the molecules that can absorb or emit lights but to the periodic structures in the media in which light is propagating. When a material has some kinds of periodic structures in a certain direction, it reflects an electromagnetic wave propagating in the same direction with a wavelength corresponding to the pitch of the periodic structure. This is because the periodic structure strongly affects the propagation of the electromagnetic wave as a periodically varying refractive index [67]. In particular, when the electromagnetic wave reflected by the material is light, the materials and the reflection band are called photonic crystals and photonic bandgap (PBG), respectively. Natural photonic crystals such as opals are solid with periodical structures and exhibit beautiful iridescent colors.

Photonic crystals can also be artificially fabricated: colloidal crystals consisting of particles with the diameter of several hundreds of nanometers [68] and periodically created holes on substrates [69]. If the periodical structure is circularly twisted and its refractive index is anisotropic, it could affect circularly polarized light when the pitch of the structure should be several hundreds of nanometers. In fact, as the most popular molecular materials with such twisted structure materials showing chiral liquid-crystalline phases like cholesteric LC phases and LC blue phases are well known [70]. In these chiral LC phases, the twisted structures of molecules are self-assembled, and one of the circularly polarized lights propagating in the same direction as the periodic structure is reflected, whereas the other is not affected and transmits. The photonic structures of the chiral LC phases work as circularly polarized light sources because of their circularly polarized laser action as described in the following.

Moreover, these chiral LC phases are flexible and fluid. Thus, these photonic properties could be controllable by external stimuli, and various three-dimensional (3D) chiral photonic structures showing 3D laser actions could be self-assembled. As these properties are unique to the chiral LC phases, they are promising materials to control the circular polarization of light both spatially and temporally. In fact, there have been a lot of reports on the self-assembled photonic structures showing 3D lasing and on the external-stimuli-induced control of the laser action.

Here, the origins of the asymmetric PBG in the chiral LC phases are explained, and the examples of the molecular technologies to control the reflection, absorption, and emission in the chiral LC phases are introduced.

6.3.1 Chiral Liquid Crystalline Phases with Self-assembled Helical Structures

LC phases are the mesophases between liquid and crystalline phases that mainly molecular materials exhibit. Thus, they show both fluidity like liquids and optical anisotropy like crystals. The optical anisotropy is attributed to the anisotropy of the molecular dielectric constant. If the light propagates in the direction perpendicular to the molecular long axis of the rod-like molecules in chiral LC phases with helical pitches corresponding to the visible wavelength range, selective reflection of visible light occurs [71]. Interestingly, the selectively reflected light has one of circularly polarized lights with the same handedness as the helical superstructure and with the same wavelength range as the helical pitch.

Calculation methods based on Maxwell's equations like 4×4 matrix method [72, 73] and finite-difference time domain (FDTD) method lead to the dispersion relation in the chiral LC materials in detail (Figure 6.5) [74, 75]. In particular, the FDTD method is widely applicable to the light propagation and polarization in chiral LC materials with curved interfaces, which have highly complicated 3D photonic structures. The topic in the next section is about the laser action in simple one-dimensional (1D) and complicated 3D photonic structures consisting of chiral LC materials.

6.3.2 Strong CPL of CLC Laser Action

For one of the circularly polarized light sources, the helical structure of a certain CLC material works as a 1D laser resonator [71, 76]. CLC materials show two lasing modes, distributed feedback (DFB) and defect modes. In both cases, CLC materials dissolving a laser dye emit laser in the same direction as the helical axis,

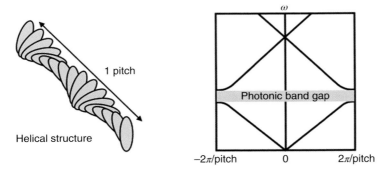

Figure 6.5 Schematic diagram of propagation of circularly polarized light in helical superstructure and photonic bandgap arising from the helical superstructure.

and the emitted laser light is circularly polarized with the same twisting direction as the CLC helices. The emitted laser light has a wavelength at the edge of the PBG for the DFB modes. In contrast, CLC materials with a defect in the photonic structure like a discontinuity of the twisted structure show laser action with a certain wavelength inside the PBG because the defect induces a resonant mode inside the PBG [77, 78]. Such kind of dye lasers has a threshold on the excitation light energy. As the lasing threshold of defect mode is lower than that of DFB mode of the same material, defect mode lasing has been vigorously investigated [70].

Threshold should be reduced to realize continuous wave lasing. A lot of strategies have been tried to reduce lasing threshold. The structure of cavity, the condition of excitation, and materials are mainly important for the lasing threshold [70]. As the molecular technology to control circular polarization of light, here the structure of cavity and the materials are worth noting. As mentioned earlier, defect structure in the CLC materials could have a variety of morphologies. In addition, the combination of right- and left-handed CLC materials has also been effective to reduce the lasing threshold [79]. In contrast, molecular structure should also be modified for lowering lasing threshold; orientational ordering and mechanisms of luminescence are important [70].

As the LC materials are flexible and fluid enough to fabricate 3D photonic structures easily, 3D lasing in LC materials is one of the hot topics [70, 80]. NLC droplets show WGM lasing [81]. CLC droplets and shells with helical axes aligning in a radial fashion with 3D omnidirectional photonic structures can be easily fabricated, and omnidirectional laser action has been reported [82, 83]. DFB laser action in CLC shells emits circularly polarized light, whereas polarization of the laser light from CLC droplets has not been clarified. In addition, as the polymer-stabilization of LC blue phases is possible [84], blue phases with appropriate dyes showing 3D laser actions are one of the powerful candidates of 3D circularly polarized light sources [84].

Tunability is important. As the PBGs of CLC materials are controllable by external stimuli, temperature, electric and magnetic fields, light irradiation, mechanical strain, and so on, CLC materials should be useful as switchable circularly polarized laser resonators [82, 83]. Recently, NLC materials showing ferromagnetism have been reported [85]. NLC droplets with the ferromagnetic NLC materials with laser dye show magnetically controllable WGM laser action [86]. If ferromagnetic CLC droplets are possible, they could be expected to be magnetically switchable circularly polarized laser resonators.

6.4 Optical Activity Caused by Mesoscopic Chiral Structures and Microscopic Analysis of the Chiroptical Properties

When a sample is illuminated with light, one can observe reflected and scattered light as well as absorbed and transmitted light. The optical outputs, which are a result from the light–matter interaction, are dependent on material properties of the sample such as composition and structure. The light–matter interaction obviously relates to the condition of incident light of wavelength, polarization, and so on. This suggests that detailed material properties can be evaluated with well-controlled light illumination, which is commonly used for analysis purpose.

Chiral molecules do not have improper rotation axis and show optical activity typified by optical rotation or CD. The optical activity is caused by differential refractive index between LCP and RCP illuminations. The real part of the differential refractive index represents optical rotation, whereas imaginary one corresponds to CD. Contribution of the optical activity is tiny, which is 100–1000 times smaller than that for orthogonal linear polarizations [87, 88]. For this reason, high-sensitive detection technique is usually adopted to analyze the tiny contribution of the optical activity [89].

Origin of optical activity in molecular scales, which is associated with individual molecules, is explained with rotational strength described in Eq. (6.2) [88]. Therefore, chiroptical property of the molecule is strongly influenced by relative configuration of the transition dipole moments.

These years, optical activity distinct from that observed in the molecular scales attracts a lot of interest. As shown in Figure 6.6, the novel optical activity originates from the interaction occurred at the spatial scales, which is comparable with or larger than optical wavelength [90, 91]. In samples showing optical activity, cause of chirality should be involved not in individual molecules but also in structured molecular assembly and/or its shape itself with the scales from nanometers to micrometers. The optical activity caused by the larger scale samples occasionally shows singular properties (nonreciprocal behavior [92, 93], etc.), which is not observed for the individual molecules. This indicates that optical activity could

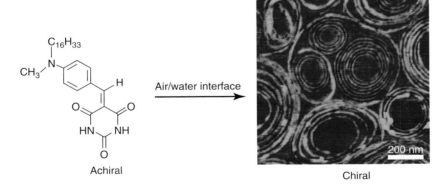

Figure 6.6 This is the first report on the chirality of molecular assemblies and spiral nanostructures formed through the air/water interface by achiral molecules. An amphiphilic barbituric acid derivative was found to form stable monolayers showing a clear phase transition at the air/water interface. It is interesting to find that the deposited Langmuir–Blodgett (LB) films of the compound showed CD although the molecule itself was achiral. Atomic force microscope (AFM) measurements on the transferred one-layer LB film revealed that spiral nanoarchitectures were formed. Source: Huang et al. 2004 [90]. Reprinted with permission of American Chemical Society.

not be dominated by usual physical mechanism described with the rotational strength, described in Eq. (6.2). The larger chiral materials can be tuned through modifying their composition and shape. It must be beneficial to consider tuning as alternative degrees of freedom to artificially manipulate chiroptical property. For relatively simple model cases, optical properties and physical mechanisms behind optical activity are surveyed with the help of electromagnetic simulation [94].

The optical activity of the larger chiral materials has been experimentally investigated by measuring changes in macroscopic polarization, which is classically done by CD spectrometer for molecules. Size of structural origin for the optical activity is larger than molecular size, which is roughly enough to employ microscopic technique. This provides us a chance to experimentally specify dominant site and structure for the optical activity, which should strongly promote understanding unrevealed mechanism. If spatial information on the optical activity is available, it must be advantageous to explore the microscopic origin as well as to examine whether desired optical property is realized. Therefore, visualizing spatial distribution of the optical activity will have an impact on fields of chiroptical control and molecular technology.

Recent progress on two types of CD microscopic technique (far-field and near-field detection) to investigate chiral materials with sizes from nanometers to micrometers will be described in the following.

6.4.1 Microscopic CD Measurements via Far-field Detection

Photo-elastic modulator (PEM) provides a periodically modulated circular polarization. In the polarization modulation, linear polarization component is partially involved at any time because of the properties of PEM. Isotropic samples or samples with random orientation like solution should not suffer from the linear polarization component. However, CD detection for ordered structures and crystalline samples is affected. As a result, the false signals arising from the linear polarization component (linear birefringence and linear dichroism) considerably commingles in actual CD measurement because various nonideal properties of the apparatus (i.e. parasitic ellipticities), such as residual strain in a PEM element, and higher harmonic responsiveness of the lock-in amplifier, prevent from exact synchronization of polarization modulation [95, 96]. Thus, for samples with ordered structure or anisotropy for linear polarization, detecting genuine signals of optical rotation and CD is extremely a difficult task.

As, in general, the difficulty is similarly associated with microscopic CD measurements, CD observation with high reliability is not easy for samples such as crystals and liquid crystals. Although microscopic CD measurement was attempted since a long time ago [97, 98], it has not been commonly used because of the aforementioned difficulties. Many ideas to remove the artifacts arising from linear polarization anisotropy have recently been proposed [99–104], which is currently pushing forward optical activity studies based on microscopic measurements.

6.4.2 Optical Activity Measurement Based on Improvement of a PEM Technique

A PEM technique, which is adopted in most of the conventional CD spectrometers, is not simply applicable to measurements for samples such as crystals and liquid crystals and its microscopic observation, as described above. Therefore, CD measurements for solid-state samples have been conducted by carefully selecting a photomultiplier tube with least polarization characteristics and a PEM element with least residual strain [99]. Alternatively, elimination of artifact signals due to the linear polarization anisotropy was proposed with the use of two PEMs at different modulation frequencies [101]. In another attempt to suppress commingling of linear polarization components from the PEM device, an optical gate method synchronized with the PEM was used [100] and actual CD imaging has been recently demonstrated [105]. As a different approach with the PEM device, CPL from a sample was microscopically detected. The CPL also provides information to analyze chirality in the sample, which is similar to the CD [106, 107].

6.4.3 Discrete Illumination of Pure Circularly Polarized Light

In order to remove the artifact due to the linear polarization anisotropy, many attempts have been proposed by extending the conventional PEM technique. This is basically because the PEM technique has been commonsensically accepted and adopted to realize high-sensitive CD detection. As an essentially different approach apart from the stereotyped idea, it has been reported that samples are discretely illuminated with uncontaminated circularly polarized light. In the first report based on this idea, respective transmission images under LCP and RCP illuminations were observed using an imaging sensor.

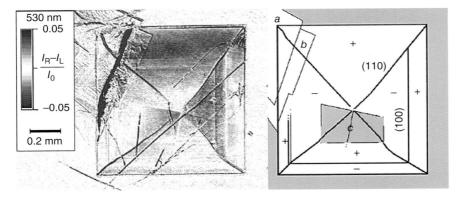

Figure 6.7 CD micrograph of a thin (<50 mm) DHA crystal showing the (110) and (100) twin planes between enantiomorphous domains. The wavelength of observation was 530 nm. *Source*: Claborn et al. 2003 [102]. Reprinted with permission of American Chemical Society.

Then, a CD image was constructed by calculating the difference in transmitted signal at each pixel between these images [102]. In the CD image of the crystal of 1,8-dihydroxyanthraquinone (DHA), enantiomorphous twinning was clearly observed, which can be recognized as red- and blue-colored regions in Figure 6.7. In the CD imaging method, S/N ratio was improved by integrating many transmission images under LCP or RCP illuminations.

For further sensitivity improvement, the discrete circularly polarized light illumination has been recently combined with high-sensitive lock-in detection [103, 104]. The lock-in detection is not easily realized with the imaging sensors because signal acquisition of the usual imaging sensors is not sufficiently fast. In this approach, CD image was obtained with a sample scanning. Consequently, the approach demonstrated that highly sensitive and reliable microscopic CD measurement that is less influenced by linear polarization anisotropy is feasible. Simultaneously, nearly diffraction-limited spatial resolution was realized in the practical microscopic CD imaging. Figure 6.8 is observed images based on the approach for a sample of arrayed chiral nanostructures. The transmission images (Figure 6.8a,c) show the lattice patterns that arise from the arrayed configuration of the nanostructures. In the CD images (Figure 6.8b,d), local maximal CD signals are found at four sites of left, right, top, and bottom of the unit nanostructures, rather than at the dark (large extinction) regions in Figure 6.8a,c. This fine structure of the CD image reveals that CD is spatially not uniform even in the individual unit chiral nanostructure.

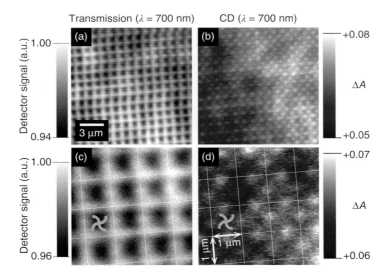

Figure 6.8 Magnified transmission (a, c) and CD (b, d) images of the two-dimensional array of chiral (swirl-shaped) gold nanostructures. The wavelength of observation both for the transmission and CD images was 700 nm. *Source*: Narushima and Okamoto 2016 [104]. Reproduced with permission of Nature Publishing Group.

6.4.4 Complete Analysis of Contribution From All Polarization Components

In the approaches described earlier, genuine CD imaging was tried by eliminating the artifact from the linear polarization anisotropy. In contrast, an absolutely accurate method has been developed to measure not merely CD but all polarization components contributed from a sample [108]. With this method, for example, optical activity of D- and L-alanine crystals was accurately measured. Furthermore, absolute chirality of alanine was determined: the crystal of D-alanine is dextrorotatory [109]. Visualization of all polarization components was examined by several groups. Depolarization (partial polarization) can also be evaluated because this visualization method is based on Mueller matrix. High-sensitive visualization has been reported with the use of four PEM devices [110, 111]. The visualization method is also realized even with mechanically rotated polarizer and analyzer. As it suppresses modulation frequency, imaging sensors become available. Actually, all images corresponding to each of 16 elements of the Mueller matrix were readily acquired within several minutes [112]. As discussed above, imaging method for optical activity has been gradually established these years. As a new tool to analyze chiral materials with micrometer scales, these imaging approaches will make a solid progress in many fields such as chemistry and biology.

6.4.5 Near-field CD Imaging

Nanostructures with chiral shapes may behave like chiral molecules. Actually, artificially fabricated chiral nanostructures exhibited giant optical activity [113], which was much larger than that shown by molecular system. Regarding the exceptional case, microscopic CD detection with sufficiently high spatial resolution of nanometer scale should yield useful information on physical mechanism behind the giant optical activity. Thus, methodology of near-field optics was employed for the realization of CD microscopy with the nanometer resolution [114]. As contribution of the optical activity is huge, the linear polarization anisotropy can be considered as relatively small influence. A sample was illuminated with modulated circularly polarized light through an ordinary objective lens and then an aperture of a near-field probe picked up local optical response from the sample. Thus, a near-field CD image is constructed by converting the local optical response into CD signal.

Figure 6.9 shows CD spectra macroscopically obtained for chiral gold nanostructures an "S" shape and near-field CD distribution observed in the individual nanostructure. S-shaped structures showed CD spectrum with positive values, whereas negative CD was obtained for their mirrored structures, as shown in Figure 6.9a. Maximum value of the CD spectra was one order of magnitude larger than that observed for a typical chiral molecule. Near-field CD images Figure 6.9b showed that both positive and negative CD signals were locally distributed in each individual nanostructure. Spatial distribution of the local CD

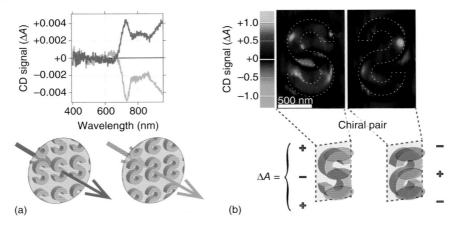

Figure 6.9 Macroscopic and local CD of gold nanostructures. (a) CD spectra macroscopically measured with the use of arrayed nanostructures Source: Narushima and Okamoto 2013 [114]. Reproduced with permission of Royal Society of Chemistry. (b) Distributions of local CD signals in the individual nanostructures observed with a near-field CD imaging technique. The CD signal was defined as $\Delta A = A_L - A_R$ as described above. Source: Narushima et al. 2014 [115]. Reprinted with permission of American Chemical Society.

signals exhibited antisymmetric relation between the S-shaped and its mirrored structures, which correctly reproduces chiral symmetry. More specifically, negative CD was observed at the central part of the S-shaped structure, whereas positive CD was seen for the mirrored structure. At bottom and top regions in the nanostructures, CD signals with opposite sign were distributed, respectively. Another interesting point is that absolute values of the CD signals localized in the nanostructures was two orders of magnitude stronger than that macroscopically observed in the CD spectra [92]. This indicates that macroscopic CD was observed as a result of balancing both positive and negative local CD. In a similar way, visualizing local CD distribution with the near-field CD microscopy has revealed developing optical activity via association of achiral nanostructure [115] and explained why achiral nanostructure does not show optical activity [116].

6.5 Conclusions

Here, we introduced the latest developments on molecular technologies to control and detect the interactions between materials and circular polarization of light. Chiral symmetry of materials can be transferred to light, i.e. localized optical field becomes chiral around the chiral molecules, chiral superstructures, and materials in magnetic fields under light illumination.

Molecular chirality can directly induce CD and CPL. However, there is no still rational strategy to design chiral lanthanide(III) complexes achieving excellent CPL performance due to the lack of conventionally accepted theory to rationalize the CPL activity of lanthanide(III). Rational planning of the CPL materials from molecular level should be an important objective in "Molecular Technology," enabling us to develop smarter photonic materials.

Magnetic field can also induce CD and CPL. Although the intensity of MCD and MCPL is generally lower than that of natural CD and CPL, these phenomena are nonreciprocal and they could be useful for magnetically switching circular polarization of light.

Chiral superstructures of molecules such as liquid crystals and chiral nanostructures under light illumination can greatly improve sensitivity of chiral molecules because huge scale mismatch between the molecular size and spatial scale of light (wavelength) may be reduced via the localized chiral optical field [117]. In a similar way, it will open the way to future applications such as asymmetric reaction fields and spin devices coupled with the chiral optical field.

Furthermore, spectroscopic analyses based on Raman scattering or CD provide information on chemical bonding or structures in a molecule. In contrast, CD microscopy may offer opportunities to investigate chiral materials in relatively large scales from nanometers to micrometers as well as chirality in molecular scale. The methodology must be advantageous to comprehend mesoscopic chiral structures hierarchically existing in various scales. If visualization of spatial distribution of chirality and its structural change in time becomes readily available, vital information will be extracted to reveal intermolecular coupling and cooperation for expression of functions in macromolecular and biological systems.

References

1 Andrews, D.L. (2015). *Front Matter*, 1–14. Hoboken, NJ: Wiley.
2 Wagniere, G.H. (2012). *On the Interaction of Light with Molecules: Pathways to the Theoretical Interpretation of Chiroptical Phenomena*, 1–34. Hoboken, NJ: Wiley.
3 Brittain, H.G. (1996). Excited-state optical activity, 1987–1995. *Chirality* 8 (5): 357–363.
4 Lu, H., Mack, J., Nyokong, T. et al. (2016). Optically active {BODIPYs}. *Coord. Chem. Rev.* 318: 1–15.
5 Sanchez-Carnerero, E.M., Agarrabeitia, A.R., Moreno, F. et al. (2015). Circularly polarized luminescence from simple organic molecules. *Chem. Eur. J.* 21 (39): 13 488–13 500.
6 Roose, J., Tang, B.Z., and Wong, K.S. (2016). Circularly-polarized luminescence (CPL) from chiral AIE molecules and macrostructures. *Small* 12 (47): 6495–6512.
7 Riehl, J.P. and Muller, G. (2004). Circularly polarized luminescence spectroscopy from lanthanide systems. In: *Handbook on the Physics and Chemistry of Rare Earths*, vol. 34 (ed. V.K. Bunzli and J.-C.G. Pecharsky), 289–357. Elsevier.
8 Zinna, F. and Di Bari, L. (2015). Lanthanide circularly polarized luminescence: bases and applications. *Chirality* 27 (1): 1–13.
9 Bari, L.D. and Salvadori, P. (2012). *Chiroptical Properties of Lanthanide Compounds in an Extended Wavelength Range*, 221–246. Hoboken, NJ: Wiley.

10 Grynberg, G., Aspect, A., and Fabre, C. (2010). *Introduction to Quantum Optics: From the Semi-Classical Approach to Quantized Light*. Cambridge: Cambridge University Press.

11 Latal, H. (1991). *Parity Violation in Atomic Physics*, 1–17. Berlin, Heidelberg: Springer.

12 Zhang, Y.J., Oka, T., Suzuki, R. et al. (2014). Electrically switchable chiral light-emitting transistor. *Science* 344 (6185): 725–728.

13 Zinna, F., Giovanella, U., and Bari, L.D. (2015). Highly circularly polarized electroluminescence from a chiral europium complex. *Adv. Mater.* 27 (10): 1791–1795.

14 Riehl, J.P. and Richardson, F.S. (1986). Circularly polarized luminescence spectroscopy. *Chem. Rev.* 86 (1): 1–16.

15 Carr, R., Evans, N.H., and Parker, D. (2012). Lanthanide complexes as chiral probes exploiting circularly polarized luminescence. *Chem. Soc. Rev.* 41: 7673–7686.

16 Wu, T., You, X.Z., and Bour, P. (2015). Applications of chiroptical spectroscopy to coordination compounds. *Coord. Chem. Rev.* 284: 1–18.

17 Muller, G. (2009). Luminescent chiral lanthanide(III) complexes as potential molecular probes. *Dalton Trans.*: 9692–9707.

18 Coruh, N. and Riehl, J.P. (1992). Circularly polarized luminescence from terbium(III) as a probe of metal ion binding in calcium-binding proteins. *Biochemistry* 31 (34): 7970–7976.

19 Abdollahi, S., Harris, W.R., and Riehl, J.P. (1996). Application of circularly polarized luminescence spectroscopy to Tb(III) and Eu(III) complexes of transferrins. *J. Phys. Chem.* 100 (5): 1950–1956.

20 Frawley, A.T., Pal, R., and Parker, D. (2016). Very bright, enantiopure europium(III) complexes allow time-gated chiral contrast imaging. *Chem. Commun.* 52: 13 349–13 352.

21 Montgomery, C.P., New, E.J., Parker, D., and Peacock, R.D. (2008). Enantioselective regulation of a metal complex in reversible binding to serum albumin: dynamic helicity inversion signalled by circularly polarised luminescence. *Chem. Commun.*: 4261–4263.

22 Yuasa, J., Ohno, T., Tsumatori, H. et al. (2013). Fingerprint signatures of lanthanide circularly polarized luminescence from proteins covalently labeled with a [small beta]-diketonate europium(III) chelate. *Chem. Commun.* 49: 4604–4606.

23 Schauerte, J.A., Steel, D.G., and Gafni, A. (1992). Time-resolved circularly polarized protein phosphorescence. *Proc. Natl. Acad. Sci. U.S.A.* 89 (21): 10 154–10 158.

24 Schauerte, J.A., Schlyer, B.D., Steel, D.G., and Gafni, A. (1995). Nanosecond time-resolved circular polarization of fluorescence: study of NADH bound to horse liver alcohol dehydrogenase. *Proc. Natl. Acad. Sci. U.S.A.* 92 (2): 569–573.

25 Okutani, K., Nozaki, K., and Iwamura, M. (2014). Specific chiral sensing of amino acids using induced circularly polarized luminescence of bis(diimine)dicarboxylic acid europium(III) complexes. *Inorg. Chem.* 53 (11): 5527–5537.

26 Iwamura, M., Kimura, Y., Miyamoto, R., and Nozaki, K. (2012). Chiral sensing using an achiral europium(III) complex by induced circularly polarized luminescence. *Inorg. Chem.* 51 (7): 4094–4098.

27 Mahajan, R.K., Kaur, I., Kaur, R. et al. (2004). Lipophilic lanthanide tris(β-diketonate) complexes as an ionophore for Cl- anion-selective electrodes. *Anal. Chem.* 76 (24): 7354–7359.

28 Claborn, K., Isborn, C., Kaminsky, W., and Kahr, B. (2008). Optical rotation of achiral compounds. *Angew. Chem. Int. Ed.* 47 (31): 5706–5717.

29 Richardson, F.S. (1980). Selection rules for lanthanide optical activity. *Inorg. Chem.* 19 (9): 2806–2812.

30 Heffern, M.C., Matosziuk, L.M., and Meade, T.J. (2014). Lanthanide probes for bioresponsive imaging. *Chem. Rev.* 114 (8): 4496–4539.

31 Binnemans, K. (2015). Interpretation of europium(III) spectra. *Coord. Chem. Rev.* 295: 1–45.

32 Schippers, P.H. (1982). Optical activity in chemiluminescence. PhD thesis. The Netherlands: University of Leiden.

33 Schippers, P.H., van den Buekel, A., and Dekkers, H.P.J.M. (1982). An accurate digital instrument for the measurement of circular polarisation of luminescence. *J. Phys. E* 15 (9): 945.

34 Bonsall, S.D., Houcheime, M., Straus, D.A., and Muller, G. (2007). Optical isomers of n,n[prime or minute]-bis(1-phenylethyl)-2,6-pyridinedicarboxamide coordinated to europium(III) ions as reliable circularly polarized luminescence calibration standards. *Chem. Commun.*: 3676–3678.

35 Lunkley, J.L., Shirotani, D., Yamanari, K. et al. (2008). Extraordinary circularly polarized luminescence activity exhibited by cesium tetrakis(3-heptafluoro-butylryl-(+)-camphorato) Eu(III) complexes in EtOH and CHCl3 solutions. *J. Am. Chem. Soc.* 130 (42): 13 814–13 815.

36 Harada, T., Nakano, Y., Fujiki, M. et al. (2009). Circularly polarized luminescence of Eu(III) complexes with point- and axis-chiral ligands dependent on coordination structures. *Inorg. Chem.* 48 (23): 11 242–11 250.

37 Harada, T., Tsumatori, H., Nishiyama, K. et al. (2012). Nona-coordinated chiral Eu(III) complexes with stereoselective ligand-ligand noncovalent interactions for enhanced circularly polarized luminescence. *Inorg. Chem.* 51 (12): 6476–6485.

38 Yuasa, J., Ohno, T., Miyata, K. et al. (2011). Noncovalent ligand-to-ligand interactions alter sense of optical chirality in luminescent tris(β-diketonate) lanthanide(III) complexes containing a chiral bis(oxazolinyl) pyridine ligand. *J. Am. Chem. Soc.* 133 (25): 9892–9902.

39 Di Pietro, S. and Di Bari, L. (2012). The structure of MLn(hfbc)4 and a key to high circularly polarized luminescence. *Inorg. Chem.* 51 (21): 12 007–12 014.

40 Kumar, J., Marydasan, B., Nakashima, T. et al. (2016). Chiral supramolecular polymerization leading to eye differentiable circular polarization in luminescence. *Chem. Commun.* 52: 9885–9888.

41 Schatz, P.N. and McCaffery, A.J. (1969). The faraday effect. *Q. Rev. Chem. Soc.* 23: 552–584.

42 Kemp, J.C., Swedlund, J.B., and Evans, B.D. (1970). Magnetoemission from incandescent bodies. *Phys. Rev. Lett.* 24: 1211–1214.

43 Wagniere, G.H. (2008). *Light, Magnetism, and Chirality*, 49–74. Wiley-VCH.

44 Schatz, P., Mowery, R., and Krausz, E. (1978). M.C.D./M.C.P.L. saturation theory with application to molecules in $D_{\infty h}$ and its subgroups. *Mol. Phys.* 35 (6): 1537–1557.

45 Krausz, E.R., Mowery, R.L., and Schatz, P.N. (1978). Magnetic circular dichroism (MCD) and magnetic circularly polarized luminescence (MCPL) studies of transient and stable matrix isolated species. *Ber. Bunsen Ges. Phys. Chem.* 82 (1): 134–136.

46 Fiederling, R., Keim, M., Reuscher, G. et al. (1999). Injection and detection of a spin-polarized current in a light-emitting diode. *Nature* 402 (6763): 787–790.

47 Ohno, Y., Young, D.K., Beschoten, B. et al. (1999). Electrical spin injection in a ferromagnetic semiconductor heterostructure. *Nature* 402 (6763): 790–792.

48 Kobayashi, N. and Muranaka, A. (2012). *Circular Dichroism and Magnetic Circular Dichroism Spectroscopy for Organic Chemists*. The Royal Society of Chemistry.

49 Shikoh, E., Fujiwara, A., Ando, Y., and Miyazaki, T. (2006). Spin injection into organic light-emitting devices with ferromagnetic cathode and effects on their luminescence properties. *Jpn. J. Appl. Phys.* 45 (9R): 6897.

50 Stephens, P.J. (1970). Theory of magnetic circular dichroism. *J. Chem. Phys.* 52 (7): 3489–3516.

51 Neese, F. and Solomon, E.I. (1999). MCD C-term signs, saturation behavior, and determination of band polarizations in randomly oriented systems with spin $S \geq 1/2$. Applications to $S = 1/2$ and $S = 5/2$. *Inorg. Chem.* 38 (8): 1847–1865.

52 Stephens, P.J., Schatz, P.N., Ritchie, A.B., and McCaffery, A.J. (1968). Magnetic circular dichroism of benzene, triphenylene, and coronene. *J. Chem. Phys.* 48 (1): 132–138.

53 Serber, R. (1932). The theory of the faraday effect in molecules. *Phys. Rev.* 41: 489–506.

54 Lehnert, N., George, S.D., and Solomon, E.I. (2001). Recent advances in bioinorganic spectroscopy. *Curr. Opin. Chem. Biol.* 5 (2): 176–187.

55 Laguta, O., Hamzaoui, H.E., Bouazaoui, M. et al. (2015). Magnetic circular polarization of luminescence in bismuth-doped silica glass. *Optica* 2 (8): 663–666.

56 Ganyushin, D. and Neese, F. (2008). First-principles calculations of magnetic circular dichroism spectra. *J. Chem. Phys.* 128 (11): 114 117.

57 Wagniere, G. and Meier, A. (1982). The influence of a static magnetic field on the absorption coefficient of a chiral molecule. *Chem. Phys. Lett.* 93 (1): 78–81.

58 Barron, L. and Vrbancich, J. (1984). Magneto-chiral birefringence and dichroism. *Mol. Phys.* 51 (3): 715–730.

59 Rikken, G.L.J.A. and Raupach, E. (1997). Observation of magneto-chiral dichroism. *Nature* 390 (6659): 493–494.

60 Train, C., Gheorghe, R., Krstic, V. et al. (2008). Strong magneto-chiral dichroism in enantiopure chiral ferromagnets. *Nat. Mater.* 7 (9): 729–734.
61 Rikken, G.L.J.A. and Raupach, E. (1998). Pure and cascaded magnetochiral anisotropy in optical absorption. *Phys. Rev. E* 58: 5081–5084.
62 Rikken, G.L.J.A. and Raupach, E. (2000). Enantioselective magnetochiral photochemistry. *Nature* 405 (6789): 932–935.
63 Koerdt, C., Düchs, G., and Rikken, G.L.J.A. (2003). Magnetochiral anisotropy in Bragg scattering. *Phys. Rev. Lett.* 91: 073 902.
64 Kitagawa, Y., Segawa, H., and Ishii, K. (2011). Magneto-chiral dichroism of organic compounds. *Angew. Chem. Int. Ed.* 50 (39): 9133–9136.
65 Frackowiak, D., Bauman, D., and Stillman, M.J. (1982). Circular dichroism and magnetic circular dichroism spectra of chlorophylls in nematic liquid crystals. I. Electric and weak magnetic field effects on the dichroism spectra. *Biochim. Biophys. Acta* 681 (2): 273–285.
66 Frackowiak, D., Bauman, D., Manikowski, H. et al. (1987). Circular dichroism and magnetic circular dichroism spectra of chlorophylls a and b in nematic liquid crystals. *Biophys. Chem.* 28 (2): 101–114.
67 Yablonovitch, E. (1987). Inhibited spontaneous emission in solid-state physics and electronics. *Phys. Rev. Lett.* 58 (20): 2059.
68 Luck, W., Klier, M., and Wesslau, H. (1963). Uber bragg-reflexe mit sichtbarem licht an monodispersen kunststofflatices. II. *Ber. Bunsen Ges. Phys. Chem.* 67 (1): 84–85.
69 Noda, S., Chutinan, A., and Imada, M. (2000). Trapping and emission of photons by a single defect in a photonic bandgap structure. *Nature* 407: 608–610.
70 Takezoe, H. (2012). *Liquid Crystal Lasers*, 1–27. Hoboken, NJ: Wiley.
71 Kopp, V.I., Fan, B., Vithana, H.K.M., and Genack, A.Z. (1998). Low-threshold lasing at the edge of a photonic stop band in cholesteric liquid crystals. *Opt. Lett.* 23 (21): 1707–1709.
72 Berreman, D.W. and Scheffer, T.J. (1970). Bragg reflection of light from single-domain cholesteric liquid-crystal films. *Phys. Rev. Lett.* 25: 577–581.
73 Berreman, D.W. (1972). Optics in stratified and anisotropic media: 4×4-matrix formulation. *J. Opt. Soc. Am.* 62 (4): 502–510.
74 Taflove, A. and Hagness, S. (2005). *Computational Electrodynamics: The Finite-Difference Time-Domain Method*, Artech House Antennas and Propagation Library. Artech House.
75 Ogawa, Y., Fukuda, J., Yoshida, H., and Ozaki, M. (2013). Finite-difference time-domain analysis of cholesteric blue phase II using the landau-de Gennes tensor order parameter model. *Opt. Lett.* 38 (17): 3380–3383.
76 Il'Chishin, I.P., Tikhonov, E.A., Tishchenko, V.G., and Shpak, M.T. (1980). Generation of a tunable radiation by impurity cholesteric liquid crystals. *JETP Lett.* 32: 24.
77 Schmidtke, J., Stille, W., and Finkelmann, H. (2003). Defect mode emission of a dye doped cholesteric polymer network. *Phys. Rev. Lett.* 90: 083 902.
78 Ozaki, R., Matsui, T., Ozaki, M., and Yoshino, K. (2003). Electrically color-tunable defect mode lasing in one-dimensional photonic-band-gap system containing liquid crystal. *Appl. Phys. Lett.* 82 (21): 3593–3595.

79 Song, M., Ha, N., Amemiya, K. et al. (2006). Defect-mode lasing with lowered threshold in a three-layered hetero-cholesteric liquid-crystal structure. *Adv. Mater.* 18 (2): 193–197, 1521–4095.

80 Wang, L. and Li, Q. (2016). Stimuli-directing self-organized 3D liquid-crystalline nanostructures: from materials design to photonic applications. *Adv. Funct. Mater.* 26 (1): 10–28.

81 Humar, M., Ravnik, M., Pajk, S., and Musevic, I. (2009). Electrically tunable liquid crystal optical microresonators. *Nat. Photonics* 3 (10): 595–600.

82 Humar, M. and Muševič, I. (2010). 3D microlasers from self-assembled cholesteric liquid-crystal microdroplets. *Opt. Express* 18 (26): 26 995–27 003.

83 Uchida, Y., Takanishi, Y., and Yamamoto, J. (2013). Controlled fabrication and photonic structure of cholesteric liquid crystalline shells. *Adv. Mater.* 25: 3234–3237.

84 Cao, W., Munoz, A., Palffy-Muhoray, P., and Taheri, B. (2002). Lasing in a three-dimensional photonic crystal of the liquid crystal blue phase II. *Nat. Mater.* 1 (2): 111–113.

85 Mertelj, A., Lisjak, D., Drofenik, M., and Copic, M. (2013). Ferromagnetism in suspensions of magnetic platelets in liquid crystal. *Nature* 504 (7479): 237–241.

86 Mur, M., Sofi, J.A., Kvasić, I. et al. (2017). Magnetic-field tuning of whispering gallery mode lasing from ferromagnetic nematic liquid crystal microdroplets. *Opt. Express* 25 (2): 1073–1083.

87 Jensen, H.P., Schellman, J.A., and Troxell, T. (1978). Modulation techniques in polarization spectroscopy. *Appl. Spectrosc.* 32 (2): 192–200.

88 Berova, N., Nakanishi, K., and Woody, R.W. ed. (2000). *Circular Dichroism: Principles and Applications*, 2e. New York: Wiley-VCH.

89 Drake, A.F. (1986). Polarisation modulation-the measurement of linear and circular dichroism. *J. Phys. E* 19 (3): 170.

90 Huang, X., Li, C., Jiang, S. et al. (2004). Self-assembled spiral nanoarchitecture and supramolecular chirality in Langmuir-Blodgett films of an achiral amphiphilic barbituric acid. *J. Am. Chem. Soc.* 126 (5): 1322–1323.

91 Valev, V.K., Baumberg, J.J., Sibilia, C., and Verbiest, T. (2013). Chirality and chiroptical effects in plasmonic nanostructures: fundamentals, recent progress, and outlook. *Adv. Mater.* 25 (18): 2517–2534.

92 Narushima, T. and Okamoto, H. (2013). Strong nanoscale optical activity localized in two-dimensional chiral metal nanostructures. *J. Phys. Chem. C* 117 (45): 23 964–23 969.

93 Fedotov, V.A., Mladyonov, P.L., Prosvirnin, S.L. et al. (2006). Asymmetric propagation of electromagnetic waves through a planar chiral structure. *Phys. Rev. Lett.* 97: 167 401.

94 Schäferling, M., Dregely, D., Hentschel, M., and Giessen, H. (2012). Tailoring enhanced optical chirality: design principles for chiral plasmonic nanostructures. *Phys. Rev. X* 2: 031 010.

95 Shindo, Y. and Ohmi, Y. (1985). Problems of CD spectrometers. 3. Critical comments on liquid crystal induced circular dichroism. *J. Am. Chem. Soc.* 107: 91–97.

96 Shindo, Y. and Nakagawa, M. (1985). Circular dichroism measurements. I. Calibration of a circular dichroism spectrometer. *Rev. Sci. Instrum.* 56 (1): 32–39.

97 Maestre, M.F. and Katz, J.E. (1982). A circular dichroism microspectrophotometer. *Biopolymers* 21 (9): 1899–1908.

98 Livolant, F., Mickols, W., and Maestre, M.F. (1988). Differential polarization microscopy (CD and linear dichroism) of polytene chromosomes and nucleoli from the dipteran sarcophaga footpad. *Biopolymers* 27 (11): 1761–1769.

99 Kuroda, R., Harada, T., and Shindo, Y. (2001). A solid-state dedicated circular dichroism spectrophotometer: development and application. *Rev. Sci. Instrum.* 72 (10): 3802–3810.

100 Satozono, H. (2012). Measurement technique and system of circular dichroism. P2013-231707a, Patent pending (Japan).

101 BioTools, I. (2002). Dual circular polarization modulation spectrometer. US Patent 6,480,277.

102 Claborn, K., Puklin-Faucher, E., Kurimoto, M. et al. (2003). Circular dichroism imaging microscopy: application to enantiomorphous twinning in biaxial crystals of 1,8-dihydroxyanthraquinone. *J. Am. Chem. Soc.* 125 (48): 14 825–14 831.

103 Narushima, T. and Okamoto, H. (2015). Illumination, analysis and microscope systems by circularly polarized light. P2015-257226, Patent pending (Japan).

104 Narushima, T. and Okamoto, H. (2016). Circular dichroism microscopy free from commingling linear dichroism via discretely modulated circular polarization. *Sci. Rep.* 6: 35 731.

105 Kushida, Y., Sawato, T., Saito, N. et al. (2016). Spatially heterogeneous nature of self-catalytic reaction in hetero-double helix formation of helicene oligomers. *ChemPhysChem* 17 (20): 3283–3288.

106 Tsumatori, H. and Kawai, T. (2010). Chirality of molecular aggregates evaluated by circularly polarized luminescence microscope. *Jpn. J. Opt.* 39 (5): 230–234.

107 Tsumatori, H., Nakashima, T., and Kawai, T. (2010). Observation of chiral aggregate growth of perylene derivative in opaque solution by circularly polarized luminescence. *Org. Lett.* 12 (10): 2362–2365.

108 Asahi, T. and Kobayashi, J. (2003). Polarimeter for anisotropic optically active materials. In: *Introduction to Complex Mediums for Optics and Electromagnetics, Part VII*, 645–676. SPIE Press.

109 Ishikawa, K., Terasawa, Y., Tanaka, M., and Asahi, T. (2017). Accurate measurement of the optical activity of alanine crystals and the determination of their absolute chirality. *J. Phys. Chem. Solids* 104: 257–266.

110 Arteaga, O., Freudenthal, J., Wang, B., and Kahr, B. (2012). Mueller matrix polarimetry with four photoelastic modulators: theory and calibration. *Appl. Opt.* 51 (28): 6805–6817.

111 Arteaga, O., Sancho-Parramon, J., Nichols, S. et al. (2016). Relation between 2D/3D chirality and the appearance of chiroptical effects in real nanostructures. *Opt. Express* 24 (3): 2242–2252.

112 Arteaga, O., Baldrís, M., Antó, J. et al. (2014). Mueller matrix microscope with a dual continuous rotating compensator setup and digital demodulation. *Appl. Opt.* 53 (10): 2236–2245.

113 Kuwata-Gonokami, M., Saito, N., Ino, Y. et al. (2005). Giant optical activity in quasi-two-dimensional planar nanostructures. *Phys. Rev. Lett.* 95: 227 401.

114 Narushima, T. and Okamoto, H. (2013). Circular dichroism nano-imaging of two-dimensional chiral metal nanostructures. *Phys. Chem. Chem. Phys.* 15: 13 805–13 809.

115 Narushima, T., Hashiyada, S., and Okamoto, H. (2014). Nanoscopic study on developing optical activity with increasing chirality for two-dimensional metal nanostructures. *ACS Photonics* 1 (8): 732–738.

116 Hashiyada, S., Narushima, T., and Okamoto, H. (2014). Local optical activity in achiral two-dimensional gold nanostructures. *J. Phys. Chem. C* 118 (38): 22 229–22 233.

117 Hendry, E., Carpy, T., Johnston, J. et al. (2010). Ultrasensitive detection and characterization of biomolecules using superchiral fields. *Nat. Nanotechnol.* 5 (11): 783–787.

7

Molecular Technology of Excited Triplet State

Yuki Kurashige[1], *Nobuhiro Yanai*[2], *Yong-Jin Pu*[3,4], *and So Kawata*[3]

[1] *Institute for Molecular Science (IMS), Department of Theoretical and Computational Molecular Science, 38 Nishigo-Naka, Myodaiji, Okazaki 444-8585, Japan*
[2] *Kyushu University, Center for Molecular Systems (CMS), Graduate School of Engineering, Department of Chemistry and Biochemistry, 744 Moto-oka, Nishi-ku, Fukuoka 819-0395, Japan*
[3] *Yamagata University, Department of Organic Materials Science, 4-3-16 Jonan, Yonezawa, Yamagata 92-8510, Japan*
[4] *RIKEN, Emergent Supramolecular Materials Research Team, Center for Emergent Matter Science (CEMS), 2-1 Hirosawa, Wako 351-0198, Japan*

7.1 Properties of the Triplet Exciton and Associated Phenomena for Molecular Technology

7.1.1 Introduction: The Triplet Exciton

One of the most important requirements for high-performance molecular electronic devices is the ability to transfer carriers, e.g. electrons, holes, and singlet and triplet excitons, "faster" and "farther" as much as possible without losing energy. Unfortunately, in terms of the "faster," the use of triplet excitons as carriers of energy is not a good idea because the transfer process of triplet excitons, which requires two electrons exchange, is in general much slower than that of singlet excitons, which can be transferred without the electron migration process. Nevertheless, triplet excitons are still promising carriers that can transfer energy "farther" because the lifetime of triplet excitons are, in general, much longer than that of singlet excitons because the decay to the singlet ground state should involve spin-forbidden transitions. In fact, as shown in Table 7.1, the diffusion coefficients D of triplet excitons are smaller by four orders of magnitude than those of singlet excitons, but the diffusion lengths L of triplet excitons are comparable with or longer than those of singlet excitons in CBP (4,4′-bis(9-carbazolyl)-biphenyl), although the values can differ among the measurements and sample preparations.

7.1.2 Molecular Design for Long Diffusion Length

From a theoretical point of view, the diffusion length L can be defined as a square root of a product of the diffusion coefficient D and the exciton lifetime τ

$$L = \sqrt{D\tau},$$

Molecular Technology: Energy Innovation, Volume 1, First Edition.
Edited by Hisashi Yamamoto and Takashi Kato.
© 2018 Wiley-VCH Verlag GmbH & Co. KGaA. Published 2018 by Wiley-VCH Verlag GmbH & Co. KGaA.

Table 7.1 Comparison of the diffusion coefficient (D) and length (L) between singlet and triplet exciton diffusions.

Material	D (cm² s⁻¹)		L (nm)	
	Singlet[a]	Triplet[b]	Singlet[a]	Triplet[b]
CBP	$28–52 \times 10^{-4}$	$1.4–770 \times 10^{-8}$	16.8	8.3–300
NPD	$0.5–0.9 \times 10^{-4}$	—	5.1	6–87
PCBM	—	4.2×10^{-6}	5	21

a) Reference [1].
b) Reference [2].

thus, we should design a material so that it has a large diffusion coefficient and a long lifetime. However, the two parameters, D and L, are not independent of each other. In the case of the singlet excitons, if we assume that the transfer occurs only by the hopping between the nearest-neighbor molecules and an equal separation between molecules and an equal hopping rate, the diffusion coefficient can be approximately estimated by

$$D = \frac{1}{2n}kr^2,$$

where n is the dimensionality of the material ($n = 1, 2,$ or 3), k is the hopping rate constant between the nearest-neighbor molecules, and r is the separation distance between the molecules. Although it seems that large r, i.e. a long length of the exciton's stride, enhances the diffusion, the rate k for the singlet exciton transfer exhibits r^{-6} dependence based on the famous Förster energy transfer model because k is quadratic in the electronic coupling V, which decays as r^{-3} in the model. Consequently, the diffusion length decreases with increase in the separation distance between the molecules. The situation is much worse in the case of triplet exciton transfer; the rate k decays exponentially with the separation distance because the triplet exciton hopping requires two-electron exchange coupling, which depends on the overlap of the wave function of the nearest-neighbor molecules.

The electronic coupling V is one of the parameters that governs the rate of the transition processes and will be explained in detail in the next subsection. Figure 7.1 shows the electronic couplings associated with the various transition processes, singlet-exciton transfer, hole transfer, triplet exciton transfer, against the distance of two anthracene molecules, which are placed in face-to-face orientation with slip parallel to the long axis and short axis by 1.0 and 0.5 Å, respectively. The corresponding diabatic states were obtained by unitary transformations from the complete active-space self-consistent field wave functions [3, 4]. The augmented-cc-pVDZ basis sets [5] were adapted in the calculations. Apparently, the coupling in the triplet exciton transfer decays rapidly and the transition should occur only in well-packed aggregates, in which the molecules are not separated larger than 4–5 Å. In contrast, the singlet exciton transfer should occur even when the molecules are far apart, e.g. the fluorescence

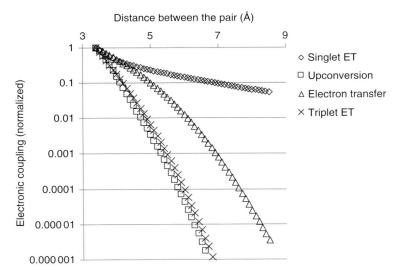

Figure 7.1 Electronic couplings associated with the various transition processes against the distance of two anthracene molecules, which are placed in face-to-face orientation with slip parallel to the long-axis and short-axis by 1.0 and 0.5 Å, respectively.

resonance energy transfer (FRET) measurements, which rely on the Förster resonance energy transfer between two fluorophores and determine whether the two molecules are within a certain distance in typically biosystems, are applicable even when the molecules are 10 nm away from each other.

Another important parameter that governs the diffusion length of the singlet excitons through D and τ is the transition dipole moment μ on the molecule, which appears in a leading term of the multipole expansion of the Coulomb coupling,

$$V_{\text{Coulomb}} = \iint d\vec{r}_1 d\vec{r}_2 \frac{\rho_D(\vec{r}_1) - \rho_A(\vec{r}_2)}{|\vec{r}_1 - \vec{r}_2|}$$

$$\approx \frac{\vec{\mu}_D \cdot \vec{\mu}_A - 3(\vec{\mu}_D \cdot \vec{R}_{DA})(\vec{\mu}_A \cdot \vec{R}_{DA})}{|\vec{R}_{DA}|^3},$$

where ρ_D and ρ_A are the transition density of the donor and accepter molecules, respectively, and $\vec{\mu}_D$ and $\vec{\mu}_A$ are their transition dipole moment, and $\vec{R}_{DA}$ is intermolecular distance between the center of the charge distributions; thus, the V, namely k and D, is maximized when the transition dipoles are as large as possible and oriented in parallel to each other. However, the large transition dipoles result in very short lifetime of the singlet excitons; the major deactivate processes of the singlet excitons are nonradiative and radiative transition, and the transition rate of the latter is determined by the magnitude of the transition dipole, which appears in the leading term of the matter–light interaction Hamiltonian. In Ref. [6], it is shown that the diffusion length curves of the singlet exciton against the transition dipole moment have a maximal point at an intermediate size of the transition dipole moment, and the theoretical maximum of the singlet diffusion length is estimated to be 230 nm, which is already realized by existing materials, e.g. tetracene crystals where $L = 115$ nm. In the case of the triplet exciton transfer, the deactivation processes, both radiative and nonradiative transitions, are spin-forbidden process, and the lifetime should be determined by the strength of the spin–orbit coupling, which is usually small in organic molecules. There is, therefore, no interdependence between the electronic coupling and lifetime, and thus, these parameters can be optimized without a restriction in contrast to the case of the singlet exciton transfer.

7.1.3 Theoretical Analysis for the Electronic Transition Processes Associated with Triplet

Triplet excitons run around in the material following quantum mechanics, i.e. the Schrödinger equation, under the effect of various fluctuations from the environments such as molecular vibrations and polarizations. A lot of effort has been made for analyzing quantum transport theoretically in condensed phases, particularly molecular crystals, and various theories have been developed to predict the efficiency of the quantum transfer in the molecular materials, from simple models to sophisticated quantum dynamics simulations. In the case of the triplet exciton transfer, it should be well described by the hopping model based on the Marcus theory because the excitons cannot be delocalized over several molecules, that is, localized on a molecule, and as mentioned above, the transitions are generally very slow, typically sub-nanosecond, because of their large reorganization energies and small electronic coupling.

$$k = \frac{2\pi}{\hbar} V^2 \frac{1}{\sqrt{4\pi \lambda k_b T}} \exp\left(-\frac{(\lambda + \Delta G)^2}{4\lambda k_b T}\right),$$

where V is the electronic coupling associated with the transition, λ is the reorganization energy, ΔG is the driving force, k_b is the Boltzmann constant, and T is the absolute temperature.

In contrast to the triplet exciton transfer, the triplet–triplet annihilation (TTA) process in Figure 7.2, in which two triplet excitons collide in solutions or materials by diffusion of the molecules and energies, respectively, and generate a singlet excited state that have equal energy with two triplets, i.e. $E(S_1) = 2 \times E(T)$, or decay to the ground state, exhibits a relatively large electronic coupling and thus can occur in short time, typically sub-picosecond, in certain situations. The TTA is an important process for the efficient use of the triplet excitons in the materials; e.g. the photon upconversion (UC) materials utilize the TTA process to convert long-wavelength photon energies to short-wavelength energies (see

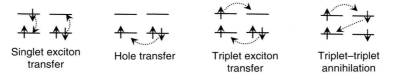

Figure 7.2 Schematic diagrams for various transition processes in the materials.

Section 7.2). The reverse process of the TTA is the singlet-fission process, in which one singlet exciton, which is usually created by sunlight absorption, is split into two triplet excitons, which may have the energy again $E(T) = 0.5 \times E(S)$. The singlet-fission process has recently drawn much attention because it potentially enables us to surpass the Shockley–Queisser limit (a theoretical efficiency limit of 33.7%) for a single-junction photovoltaic cells by splitting the excess energies of the UV of sunlight above the bandgap of the semiconductor into the utilizable two triplet excitons (see Section 7.3). The electronic coupling associated with the UC and singlet fission are identical, i.e. $\langle S_1 | V | {}^1(T_1T_1) \rangle$ where ${}^1(T_1T_1)$ represents two triplet states are spin coupled to form a singlet state in total; therefore, they are uniformly indicated by V_{TTA} in this chapter. As mentioned above, the time scale of the TTA is sometimes short enough and there is an argument over the validity of the application of the simple hopping model as the Marcus model to the TTA process [7–9]. In many previous studies, sophisticated quantum dynamics simulations that consider the coherence of the electronic states between S_1 and ${}^1(T_1T_1)$ states have been adapted to describe the TTA process.

If one adapts the hopping model, the transition rate can be predicted accurately in an *ab initio* manner; i.e. the parameters are determined by the *ab initio* quantum chemical calculations. The reorganization energies λ and driving force ΔG are obtained from the potential energy surfaces of the reactor and product states of the transitions, and the electronic coupling is defined as the energy gap of the two adiabatic potential energy surfaces at the crossing point. Unfortunately, the π-conjugated molecules used as building blocks for the organic electronic devices often have complex low-lying electronic states and predictive molecular electronic structure theories are not well established as for the ground state. Table 7.2 shows the excitation energies and reorganization energies with various quantum chemical methods. The vertical excited energies of L_a, L_b, and $2A_g$ states were calculated at the ground-state geometry and that of T_1 state were calculated at the T_1 optimized geometry in order to directly compare with the experiments. The reorganization energies are calculated from the ground-state energy at the ground and excited state geometries and the excited-state energy at

Table 7.2 Comparison of the vertical excitation energies and reorganization energies obtained by DFT and ab initio theories for the low-lying states of the pentacene molecule (eV).

	BP86	B3LYP	CAM-B3LYP	CC2	DMRG-CASPT2[a]	Expr.[b]
$E(L_a)$	1.56	1.81	2.12	2.27	2.17	2.37
$E(L_b)$	2.34	2.88	3.81	3.24	3.23	3.12
$E(2A_g)$	3.83	4.32	4.65	4.37	2.91	—
$E(T_1)$[c]	0.58	0.54	0.46	0.80	0.86	0.86
λ_{La}	0.18	0.20	0.24	0.19	0.20	—
λ_{T1}	0.28	−0.12	0.48	0.35	0.36	—

a) Reference [10].
b) Reference [11, 12].
c) The vertical excitation energies for the T_1 state were calculated at the T_1-optimized geometry.

the ground and excited state geometries by using the four-point rule [13]; thus, only the reorganization of the inner sphere is included. In the calculations of the time-dependent density functional theory (TD-DFT) and the second-order approximate coupled-cluster theory (CC2), def2-TZVP basis sets were adapted.

In most of the studies, the TD-DFT is frequently used for computing the excitation energies of the π-conjugated molecules due to the good balance between the accuracy and the computational costs. Nevertheless, it is well known that the DFT calculations with conventional functionals give quantitatively incorrect results, even qualitatively incorrect sometimes, for the low-lying states of polycyclic aromatic hydrocarbons. In Table 7.2, the excitation energy for the L_a and L_b states of the pentacene molecule is considerably underestimated by the BP86 (by 0.81 and 0.78 eV for the L_a and L_b states, respectively) and B3LYP (by 0.56 and 0.24 eV for the L_a and L_b states, respectively). Although the excitation energy for the L_a state is improved by the CAM-B3LYP [14] calculation, that of the L_b state is overestimated by 0.5 eV. In addition, the excitation energy of the T_1 state is significantly underestimated in all the DFT calculations, which should cause large errors for evaluating the triplet transfer rate between the molecules. The CC2 theory [15], which is an *ab initio* method with moderate computational cost, gives accurate results for the excitation energies of the L_a, L_b, and T_1 states. It is, however, based on the single-reference picture and cannot describe double excitations, and the excitation energy for the $2A_g$ state, which is a *dark* state and difficult to be observed with experiments, is largely overestimated. The DMRG-CASPT2 theory [16, 17], which is based on the multireference picture and gives a balanced description for various complex excited states, gives very accurate excitation energies regardless of the character of the excited states. The development of the DMRG-CASPT2 theory, therefore, will pave the way for reliable theoretical predictions of the transition processes in the molecular materials.

The *V* is so-called "electronic coupling" between the reactant state and the product state. Many approximated methods for evaluating *V* for various transitions have been developed so far; e.g. the transfer integral approach is famous for the hole and electron transfer analysis, and the dipole–dipole interaction approach is frequently used for the singlet energy transfer analysis. Although these approaches are based on the one-electron theories and the entanglement within and between the molecules, i.e. intra- and intermolecular entanglements, are neglected, it is difficult to assess the accuracy of the approximations to *V* because the electronic coupling is not an observable and cannot be directly measured by experiments. Comparisons between theoretical calculations and experiments are often made through the transition rate *k*, for which the errors in theoretical calculations will be caused not only by *V* but also by the parameters λ and ΔG associated with the potential energy surfaces, although the accuracy of those with the conventional quantum chemical methods are not always reliable as described above. The errors in the several parameters can therefore be sometimes canceled or affected by each other in the result for *k*, i.e. the diffusion coefficient and carrier mobility.

In theory, the electronic coupling is defined as the nondiagonal elements of the electronic Hamiltonian between the so-called diabatic states representing the reactant and product states, for which the adiabatic coupling is completely eliminated, although as already described in the literature, the strictly diabatic states cannot be obtained in general [18]. Most of the diabatization methods including the abovementioned approaches can be categorized to the constructive approaches in which a diabatic state is approximated as a composite of the fragment states and directly constructed with some constraints, implicitly or explicitly, to the character of the fragment states. It is based on the idea that wave functions of the diabatic states do not change when the nuclei move and should maintain the same character along the dissociation to the fragments. The diabatic dimer states therefore usually described as a simple direct product of well-characterized monomer state, e.g. the reactant and product diabatic states for the singlet energy transfer are often assumed as $|S_1\rangle|S_0\rangle \rightarrow |S_0\rangle|S_1\rangle$ where $|S_0\rangle$, $|S_1\rangle$ are monomer wave functions of the ground and excited states, respectively. Even if the monomer wave functions are described by highly accurate electron correlation theories, i.e. intra-molecular entanglements are taken into account, the direct product dimer states are not sufficient, in particular when the molecules are strongly interacting, because there is a lack of effect of intermolecular entanglement. In such a case, the intermolecular entanglement can be accounted by using the augmented diabatic states as described in the following.

For the transition processes in which the triplet excitons are involved, such as triplet energy transfer and TTA, effect of the entanglements between the molecules cannot be neglected for the quantitative, or even qualitative, prediction of their electronic coupling. A pair of molecules is called entangled if the dimer wave function cannot be written as a direct product of the two monomer states; i.e. even the diabatic states have to be expressed as linear combinations of the several states that have different electronic characters. In an early work of Harcourt et al. [19], the diabatic states for the triplet exciton transfer were expressed in the four-configuration model

$$|\Psi_{T_1 S_0}\rangle = N(|T_1\rangle|S_0\rangle + \lambda|D\rangle|A\rangle + \mu|A\rangle|D\rangle),$$
$$|\Psi_{S_0 T_1}\rangle = N(|S_0\rangle|T_1\rangle + \mu|D\rangle|A\rangle + \lambda|A\rangle|D\rangle),$$

where N is the normalization factor and $|A\rangle$ indicates the monomer acceptor state that has been accepted one electron from the monomer donor state $|D\rangle$; that is, the $|D\rangle|A\rangle$ and $|A\rangle|D\rangle$ indicate the two lowest charge transfer (CT) states. The coefficient μ and λ will be determined by the perturbative corrections. It was reported that the contributions of the CT states were particularly significant for the evaluation of the electronic couplings for the triplet energy transfer. There are related approaches that consider the contributions of the CT states [20–22].

In molecular materials, the triplet excitons have not been effectively used or even regarded as undesired byproducts that cause a leak of the energy. Recently, however, they have attracted renewed interest in the context of the effective use of the solar energy. The TTA process, which, of course, involves triplet excitons, is a key process for the efficient conversion of the wavelength, i.e. energy level, of excitons; in the photon UC materials (Section 7.2), two triplet excitons collide and are fused into one singlet exciton, and in the singlet-fission photovoltaic materials (Section 7.3), one singlet exciton is split up into two triplet excitons. The macroscopic functions of these molecules are delivered not only through one kind of microscopic transition process but by the cooperation of several kinds of microscopic transition processes; thus, the molecule design for efficient materials is so complicated like as a multivariable optimization problem and a molecular theory that reveals the underlying principles that connect the microscopic properties and the macroscopic functions is desired. Although theoretical analysis for the efficient molecular design has been hampered by the absent of predictive theories for the microscopic transition processes occurred in the materials due to the complex electronic–structure of the π-conjugated molecules that involves the significant intra- and inter-molecular entanglements, the developments of the theoretical approaches based on the multiconfigurational wave function theories, in particular those based on the density matrix renormalization group theory, will pave the way for theoretical molecular technology for the design of molecular electronic and energy devices.

7.2 Near-infrared-to-visible Photon Upconversion: Chromophore Development and Triplet Energy Migration

7.2.1 Introduction

The wavelength conversion from longer wavelength (lower energy photons) to shorter wavelength (higher energy photons) is called photon UC [23–26]. There has been an increasing attention for UC technologies since all the solar energy devices can be improved by implementing UC materials. For example, the UC from near-infrared (NIR) light to visible light has a potential to overcome the Shockley–Queisser limit of single-junction photovoltaic cells [27–29]. For example, metal halide perovskites have recently emerged as promising materials for photovoltaics; however, their absorption is limited to the visible range below 800 nm [30–32]. The recovery of sub-bandgap NIR photons through UC should largely enhance the efficiency of perovskite solar cells. Another application of UC can be found in photocatalysis. Photocatalytic reactions including hydrogen production and carbon dioxide reduction are regarded as next-generation energy production routes. Efficient photocatalytic systems under UV light or high-energy visible light (<500 nm) have been reported; however, efficient photocatalysis under low-energy visible light is still a challenging issue [33]. The utilization of such low-energy visible light and NIR light via UC would contribute to enhance the photocatalytic performance. Furthermore, the UC has a potential to solve major issues in photobiology. The poor permeability of visible light hampers the further development of photodynamic therapy, photoinduced drug delivery, and optogenetics. Taking advantage of high permeability of NIR light in biological tissues, materials with NIR-to-visible UC ability would offer in-body light sources under NIR excitation.

Triplet–triplet annihilation-based photon upconversion (TTA-UC) has recently attracted much attention for its occurrence with noncoherent weak excitation such as sunlight, which is difficult to be achieved by other UC

7.2 Near-infrared-to-visible Photon Upconversion

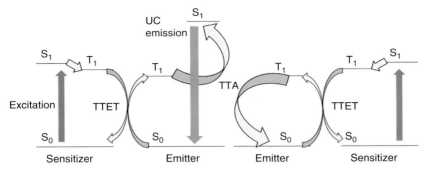

Figure 7.3 Scheme for the common mechanism of TTA-based UC. The system includes triplet sensitizer and emitter molecules (S = singlet, T = triplet).

mechanisms including two-photon absorption and multistep excitation of lanthanide nanoparticles [23–26]. The finding of TTA-UC phenomenon dates back to the 1960s [34], and there has been renewed attention in this decade by the demonstration of efficient TTA-UC under weak solar excitation [23].

The common scheme of TTA-UC is shown in Figure 7.3. The system consists of donor (triplet sensitizer) and acceptor (emitter) molecules. First, sensitizer molecules absorb light, and triplet excited state $T_{1,D}$ is formed through intersystem crossing (ISC) from the singlet excited state $S_{1,D}$. This is followed by triplet–triplet energy transfer (TTET) from sensitizer to emitter. Two sensitized emitter triplets $T_{1,A}$ diffuse and collide to undergo TTA and consequently formed excited singlet $S_{1,A}$ produces upconverted delayed fluorescence. Both of TTET and TTA processes take place via Dexter-type electron exchange mechanism, and thus, the molecules should come within a close distance, usually less than 1 nm.

Basic requirements for sensitizer and emitter are as follows. The sensitizer should have large excitation coefficients at the excitation wavelength, and its ISC efficiency should be high. For efficient TTET, microsecond-scale long triplet lifetime is desired for the sensitizer. As for the emitter, its triplet lifetime needs to be longer than ~100 μs for TTA, the $S_{1,A}$ energy level should be lower than twice of $T_{1,A}$ energy, and the emitter shows a high fluorescence quantum yield. The most common combination of the sensitizer and emitter is metalloporphyrins and polyaromatic compounds, respectively. In particular, the benchmark combination is Pt(II) octaethylporphyrin (PtOEP) and 9,10-diphenyl anthracene (DPA), which shows efficient green-to-blue TTA-UC [24, 35, 36].

Although many sensitizer and emitter molecules have been recently developed for various wavelength UCs, it still remains a challenge to achieve efficient NIR-to-visible TTA-UC. This is partly due to the energy loss during the ISC of sensitizer, which limits the use of emitters with high $T_{1,A}$ and $S_{1,A}$ energy levels [37]. To solve this problem, there have been recent efforts to develop new types of sensitizer that can minimize the energy loss for triplet sensitization, which is introduced later.

It should be noted that the excited state of these new sensitizer materials often has short lifetime (<μs) [38]. Most efficient TTA-UC systems using conventional triplet sensitizers have been reported in solution; however, in the case of new sensitizers with short lifetime, the molecular diffusion in solution is not fast enough to achieve an efficient energy transfer from sensitizer to emitter. Therefore, it is important to integrate the concept of triplet energy migration-based photon upconversion (TEM-UC), which utilizes the fast TEM in dense co-assemblies of sensitizer and emitter [26]. Even though the sensitizer excited state is short-lived, the close contact between sensitizer and emitter allows the efficient energy transfer.

7.2.2 Evaluation of TTA-UC Properties

Along with the conversion wavelength, excitation intensity and quantum yield are the key parameters to characterize a TTA-UC system. The efficiency of multiexciton TTA process depends on concentration of excited species, namely, the excitation intensity. Highly efficient TTA-UC systems under weak excitation intensity such as sunlight are strongly desired. Monguzzi et al. proposed the figure-of-merit parameter called threshold excitation intensity I_{th}, at which half of produced T_1 is used for TTA [39]. This means half of maximum UC quantum yield Φ_{UC} of the system is obtained at I_{th}. Generally, at low-incident light intensity, the emission from bimolecular annihilation process exhibits the quadratic dependence to the excitation intensity. At the high excitation regime, the TTA process becomes dominant for emitter triplet decay, resulting in quasi-linear dependence [39–41]. The I_{th} value can be experimentally determined as a crossing point between two fitting lines for these quadratic and linear regimes. In the case of three-dimensional diffusion for triplet exciton, the threshold is expressed as a function of the system's fundamental parameters using the equation

$$I_{th} = (\alpha \Phi_{ET} 8\pi D_T a_0)^{-1} (\tau_T)^{-2}, \tag{7.1}$$

where α is the absorption coefficient at the excitation wavelength, Φ_{ET} is the sensitizer-to-emitter TTET efficiency, D_T is the diffusion constant of emitter triplet, a_0 is the annihilation distance between emitter triplets (usually around 1 nm), and τ_T is the lifetime of the emitter triplet [39]. To achieve a low I_{th} value, the TTET efficiency needs to be high, and the triplet diffusion should be fast.

In general, the quantum yield is defined as the ratio of absorbed photons to emitted photons, and thus the maximum TTA-UC quantum yield Φ_{UC} is 50%. However, many reports multiply this value by 2 to normalize the maximum quantum yield at 100%. To avoid the confusion between these different definitions, the UC quantum yield is written as Φ_{UC}' ($=2\Phi_{UC}$) when the maximum efficiency is set as 100%. The Φ_{UC}' can be expressed as follows:

$$\Phi_{UC}' = f \Phi_{ISC} \Phi_{ET} \Phi_{TTA} \Phi_{FL}, \tag{7.2}$$

where Φ_{ISC}, Φ_{ET}, Φ_{TTA}, and Φ_{FL} represent the quantum efficiencies of sensitizer ISC, sensitizer-to-emitter TTET, emitter–emitter TTA, and emitter fluorescence, and f is the statistical probability for obtaining the singlet excited state after the annihilation of two triplets [24, 25]. Besides employing a sensitizer with high Φ_{ISC} and a highly fluorescent emitter, the high triplet sensitization efficiency is important to achieve a high UC quantum yield. It is possible to optimize all the parameters for TTA-UC in the visible range. For example, the PtOEP-DPA pair in deaerated solution shows almost quantitative Φ_{ISC}, Φ_{ET}, Φ_{TTA}, and Φ_{FL} values. The Φ_{UC} was reported as high as 52%, suggesting the f value of 0.52 [24].

7.2.3 NIR-to-visible TTA-UC Sensitized by Metalated Macrocyclic Molecules

The key of upconverting NIR light over 700 nm is to develop an appropriate triplet sensitizer. Requirements for NIR sensitizer include a large absorption coefficient in the NIR region, an efficient S_1-to-T_1 ISC (large Φ_{ISC} value), and a long triplet lifetime. Recent efforts of several research groups have shown that metalated macrocyclic compounds can meet these requirements. Figure 7.4 illustrates chemical structures of employed sensitizers. Porphyrin [42–48], phthalocyanine [37, 49], and texaphyrin [50] derivatives with extended π-conjugated structures were found to effectively adsorb NIR light and sensitize emitter triplets.

Emitter molecules need to have a lower T_1 energy level in the NIR region than that of sensitizer molecules. In addition, an emitter S_1 energy level should be lower than twice of T_1 energy, and a fluorescent quantum yield Φ_{FL} from emitter S_1 should be high. Derivatives of tetracene [42–44, 46–48], perylene diimide (PDI) [45], and terrylenediimide (TDI) [37] have been employed as emitter (Figure 7.5).

Baluschev et al. have reported the NIR-to-green TTA-UC by employing a π-extended palladium tetranaphthoporphyrin **1** and a bis(tetracene) **13** as sensitizer and emitter, respectively (Figures 7.4 and 7.5) [43]. The photoluminescence spectra of a mixed solution of **1** and **13** showed an upconverted emission with local maxima at 500, 533, and 573 nm under excitation by a 700-nm laser. The fact that the phosphorescence of the sensitizer (local maxima at 916 and 942 nm) was almost completely quenched indicates the efficient sensitizer-to-emitter TTET process. When the excitation source was the NIR part of the noncoherent solar spectrum, the upconverted emission was clearly observed, and the hypsochromic shift between excitation and emitted energy was as large as 0.7 eV (Figure 7.6). The quantum yield of this NIR-to-visible UC was more than 4%. Importantly, these results demonstrate the promising potential of TTA-UC to harvest such deep-lying NIR triplets by means of visible fluorescence.

Baluschev et al. have further extended the excitation wavelength to 785 nm by expanding the sensitizer structure to palladium tetraanthraporphyrin **2** (Figure 7.4). By combining it with an emitter rubrene, a NIR-to-yellow (570 nm) UC emission with a large anti-Stokes shift of 0.6 eV was observed [44]. They further tuned the absorption band of tetraanthraporphyrin by systematic chemical modification (**3–6** in Figure 7.4). The mixture of these four sensitizers and rubrene allowed the wide-range absorption of the solar spectrum (720–840 nm) and the UC into visible emission (520–680 nm) [42].

Castellano and coworker have sensitized rubrene by employing Pd phthalocyanine **7** as a triplet sensitizer (Figure 7.4) [49]. A yellow upconverted emission was observed under excitation of 725 nm laser. In air-saturated solution, the excitation of sensitizer **7** resulted in the decomposition of rubrene to form oxygen

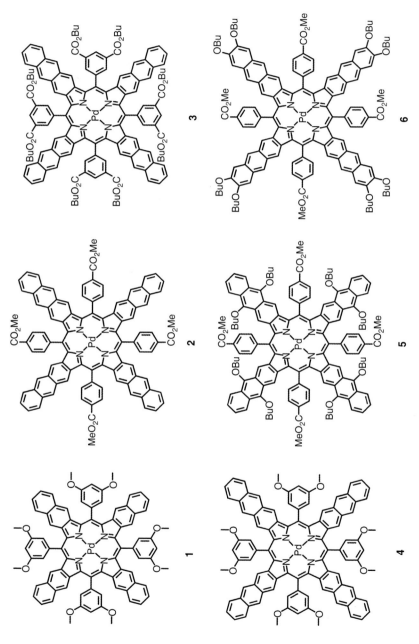

Figure 7.4 Chemical structures of sensitizers having NIR (>700 nm) absorption.

Figure 7.4 (*Continued*)

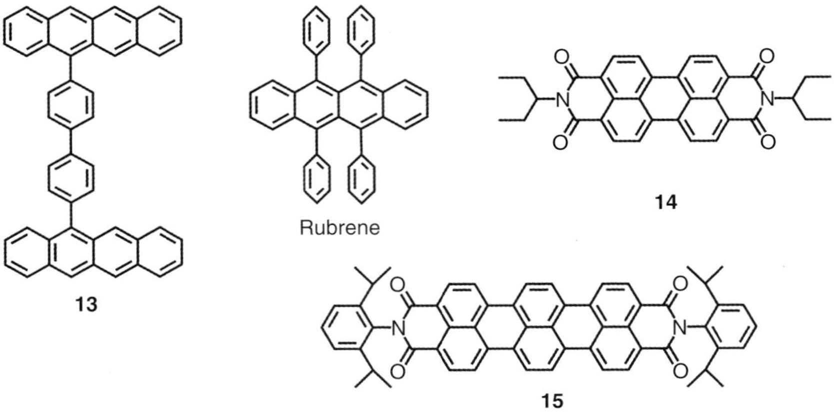

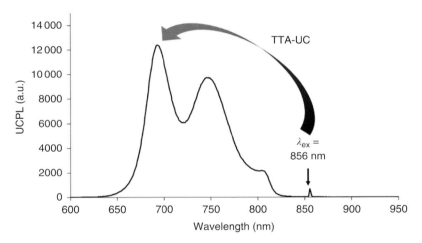

Figure 7.5 Chemical structures of emitters that show fluorescence in the visible range (<700 nm).

Figure 7.6 Upconverted emission spectra of the **12–15** pair in deaerated 1,2-dichlorobenzene under excitation of the 856 nm laser ([**12**] = 0.05 mM, [**15**] = 0.6 mM, 810 nm short-pass filter). *Source*: From Christiansen et al. [15]. Reproduced with permission of Elsevier.

adducts. This decomposition reaction was suppressed by the removal of oxygen from solution using freeze–pump–thaw degassing. This sensitizer–emitter pair was dispersed in ethyleneoxide/epichlorohydrin copolymer to make a solid film. The film showed an UC emission under Ar atmosphere; however, air exposure of the film slowly degraded over time as evident by the disappearance of rubrene absorption. This air stability issue is discussed later in this section. The Castellano group has also reported a larger anti-Stokes shift of 0.86 eV from 780 nm laser excitation to 541 nm emission by combining a sensitizer **8** with a PDI emitter **14** (Figures 7.4 and 7.5) [45]. The UC quantum yield Φ_{UC}' of this system was 0.75%. An improved UC quantum yield Φ_{UC}' of 1.54% was achieved when a Cd(II) texaphyrin **9** was coupled with rubrene in degassed dichloromethane [50].

The integration of TTA-UC into photovoltaic cells has been explored by Schmidt and coworkers [46]. The inability to harvest low-energy photons is the major efficiency loss mechanism of all higher bandgap materials including hydrogenated amorphous silicon (a-Si:H), lead halide perovskites, and organic and dye-sensitized cells. TTA-UC can be a simple and cheap method to overcome the Shockley–Queisser limit of these single-threshold photovoltaic devices. A mixed solution of sensitizer **11** and emitter rubrene was placed behind an a-Si:H cell. A part of photons in the range of 600–750 nm passed through the a-Si:H material and excited the triplet sensitizer, which resulted in returning upconverted photons of 500–600 nm. A peak efficiency enhancement of 1% was observed at an excitation intensity 48 times stronger than the solar irradiance (48 suns). The photocurrent increase of organic and dye-sensitized solar cells by TTA-UC has also been demonstrated [48].

The grand challenge of NIR-to-visible TTA-UC is to harvest photons below 850 nm. The group of Yanai and Kimizuka has employed a relatively simple triplet sensitizer, metallonaphthalocyanine **12**, which exhibits a strong Q-band absorption at 828 nm with a tail extending over 900 nm [37]. This sensitizer showed a phosphorescence peak at 1348 nm (0.92 eV). This NIR energy was accepted by a π-extended emitter **15** whose triplet energy level was estimated to be 0.85 eV by DFT calculation. The sensitizer-to-emitter TTET efficiency Φ_{ET} was 31% for the optimized concentration. The excitation of the mixed solution at 856 nm resulted in a clear upconverted emission in the visible range (Figure 7.6). The TTA-based mechanism was confirmed by microsecond-scale UC lifetime and quadratic excitation intensity dependence of UC emission intensity. The relatively low UC quantum yield Φ_{UC}' of 0.13% was obtained for this system. This is mainly because the S_1 energy level (1.89 eV) was higher than twice of T_1 and thus the formation of S_1 by TTA is thermodynamically unfavorable for **15**. This result unveils the fundamental problem of NIR-to-visible UC. The energy loss during ISC from S_1 to T_1 of sensitizer forced us to use the emitter with low triplet energy level that is too low to effectively produce S_1 in the visible range. Therefore, it is crucial to circumvent the energy loss during the ISC of sensitizer, which is discussed below.

7.2.4 TTA-UC Sensitized by Metal Complexes with S–T Absorption

Triplet sensitization without energy loss of ISC is crucial to achieve an efficient TTA-UC from NIR light over 850 nm to visible light (Figure 7.7a,b). The group of Yanai and Kimizuka has focused on direct S–T absorption of metal complexes to skip the ISC process [38]. Although the S–T absorption is the spin-forbidden process, properly designed Os complexes show relatively large excitation coefficients over thousands of $cm^{-1} M^{-1}$ due to the strong spin–orbit coupling, which has been utilized to harvest NIR photons in dye-sensitized solar cells by Segawa and coworkers [51, 52]

An Os complex **16** having branched alkyl chains was designed to be miscible with nonpolar emitter, rubrene (Figure 7.7c) [38]. Deaerated CH_2Cl_2 solution of **16** and rubrene clearly showed an upconverted emission at around 570 nm under excitation of 938 nm NIR laser. The use of S–T absorption allowed to achieve the unprecedented TTA-UC of NIR light over 900 nm. However, the UC quantum

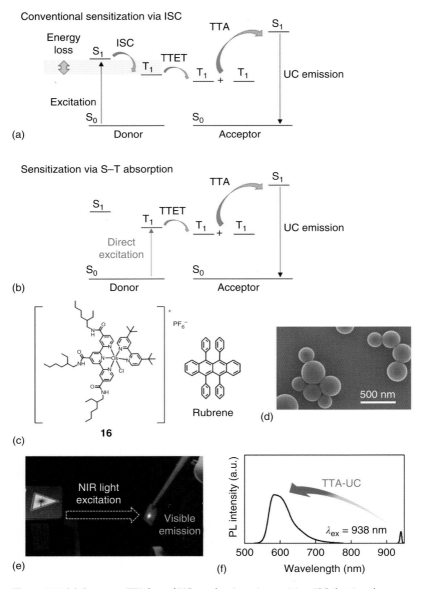

Figure 7.7 (a) Common TTA-based UC mechanism via sensitizer ISC that involves energy loss. (b) TTA-based UC mechanism utilizing S–T absorption of sensitizer. The absence of energy loss due to ISC allows the large anti-Stokes shift from NIR to visible range. (c) Chemical structures of sensitizer **16** and emitter rubrene. (d) Scanning electron microscopy image of the **16**-doped rubrene nanoparticles. (e) Photograph of upconverted yellow emission of **16**-rubrene in PVA film in air under 938 nm NIR excitation. (f) In-air upconverted emission spectrum of the **16**-doped rubrene nanoparticles dispersed in PVA film (λ_{ex} = 938, 780 nm short-pass filter). *Source*: From Amemori et al. [38]. Adapted with permission of American Chemical Society.

yield Φ_{UC}' was less than 0.01% in solution. This low efficiency is mainly because the sensitizer's triplet lifetime of 12 ns was too short to sensitizer emitter triplets by molecular diffusion and collision in solution. The strong spin–orbit coupling of the Os complex allows the S–T direct excitation; however, at the same time, the deactivation of triplet excited state becomes pronounced.

To solve this issue, the concept of TEM-UC in dense chromophore assemblies was introduced [38]. Rapid reprecipitation of **16**-rubrene mixture in water resulted in the formation of amorphous rubrene nanoparticles in which the sensitizer molecules were homogeneously doped (Figure 7.7d). The obtained nanoparticles were dispersed in poly(vinyl alcohol) (PVA) to avoid quenching the triplet state by oxygen in air. Significantly, the solid film converted NIR light (938 nm) to yellow light (580 nm) even in the ambient condition (Figure 7.7e,f). The UC quantum yield was as high as $\Phi_{UC}' = 3.1\%$, which is much higher than the one in solution ($\Phi_{UC}' = 0.01\%$). The observed difference is mainly due to more efficient TTET from the sensitizers to neighboring the emitters. Therefore, the combination between ISC-free S–T absorption and TEM-UC in dense chromophore assemblies allowed to overcome the grand challenge of efficient NIR-to-visible TTA-UC.

7.2.5 Conclusion and Outlook

In this section, we discussed the recent development of molecular sensitizers for constructing NIR-to-visible UC systems. The combination of metalated macrocycles having π-extended structures with acene and rylene dyes effectively upconvert NIR light (700–850 nm) to visible light (<700 nm). Although it remained challenging to harvest NIR light over 900 nm mainly due to the decrease in lifetime of excited states [46], it has been realized by the recent breakthrough employing S–T absorption and TEM-UC [38]. Organic UC materials can show large absorption coefficients and thus utilize weak excitation intensity such as sunlight, which is unattainable by inorganic counterparts. Further development of sensitizer materials and their integration into emitter assemblies would lead to the ideal NIR-harvesting UC systems based on energy migration [26].

7.3 Singlet Exciton Fission Molecules and Their Application to Organic Photovoltaics

7.3.1 Introduction

Singlet exciton fission (SF) [11, 53, 54] has been reported as a photophysical process, in which one singlet exciton converts into two triplet excitons through two intermediate, neighboring organic molecules. However, SF has recently attracted considerable attention since it has been reported that SF could be a means to overcome the theoretical upper limit of power conversion efficiency (PCE) in organic photovoltaics (OPVs) [55–57], known as the Shockley–Queisser limit. If the triplet excitons generated through the SF process could dissociate into free charges at the donor/acceptor interface in the OPV, the upper limit of the external quantum efficiency (EQE) of the device could reach 200%. In this regard, Baldo's group recently reported an EQE > 100% attained by utilizing the SF of pentacene as a donor material in the OPVs [58, 59].

To exhibit SF character, molecules are required to satisfy two main conditions. The first is the energy condition, wherein the exothermal conversion of one singlet exciton into two triplet excitons requires that the lowest energy of the excited singlet state (S_1) must be higher than or equal to twice the lowest energy of the triplet state (T_1), i.e. $E(S_1) \geq 2E(T_1)$. The second condition that must be satisfied is the morphology condition. The SF process is essentially a bimolecular process; therefore, the distance between the two molecules sharing the energy of either singlet exciton must be sufficiently small for this process to occur efficiently [7, 60–62]. In order to employ the SF molecules in OPV devices, in addition to satisfying the SF conditions, these molecules must satisfy the conditions required by the OPV device. These conditions include (i) broad and intense light absorption, (ii) high charge mobility, (iii) appropriate energy levels of the highest occupied molecular orbital (HOMO) and lowest unoccupied molecular orbital (LUMO), and (iv) chemical stability.

In this section, we review potential SF molecules primarily from the perspective of molecular design and categorize these into three types: (i) polycyclic π-conjugated small molecules such as acenes and rylenes, (ii) nonpolycyclic π-conjugated small molecules, and (iii) polymer-type compounds. Most SF molecules contain a polycyclic aromatic structure such as tetracene, pentacene, perylene, etc. There have been few reports on the use of nonpolycyclic SF molecules in OPV devices [63].

7.3.2 Polycyclic π-Conjugated Compounds

In oligoacenes, $E(T_1)$ decreases as the number of aromatic rings increases, due to the increased diradical character [64]. Tetracene exhibits an $E(S_1)$ of 2.32 eV and an $E(T_1)$ of 1.25 eV [11], whereas pentacene has an $E(S_1)$ of 1.83 eV and an $E(T_1)$ of 0.86 eV [65, 66]. Considering that the $E(T_1)$ of tetracene is greater than half the $E(S_1)$, the SF of tetracene should be endothermic. On the other hand, the $E(T_1)$ of pentacene is less than half of $E(S_1)$, which ensures an exothermic SF for pentacene. This energy condition, $E(S_1) \geq 2E(T_1)$, is the most important one in the molecular design of SF compounds. To tune molecules to satisfy the energy condition, the acene structure has been found to be effective in lowering the $E(T_1)$ of compounds to half of their $E(S_1)$ (Figure 7.8).

7.3.2.1 Pentacene

The SF phenomenon of pentacene was reported in the early 1970s [53, 54, 65] and has been studied within the confines of basic photophysical science. In 2009, Lee et al. reported that a photodetector consisting of pentacene and C_{60} exhibited an internal quantum efficiency (IQE) that exceeded 100% [67]. Based on the magnetic field dependence of the photocurrent, Lee et al. demonstrated that the observed high IQE was produced by the SF process of pentacene. The SF was suppressed by a magnetic field due to Zeeman splitting of T_1 state, which resulted

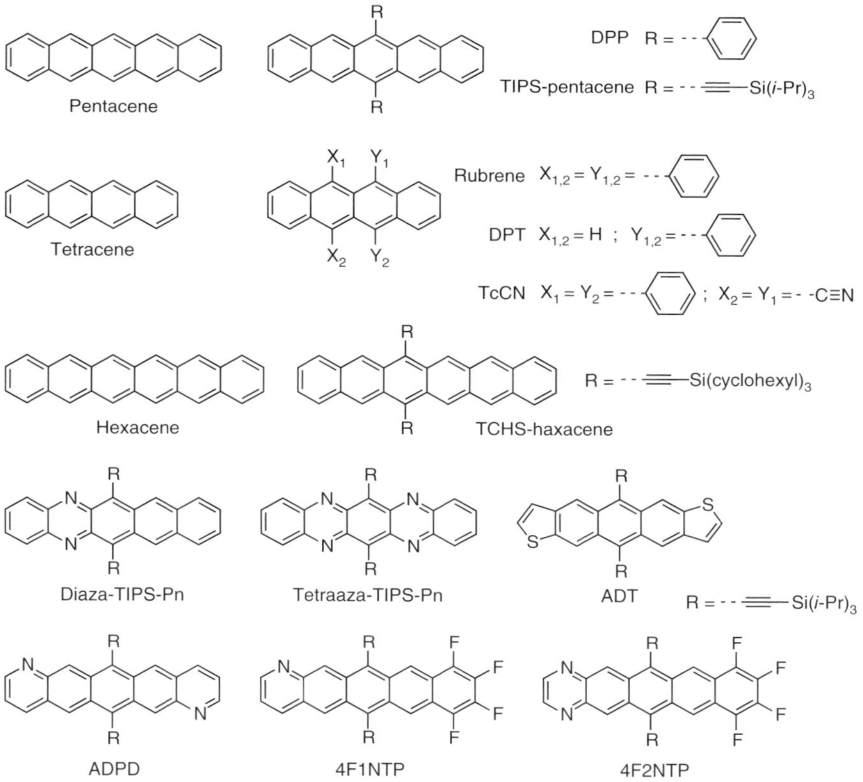

Figure 7.8 Chemical structure of the acene-type SF compounds.

in a reduction of the photocurrent. The publication of this report triggered great interest in SF, and the use of pentacene for SF has since been extensively studied. Recently, employment of the SF process of pentacene has also resulted in an EQE that exceeded 100% in a OPV that employed triplet excitons generated by pentacene/C_{60} [58].

Ultrafast transient absorption spectroscopy is a powerful technique that can be used to pursue the exciton dynamics of SF molecules. The SF mechanism of pentacene and the charge generation mechanism of pentacene/C_{60} were described by Rao, Wilson, and coworkers [66, 68]. They reported that the photoinduced absorption of the S_1 state rapidly decreased, whereas that of the T_1 state simultaneously increased. The SF of pentacenes occurs in an extremely rapid (<80 fs) time scale, comparable with the that of molecular vibration, suggesting that the SF results from a nonequilibrium delocalized state.

Pentacene derivatives substituted with specific functional groups at 6,13-positions have been reported: 6,13-bis(triisopropylsilylethynyl)pentacene (TIPS-pentacene) [61, 69–71] and 6,13-diphenyl-pentacene (DPP) [7, 72]. TIPS-pentacene is highly soluble in organic solvents due to these functional groups, and thus a film can be formed of this material using a solution-casting

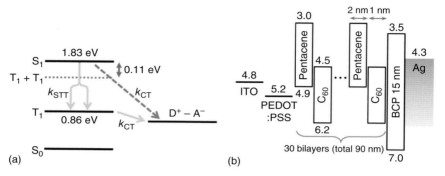

Figure 7.9 (a) Schematic of the SF process in the pentacene/C_{60} photodetector and (b) energy diagram of the pentacene/C_{60} photodetector. *Source*: From Lee et al. [67]. Reproduced with permission of American Institute of Physics.

process. The resulting solution-processed OPV, using TIPS-pentacene as a p-type material and PbS quantum dots as an n-type material, was reported to exhibit a PCE of 4.8% [71]. The CT energy estimated from the difference between the conduction band energy of the PbS QD and the HOMO of TIPS-pentacene was less than the $E(T_1)$ of the TIPS-pentacene, which enabled the charge separation of the triplet excitons that were doubled from the SF process. In DPP, two phenyl groups at 6,13-positions on the ring significantly reduced the SF rate (k_{fis}) from 12.5 ps^{-1} of unsubstituted pentacene to 0.085 ps^{-1} [7]. This was mainly because of the steric hindrance of the phenyl groups that prevented the pentacene planes from packing closely together (Figure 7.9).

7.3.2.2 Tetracene

Yost, Lee, and coworkers compared six pentacene derivatives and three tetracene derivatives and reported that a negative $\Delta E(=2E(T_1) - E(S_1))$ is a primary parameter for the design of successful SF molecules [7]. In the case of pentacene, ΔE is −0.11 eV, and this negative ΔE assures an exothermic SF process. On the other hand, the ΔE of tetracene is +0.15 eV, and this positive ΔE results in a much lower k_{fis}(=0.0091 ps^{-1}) than DPP, although tetracene exhibits a much closer molecular packing with the intermolecular distance of 3.9 Å than DPP (5.0 Å).

Rubrene (5,6,11,12-tetraphenyltetracene) is a tetracene derivative and is known to exhibit SF character similar to tetracene [73–76]. Rubrene also shows TTA, which is an UC process where two triplet excitons convert into one single exciton. In an OPV employing rubrene and C_{60}, the resulting EQE under background light exhibited a morphological dependence that was different from rubrene [76]. With amorphous rubrene, the EQE that results from the absorption of rubrene is enhanced as the background light intensity increases. This is because charge generation of the singlet exciton generated by the TTA is enhanced in the high exciton density that results from the background light. On the other hand, in the crystalline rubrene OPV device, the triplet excitons produced by the SF can be separated into charges at the interface with the C_{60}, so that the TTA resulting from the background light reduces the EQE due to the loss of triplet excitons.

The compound 5,12-diphenyltetracene (DPT) forms an amorphous film in contrast to the crystalline films formed by tetracene. However, even in the amorphous state, DPT produces triplet excitons from SF, and this occurs rapidly and slowly in time scales of 0.8 and 100 ps, respectively. There are certain bimolecular configurations of this compound that are suitable for SF in the amorphous state [77]. In contrast to tetracene, which exhibits an endothermic slow SF, the substituted tetracene, 5,11-dicyano-6,12-diphenyltetracene (TcCN), has a negative $\Delta E = -0.17$ eV and exhibits a fast subpicosecond SF [78] (Figure 7.10).

Chan et al. performed the time-resolved two-photon photoemission (TR-2PPE) spectroscopy of crystalline tetracene and observed an intermediate multiexciton (ME) state [79]. These authors suggested the formation of a coherent quantum superposition between S_1 and ME that enabled access to the higher-energy ME state and that an entropic gain is the main driving force of the decoupling of ME that produced two triplet excitons, which compensated for the negative ΔE.

7.3.2.3 Hexacene

In hexacene, $E(T_1)$ further decreases to 0.51 eV due to the large contribution of the diradical feature, and $E(S_1)$ is 1.65 eV. Therefore, the ΔE is -0.63 eV, indicating that the SF should be largely exothermic. It has been reported that this highly negative ΔE makes the SF of hexacene rather slow because of the Marcus inverted region [7, 80]. To utilize the low-energy triplet exciton of hexacene in OPVs, the LUMO level of an acceptor material must be sufficiently low to make E(CT) smaller than 0.51 eV. In a series of experiments, the compound N,N'-bis(1H,1H-perfluorobutyl)-(1,7 & 1,6)-dicyano-perylene-3,4:9,10-bis(dicarboximide) (PDIF-CN2) (LUMO energy level = 4.5 eV) was used as the acceptor, and 6,13-di-tri-cyclohexylsilylethynyl-hexacene (TCHS-hexacene) was used as a p-type hexacene derivative. The charge separation of the triplet excitons was confirmed by the negative dependence of the photocurrent on a magnetic field, which resulted from the suppressed SF [81].

7.3.2.4 Heteroacene

Incorporation of nitrogen atoms into the acene structures was accomplished by Herz et al. to yield diaza-TIPS-pentacene and tetraaza-TIPS-pentacene [82, 83]. Ultrafast transient absorption spectroscopy revealed that these materials exhibited a faster SF than TIPS-pentacene. Zhang et al. reported a systematic investigation of four heteroacenes (ADT, ADPD, 4F1NTP, and 4F2NTP) and TIPS-pentacene and the effect of ΔE on their SF efficiency [84]. Since ADT had the largest ΔE, it exhibited the lowest SF yield, whereas ADPD having the smallest ΔE showed the highest SF yield. These results suggest that excessive exoergicity reduces the SF efficiency.

7.3.2.5 Perylene and Terrylene

Since the late 1970s, perylene has also been known to show SF character similar to tetracene and pentacene [85, 86]. In the context of an OPV application, perylene has also attracted interest recently as an SF molecule. Minami

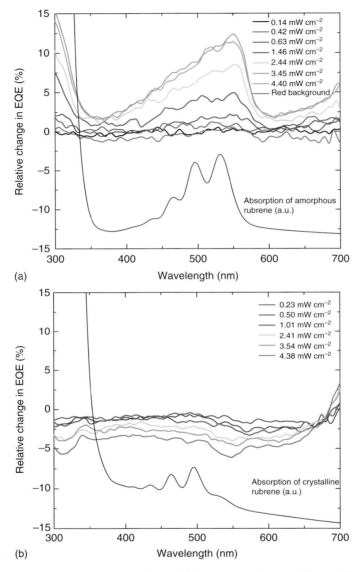

Figure 7.10 EQE under background light (a) amorphous and (b) crystalline rubrene/C_{60} OPVs, absorption spectra of (a) amorphous and (b) crystalline rubrene. *Source*: From Lin et al. [76]. Reproduced with permission of John Wiley & Sons.

et al. theoretically predicted that oligorylenes can satisfy the energy condition for SF due to their diradical character [87]. One of rylene derivatives, N,N-bis(n-octyl)-2,5,8,11-tetraphenylperylene-3,4:9,10-bis(dicarboximide) (perylenediimide), was reported to show SF character [88]. It has an $E(S_1)$ of 2.08 eV, an $E(T_1)$ of 1.14 eV, and a positive ΔE of 0.20 eV, indicating an endothermic SF. Ultrafast spectroscopy showed a fast and high-yield SF for PDI

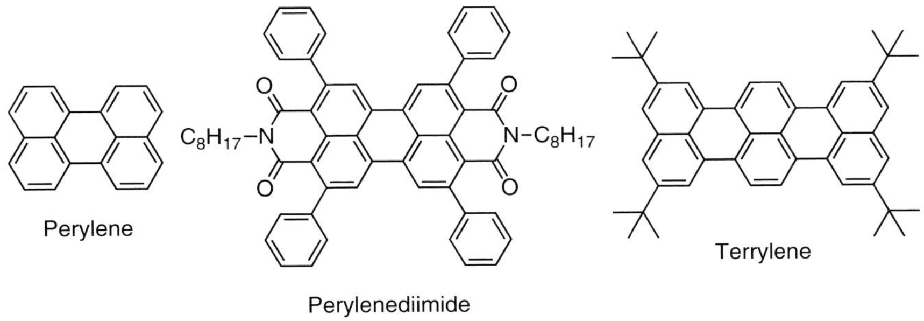

Figure 7.11 Chemical structure of the rylene-type SF compounds.

in spite of the endothermic energy condition. This was attributed to the entropic effect similar to tetracene. The compound *tert*-butyl-substituted terrylene has also been reported to be an SF molecule. This compound has an $E(S_1)$ of 1.93 eV and a $E(T_1)$ of 1.00 eV, so that the SF is slightly endothermic with a ΔE of 0.07 eV [89] (Figure 7.11).

7.3.3 Nonpolycyclic π-Conjugated Compounds

The energy level of the excited states of polycyclic aromatic compounds depends exclusively on the number of aromatic rings in their backbone; therefore, the $E(T_1)$ values of these compounds are discrete. Pentacene and tetracene exhibit $E(T_1)$ values of 0.86 and 1.25 eV, respectively, and it is difficult to obtain an intermediate T_1 value with these acene structures because of the integral number of polycyclic aromatic rings. Moreover, it is not easy to modify the polycyclic core structure of this class of compounds using general organic synthesis techniques, and the positions of the functional groups are limited to specific carbon carbons.

The $E(T_1)$ of most organic compounds, except the polycyclic π-conjugated compounds, tends to be more than half their $E(S_1)$ values, which does not satisfy the energy conditions for SF. However, there have been a few reports on nonpolycyclic π-conjugated compounds that exhibit an SF property. The basic structures that these compounds are based on can be categorized into four main types: diphenylisobenzofuran, polyene (e.g., carotenoid), diketopyrrolopyrrole, and thienoquinoid.

The SF of 1,3-diphenylisobenzofuran (DPIBF) has been reported by Johnson, Schrauben, and coworkers [90, 91] and the $E(T_1)$ of this compound is reasonably low at 1.41 eV because of its diradical character [92]. The $E(S_1)$ of this compound is 2.73 eV, so that its ΔE is 0.09 eV, indicating a slightly endothermic SF. DPIBF has two different morphologies: the more stable crystalline phase exhibits rapid excimer formation, whereas the less stable crystalline phase shows efficient singlet fission with a time constant of 10–30 ps and without thermal activation.

Zeaxanthin, one of the widely known carotenoid alcohols, was reported to exhibit an SF feature in an aggregate state. It has an $E(S_1)$ of 1.80 eV and an $E(T_1)$ of 0.87 eV, which satisfies the energy condition for SF. Wang et al. used time-resolved resonance Raman spectroscopy and revealed that the J-aggregate of zeaxanthin efficiently produces triplets in a time scale of <4 ps via an SF process [93]. The SF rate of 1,6-diphenyl-1,3,5-hexatriene (DPH) with an E(S$_1$) of 3.31 eV and E(T$_1$) of 1.54 eV is dependent on the presence of a polymorph. The monoclinic crystal showed a much higher SF rate than the orthorhombic crystal [94].

Diketopyrrolopyrrole has been commonly used as a building block of low-bandgap polymers for OPVs. The E(T$_1$) of 3,6-bis(thien-2-yl)-2,5-dihexyl-diketopyrrolopyrrole (TDPP) was measured to be 1.1 eV in toluene by Janssen et al. [95] Considering that the E(S$_1$) of 2.2 eV is twice its E(T$_1$), TDPP would be a candidate as an SF material. Actually, Hartnett, Mauck, and coworkers demonstrated the SF character of TDPP and 3,6-bis(5-phenylthien-2-yl)-2,5-bis(2-ethylhexyl)-diketopyrrolopyrrole (PhTDPP), of which E(T$_1$) was estimated to be 1.02 eV by DFT calculations, using ultrafast transient absorption spectroscopy [96, 97].

Thienoquinoid derivatives have been known to have diradical character, which lowers the E(T$_1$) of these molecules [98]. Varnavski et al. employed ultrafast spectroscopy of a diluted solution of the thienoquinoid dimer QOT2 [99] and suggested that this dimer could exhibit SF character and this behavior could possibly be intramolecular.

Kawata et al. synthesized three thienoquinoid-based nonpolycyclic aromatic SF molecules (ThBF, TThBF, and BThBF) and fabricated OPVs with them as *p*-type material and C$_{60}$ or PDIF-CN2 as n-type material [63]. The SF character of the materials was demonstrated by DFT calculations together with the magnetic field response of the photocurrent of the OPVs. The EQE spectra of the OPVs at the wavelength of the thienoquinoid compounds were largely dependent on the LUMO level of the n-type acceptor, and this demonstrated that the excitons dissociated into charges with PDIF-CN2 acceptors because the corresponding E(CT) was relatively close to E(T$_1$). However, when the excitons did not dissociate into charges with C$_{60}$ acceptors, it was found that the E(CT) was higher than E(T$_1$) (Figure 7.12).

7.3.4 Polymers

Polymeric photosensitive materials, such as polydiactylene and polyenes, which exhibit the SF property, have also been reported [100–102]. Recently, Musser et al. revealed the details of the intrachain and ultrafast SF of poly(3-dodecylthienylenevinylene) (P3VT) [103]. Busby et al. designed and synthesized a donor–acceptor-type copolymer with benzodithiophene as the donor and thiophene-1,1-dioxide as the acceptor (PBTDO1) and demonstrated its intrachain SF character that was mediated by charge-transfer states [104]. Endothermic SF through the highly excited singlet state (S$_n$) of poly(3-hexylthiophene) (P3HT) [105], poly(9,9-di-*n*-octylfluorene) (PFO) [106], and poly[4,6-(dodecyl-thieno[3,4-*b*]thiophene-2-carboxylate)-*alt*-2,6-(4,8-dioctoxylbenzo[1,2-*b*,4,5-*b*]dithiophene)] (PTB1) [107] has been reported by Ohkita and coworkers. In particular, PTB1 is a well-known low-bandgap polymer that acts as a p-type OPV material, and understanding its SF character is important for improving the performance of OPVs (Figure 7.13).

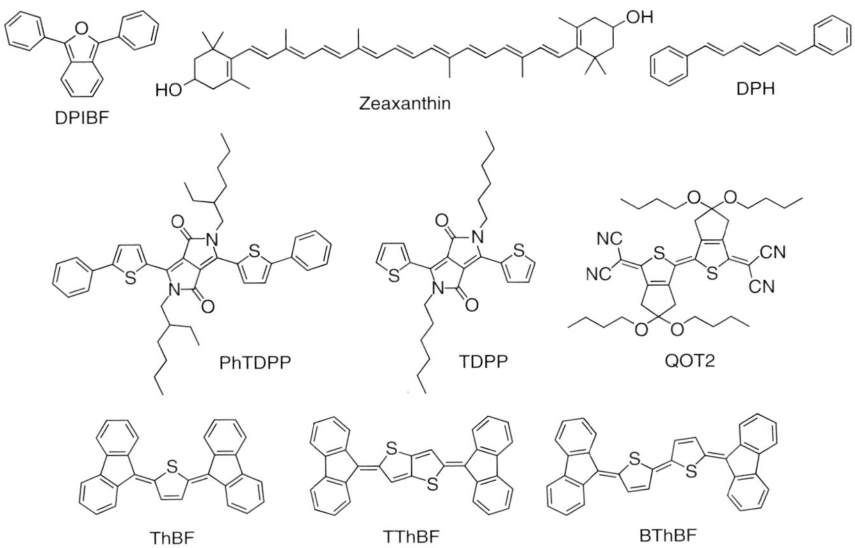

Figure 7.12 Chemical structure of the nonpolycyclic SF compounds.

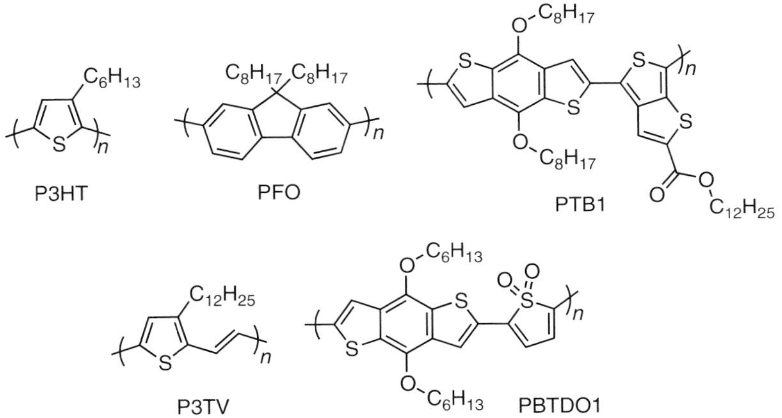

Figure 7.13 Chemical structure of the polymer-type SF compounds.

7.3.5 Perspectives

The SF phenomenon has attracted a great deal of attention as a multiexciton generating process (MEG) that can improve efficiency of photovoltaics. However, most studies on the SF remain in the realm of basic photochemistry and physics. There have been only a few reports on actual OPV devices that use SF molecules. Unfortunately, the efficiencies reported so far for these devices are still much lower than those of the state-of-the-art OPVs with non-SF organic molecules. In the SF process, the energy of the excitons decreases to half;

thus, the open-circuit voltage (V_{oc}) also drops to half, although short-circuit current (J_{sc}) doubles. Therefore, in principle, the OPV with SF cannot obtain the benefits of this energy conversion, and PCE cannot be increased, as long as the absorption wavelength of the SF materials is similar to that of conventional non-SF OPV materials. In order to increase the PCE, the SF materials must utilize the high-energy photons present in the solar light that the conventional OPV materials cannot efficiently convert to charges because of the relaxation of the hot excitons. Once this occurs, the produced triplet excitons resulting from the SF that retain relatively high energy should be transferred to the conventional OPV materials. This energy assist would be one way to use the SF materials in a practical manner.

References

1 Lunt, R.R., Giebink, N.C., Belak, A.A. et al. (2009). Exciton diffusion lengths of organic semiconductor thin films measured by spectrally resolved photoluminescence quenching. *J. Appl. Phys.* 105 (5): 053711.
2 Mikhnenko, O.V., Blom, P.W.M., and Nguyen, T.-Q. (2015). Exciton diffusion in organic semiconductors. *Energy Environ. Sci.* 8 (7): 1867–1888.
3 Matos, J., Mauricio, O., Roos, B.O., and Malmqvist, P.-Å. (1998). A CASSCF-CCI study of the valence and lower excited states of the benzene molecule. *J. Chem. Phys.* 86 (3): 1458–1466.
4 Matos, J., Mauricio, O., Malmqvist, P.-Å., and Roos, B.O. (1998). A CASSCF study of the potential curves for the X $^1\Sigma^+$, A $^1\Pi$, and C $^1\Sigma^+$ states of the AlH molecule. *J. Chem. Phys.* 86 (9): 5032–5042.
5 Dunning, T.H. (1989). Gaussian basis sets for use in correlated molecular calculations. I. The atoms boron through neon and hydrogen. *J. Chem. Phys.* 90 (2): 1007–1023.
6 Yost, S.R., Hontz, E., Yeganeh, S., and Van Voorhis, T. (2012). Triplet vs singlet energy transfer in organic semiconductors: the tortoise and the hare. *J. Phys. Chem. C* 116 (33): 17369–17377.
7 Yost, S.R., Lee, J., Wilson, M., W.B. et al. (2014). A transferable model for singlet-fission kinetics. *Nat. Chem.* 6 (6): 492–497.
8 Berkelbach, T.C., Hybertsen, M.S., and Reichman, D.R. (2013). Microscopic theory of singlet exciton fission. I. General formulation. *J. Chem. Phys.* 138 (11): 114102.
9 Berkelbach, T.C., Hybertsen, M.S., and Reichman, D.R. (2013). Microscopic theory of singlet exciton fission. II. Application to pentacene dimers and the role of superexchange. *J. Chem. Phys.* 138 (11): 114103.
10 Kurashige, Y. and Yanai, T. (2014). Theoretical study of the $\pi \to \pi^*$ excited states of oligoacenes: a full π-valence DMRG-CASPT2 study. *Bull. Chem. Soc. Jpn.* 87 (10): 1071–1073.
11 Tomkiewicz, Y., Groff, R.P., and Avakian, P. (1971). Spectroscopic approach to energetics of exciton fission and fusion in tetracene crystals. *J. Chem. Phys.* 54 (10): 4504–4507.

12 Biermann, D. and Schmidt, W. (1980). Diels–Alder reactivity of polycyclic aromatic hydrocarbons. III. New experimental and theoretical results. *J. Am. Chem. Soc.* 20 (3-4): 312–318.

13 Nelsen, S.F., Blackstock, S.C., and Kim, Y. (1987). Estimation of inner shell Marcus terms for amino nitrogen compounds by molecular orbital calculations. *J. Am. Chem. Soc.* 109 (3): 677–682.

14 Yanai, T., Tew, D.P., and Handy, N.C. (2004). A new hybrid exchange–correlation functional using the Coulomb-attenuating method (CAM-B3LYP). *Chem. Phys. Lett.* 393 (1-3): 51–57.

15 Christiansen, O., Koch, H., and Jørgensen, P. (1995). The second-order approximate coupled cluster singles and doubles model CC2. *Chem. Phys. Lett.* 243 (5-6): 409–418.

16 Kurashige, Y. and Yanai, T. (2011). Second-order perturbation theory with a density matrix renormalization group self-consistent field reference function: theory and application to the study of chromium dimer. *J. Chem. Phys.* 135 (9): 094104.

17 Kurashige, Y., Chalupský, J., Lan, T.N., and Yanai, T. (2014). Complete active space second-order perturbation theory with cumulant approximation for extended active-space wavefunction from density matrix renormalization group. *J. Chem. Phys.* 141 (17): 174111.

18 Voorhis, V., Troy, K., Tim, K., and Benjamin, et al. (2010). The diabatic picture of electron transfer, reaction barriers, and molecular dynamics. *Annu. Rev. Phys. Chem.* 61 (1): 149–170.

19 Harcourt, R.D., Scholes, G.D., and Ghiggino, K.P. (1994). Rate expressions for excitation transfer. II. Electronic considerations of direct and through–configuration exciton resonance interactions. *J. Chem. Phys.* 101 (12): 10521.

20 Thompson, A.L., Gaab, K.M., Xu, J. et al. (2004). Variable electronic coupling in phenylacetylene dendrimers: the role of Förster, Dexter, and charge-transfer interactions. *J. Phys. Chem. A* 108 (4): 671–682.

21 Fujimoto, K.J. (2012). Transition-density-fragment interaction combined with transfer integral approach for excitation-energy transfer via charge-transfer states. *J. Chem. Phys.* 137 (3): 034101.

22 Parker, S.M., Seideman, T., Ratner, M.A., and Shiozaki, T. (2013). Communication: active-space decomposition for molecular dimers. *J. Chem. Phys.* 139 (2): 021108.

23 Baluschev, S., Miteva, T., Yakutkin, V. et al. (2006). Up-conversion fluorescence: noncoherent excitation by sunlight. *Phys. Rev. Lett.* 97: 143903.

24 Monguzzi, A., Tubino, R., Hoseinkhani, S. et al. (2012). Low power, non-coherent sensitized photon up-conversion: modelling and perspectives. *Phys. Chem. Chem. Phys.* 14: 4322–4332.

25 Singh-Rachford, T.N. and Castellano, F.N. (2010). Photon upconversion based on sensitized triplet–triplet annihilation. *Coord. Chem. Rev.* 254: 2560–2573.

26 Yanai, N. and Kimizuka, N. (2016). Recent emergence of photon upconversion based on triplet energy migration in molecular assemblies. *Chem. Commun.* 52: 5354–5370.

27 de Wild, J., Meijerink, A., Rath, J.K. et al. (2011). Upconverter solar cells: materials and applications. *Energy Environ. Sci.* 4: 4835–4848.
28 Schulze, T.F. and Schmidt, T.W. (2015). Photochemical upconversion: present status and prospects for its application to solar energy conversion. *Energy Environ. Sci.* 8: 103–125.
29 Shockley, W. and Queisser, H.J. (1961). Detailed balance limit of efficiency of p–n junction solar cells. *J. Appl. Phys.* 32: 510–519.
30 Kazim, S., Nazeeruddin, M.K., Grätzel, M., and Ahmad, S. (2014). Perovskite as light harvester: a game changer in photovoltaics. *Angew. Chem. Int. Ed.* 53: 2812–2824.
31 Green, M.A., Ho-Baillie, A., and Snaith, H.J. (2014). The emergence of perovskite solar cells. *Nat. Photon.* 8: 506–514.
32 Kojima, A., Teshima, K., Shirai, Y., and Miyasaka, T. (2009). Organometal halide perovskites as visible-light sensitizers for photovoltaic cells. *J. Am. Chem. Soc.* 131: 6050–6051.
33 Kudo, A. and Miseki, Y. (2009). Heterogeneous photocatalyst materials for water splitting. *Chem. Soc. Rev.* 38: 253–278.
34 Parker, C.A. and Hatchard, C.G. (1962). Sensitised anti-stokes delayed fluorescence. *Proc. Chem. Soc. Lond.* 386–387.
35 Gray, V., Dzebo, D., Abrahamsson, M. et al. (2014). Triplet–triplet annihilation photon-upconversion: towards solar energy applications. *Phys. Chem. Chem. Phys.* 16: 10345–10352.
36 Islangulov, R.R., Kozlov, D.V., and Castellano, F.N. (2005). Low power upconversion using MLCT sensitizers. *Chem. Commun.* 3776–3778.
37 Amemori, S., Yanai, N., and Kimizuka, N. (2015). Metallonaphthalocyanines as triplet sensitizers for near-infrared photon upconversion beyond 850 nm. *Phys. Chem. Chem. Phys.* 17: 22557–22560.
38 Amemori, S., Sasaki, Y., Yanai, N., and Kimizuka, N. (2016). Near-infrared-to-visible photon upconversion sensitized by a metal complex with spin-forbidden yet strong S_0–T_1 absorption. *J. Am. Chem. Soc.* 138: 8702–8705.
39 Monguzzi, A., Mezyk, J., Scotognella, F. et al. (2008). Upconversion-induced fluorescence in multicomponent systems: steady-state excitation power threshold. *Phys. Rev. B* 78: 195112.
40 Cheng, Y.Y., Khoury, T., Clady, R.G.C.R. et al. (2010). On the efficiency limit of triplet–triplet annihilation for photochemical upconversion. *Phys. Chem. Chem. Phys.* 12: 66–71.
41 Haefele, A., Blumhoff, J., Khnayzer, R.S., and Castellano, F.N. (2012). Getting to the (square) root of the problem: how to make noncoherent pumped upconversion linear. *J. Phys. Chem. Lett.* 3: 299–303.
42 Yakutkin, V., Filatov, M.A., Ilieva, I.Z. et al. (2016). Upconverting the IR-A range of the sun spectrum using palladium tetraaryltetraanthra[2,3]porphyrins. *Photochem. Photobio. Sci.* doi: 10.1039/C1035PP00212E.
43 Baluschev, S., Yakutkin, V., Miteva, T. et al. (2007). Blue-green up-conversion: noncoherent excitation by NIR light. *Angew. Chem. Int. Ed.* 46: 7693–7696.

44 Yakutkin, V., Aleshchenkov, S., Chernov, S. et al. (2008). Towards the IR limit of the triplet–triplet annihilation-supported up-conversion: tetraanthraporphyrin. *Chem. Eur. J.* 14: 9846–9850.
45 Singh-Rachford, T.N., Nayak, A., Muro-Small, M.L. et al. (2010). Supermolecular-chromophore-sensitized near-infrared-to-visible photon upconversion. *J. Am. Chem. Soc.* 132: 14203–14211.
46 Cheng, Y.Y., Fuckel, B., MacQueen, R.W. et al. (2012). Improving the light-harvesting of amorphous silicon solar cells with photochemical upconversion. *Energy Environ. Sci.* 5: 6953–6959.
47 Nattestad, A., Cheng, Y.Y., MacQueen, R.W. et al. (2013). Dye-sensitized solar cell with integrated triplet–triplet annihilation upconversion system. *J. Phys. Chem. Lett.* 4: 2073–2078.
48 Schulze, T.F., Czolk, J., Cheng, Y.Y. et al. (2012). Efficiency enhancement of organic and thin-film silicon solar cells with photochemical upconversion. *J. Phys. Chem. C* 116: 22794–22801.
49 Singh-Rachford, T.N. and Castellano, F.N. (2008). Pd(II) phthalocyanine-sensitized triplet–triplet annihilation from rubrene. *J. Phys. Chem. A* 112: 3550–3556.
50 Deng, F., Sun, W.F., and Castellano, F.N. (2014). Texaphyrin sensitized near-IR-to-visible photon upconversion. *Photochem. Photobio. Sci.* 13: 813–819.
51 Kinoshita, T., Dy, J.T., Uchida, S. et al. (2013). Wideband dye-sensitized solar cells employing a phosphine-coordinated ruthenium sensitizer. *Nat. Photon.* 7: 535–539.
52 Kinoshita, T., Fujisawa, J., Nakazaki, J. et al. (2012). Enhancement of near-IR photoelectric conversion in dye-sensitized solar cells using an osmium sensitizer with strong spin-forbidden transition. *J. Phys. Chem. Lett.* 3: 394–398.
53 Geacintov, N.E., Burgos, J., Pope, M., and Storm, C. (1971). Heterofission of pentacene excited singlets in pentacene-doped tetracene crystals. *Chem. Phys. Lett.* 11: 504–508.
54 Burgos, J., Pope, M., Swenberg, C.E., and Alfano, R.R. (1977). Heterofission in pentacene-doped tetracene single crystals. *Phys. Stat. Solid. (b)* 77: 249–256.
55 Hanna, M.C. and Nozik, A.J. (2006). Solar conversion efficiency of photovoltaic and photoelectrolysis cells with carrier multiplication absorbers. *J. Appl. Phys.* 100.
56 Aryanpour, K., Munoz, J.A., and Mazumdar, S. (2013). Does singlet fission enhance the performance of organic solar cells? *J. Phys. Chem. C* 117: 4971–4979.
57 Smith, M.B. and Michl, J. (2010). Singlet fission. *Chem. Rev.* 110: 6891–6936.
58 Congreve, D.N., Lee, J.Y., Thompson, N.J. et al. (2013). External quantum efficiency above 100% in a singlet-exciton-fission-based organic photovoltaic cell. *Science* 340: 334–337.
59 Thompson, N.J., Congreve, D.N., Goldberg, D. et al. (2013). Slow light enhanced singlet exciton fission solar cells with a 126% yield of electrons per photon. *Appl. Phys. Lett.* 103: 263302.

60 Wang, L.J., Olivier, Y., Prezhdo, O.V., and Beljonne, D. (2014). Maximizing singlet fission by intermolecular packing. *J. Phys. Chem. Lett.* 5: 3345–3353.
61 Wu, Y.S., Liu, K., Liu, H.Y. et al. (2014). Impact of intermolecular distance on singlet fission in a series of TIPS pentacene compounds. *J. Phys. Chem. Lett.* 5: 3451–3455.
62 Piland, G.B. and Bardeen, C.J. (2015). How morphology affects singlet fission in crystalline tetracene. *J. Phys. Chem. Lett.* 6: 1841–1846.
63 Kawata, S., Pu, Y.J., Saito, A. et al. (2016). Singlet fission of non-polycyclic aromatic molecules in organic photovoltaics. *Adv. Mater.* 28: 1585–1590.
64 Bendikov, M., Duong, H.M., Starkey, K. et al. (2004). Oligoacenes: theoretical prediction of open-shell singlet diradical ground states. *J. Am. Chem. Soc.* 126: 7416–7417.
65 Jundt, C., Klein, G., Sipp, B. et al. (1995). Exciton dynamics in pentacene thin-films studied by pump-probe spectroscopy. *Chem. Phys. Lett.* 241: 84–88.
66 Rao, A., Wilson, M.W.B., Hodgkiss, J.M. et al. (2010). Exciton fission and charge generation via triplet excitons in pentacene/C-60 bilayers. *J. Am. Chem. Soc.* 132: 12698–12703.
67 Lee, J., Jadhav, P., and Baldo, M.A. (2009). High efficiency organic multilayer photodetectors based on singlet exciton fission. *Appl. Phys. Lett.* 95: 033301.
68 Wilson, M.W.B., Rao, A., Clark, J. et al. (2011). Ultrafast dynamics of exciton fission in polycrystalline pentacene. *J. Am. Chem. Soc.* 133: 11830–11833.
69 Ramanan, C., Smeigh, A.L., Anthony, J.E. et al. (2012). Competition between singlet fission and charge separation in solution-processed blend films of 6,13-bis(triisopropylsilylethynyl)pentacene with sterically-encumbered perylene-3,4:9,10-bis(dicarboximide)s. *J. Am. Chem. Soc.* 134: 386–397.
70 Walker, B.J., Musser, A.J., Beljonne, D., and Friend, R.H. (2013). Singlet exciton fission in solution. *Nat. Chem.* 5: 1019–1024.
71 Yang, L., Tabachnyk, M., Bayliss, S.L. et al. (2015). Solution-processable singlet fission photovoltaic devices. *Nano Lett.* 15: 354–358.
72 Jadhav, P.J., Brown, P.R., Thompson, N. et al. (2012). Triplet exciton dissociation in singlet exciton fission photovoltaics. *Adv. Mater.* 24: 6169–6174.
73 Ma, L., Zhang, K.K., Kloc, C. et al. (2012). Singlet fission in rubrene single crystal: direct observation by femtosecond pump-probe spectroscopy. *Phys. Chem. Chem. Phys.* 14: 8307–8312.
74 Ryasnyanskiy, A. and Biaggio, I. (2011). Triplet exciton dynamics in rubrene single crystals. *Phys. Rev. B* 84: 193203.
75 Reusswig, P.D., Congreve, D.N., Thompson, N.J., and Baldo, M.A. (2012). Enhanced external quantum efficiency in an organic photovoltaic cell via singlet fission exciton sensitizer. *Appl. Phys. Lett.* 101: 113304.
76 Lin, Y.H.L., Fusella, M.A., Kozlov, O.V. et al. (2016). Morphological tuning of the energetics in singlet fission organic solar cells. *Adv. Funct. Mater.* 26: 6489–6494.
77 Roberts, S.T., McAnally, R.E., Mastron, J.N. et al. (2012). Efficient singlet fission discovered in a disordered acene film. *J. Am. Chem. Soc.* 134: 6388–6400.

78 Margulies, E.A., Wu, Y.L., Gawel, P. et al. (2015). Sub-picosecond singlet exciton fission in cyano-substituted diaryltetracenes. *Angew. Chem. Int. Ed.* 54: 8679–8683.

79 Chan, W.L., Ligges, M., and Zhu, X.Y. (2012). The energy barrier in singlet fission can be overcome through coherent coupling and entropic gain. *Nat. Chem.* 4: 840–845.

80 Busby, E., Berkelbach, T.C., Kumar, B. et al. (2014). Multiphonon relaxation slows singlet fission in crystalline hexacene. *J. Am. Chem. Soc.* 136: 10654–10660.

81 Lee, J., Bruzek, M.J., Thompson, N.J. et al. (2013). Singlet exciton fission in a hexacene derivative. *Adv. Mater.* 25: 1445–1448.

82 Herz, J., Buckup, T., Paulus, F. et al. (2014). Acceleration of singlet fission in an aza-derivative of TIPS-pentacene. *J. Phys. Chem. Lett.* 5: 2425–2430.

83 Herz, J., Buckup, T., Paulus, F. et al. (2015). Unveiling singlet fission mediating states in TIPS-pentacene and its aza derivatives. *J. Phys. Chem. A* 119: 6602–6610.

84 Zhang, Y.D., Wu, Y.S., Xu, Y.Q. et al. (2016). Excessive exoergicity reduces singlet exciton fission efficiency of heteroacenes in solutions. *J. Am. Chem. Soc.* 138: 6739–6745.

85 Albrecht, W.G., Coufal, H., Haberkorn, R., and Michel-Beyerle, M.E. (1978). Excitation spectra of exciton fission in organic crystals. *Phys. Status Solidi B* 89: 261–265.

86 Albrecht, W.G., Michel-Beyerle, M.E., and Yakhot, V. (1978). Exciton fission in excimer forming crystal. Dynamics of an excimer build-up in α-perylene. *Chem. Phys.* 35: 193–200.

87 Minami, T., Ito, S., and Nakano, M. (2012). Theoretical study of singlet fission in oligorylenes. *J. Phys. Chem. Lett.* 3: 2719–2723.

88 Eaton, S.W., Shoer, L.E., Karlen, S.D. et al. (2013). Singlet exciton fission in polycrystalline thin films of a slip-stacked perylenediimide. *J. Am. Chem. Soc.* 135: 14701–14712.

89 Eaton, S.W., Miller, S.A., Margulies, E.A. et al. (2015). Singlet exciton fission in thin films of *tert*-butyl-substituted terrylenes. *J. Phys. Chem. A* 119: 4151–4161.

90 Johnson, J.C., Nozik, A.J., and Michl, J. (2010). High triplet yield from singlet fission in a thin film of 1,3-diphenylisobenzofuran. *J. Am. Chem. Soc.* 132: 16302–16303.

91 Schrauben, J.N., Ryerson, J.L., Michl, J., and Johnson, J.C. (2014). Mechanism of singlet fission in thin films of 1,3-diphenylisobenzofuran. *J. Am. Chem. Soc.* 136: 7363–7373.

92 Schwerin, A.F., Johnson, J.C., Smith, M.B. et al. (2010). Toward designed singlet fission: electronic states and photophysics of 1,3-diphenylisobenzofuran. *J. Phys. Chem. A* 114: 1457–1473.

93 Wang, C. and Tauber, M.J. (2010). High-yield singlet fission in a zeaxanthin aggregate observed by picosecond resonance raman spectroscopy. *J. Am. Chem. Soc.* 132: 13988–13991.

94 Dillon, R.J., Piland, G.B., and Bardeen, C.J. (2013). Different rates of singlet fission in monoclinic versus orthorhombic crystal forms of diphenylhexatriene. *J. Am. Chem. Soc.* 135: 17278–17281.

95 Karsten, B.P., Bouwer, R.K.M., Hummelen, J.C. et al. (2010). Charge separation and (triplet) recombination in diketopyrrolopyrrole-fullerene triads. *Photochem. Photobiol. Sci.* 9: 1055–1065.

96 Hartnett, P.E., Margulies, E.A., Mauck, C.M. et al. (2016). Effects of crystal morphology on singlet exciton fission in diketopyrrolopyrrole thin films. *J. Phys. Chem. B* 120: 1357–1366.

97 Mauck, C.M., Hartnett, P.E., Margulies, E.A. et al. (2016). Singlet fission via an excimer-like intermediate in 3,6-bis(thiophen-2-yl)diketopyrrolopyrrole derivatives. *J. Am. Chem. Soc.* 138: 11749–11761.

98 Ortiz, R.P., Casado, J., Hernandez, V. et al. (2007). On the biradicaloid nature of long quinoidal oligothiophenes: experimental evidence guided by theoretical studies. *Angew. Chem. Int. Ed.* 46: 9057–9061.

99 Varnavski, O., Abeyasinghe, N., Arago, J. et al. (2015). High yield ultrafast intramolecular singlet exciton fission in a quinoidal bithiophene. *J. Phys. Chem. Lett.* 6: 1375–1384.

100 Kraabel, B., Hulin, D., Aslangul, C. et al. (1998). Triplet exciton generation, transport and relaxation in isolated polydiacetylene chains: subpicosecond pump-probe experiments. *Chem. Phys.* 227: 83–98.

101 Lanzani, G., Stagira, S., Cerullo, G. et al. (1999). Triplet exciton generation and decay in a red polydiacetylene studied by femtosecond spectroscopy. *Chem. Phys. Lett.* 313: 525–532.

102 Antognazza, M.R., Luer, L., Polli, D. et al. (2010). Ultrafast excited state relaxation in long-chain polyenes. *Chem. Phys.* 373: 115–121.

103 Musser, A.J., Al-Hashimi, M., Maiuri, M. et al. (2013). Activated singlet exciton fission in a semiconducting polymer. *J. Am. Chem. Soc.* 135: 12747–12754.

104 Busby, E., Xia, J.L., Wu, Q. et al. (2015). A design strategy for intramolecular singlet fission mediated by charge-transfer states in donor–acceptor organic materials. *Nat. Mater.* 14: 426–433.

105 Guo, J.M., Ohkita, H., Benten, H., and Ito, S. (2009). Near-IR femtosecond transient absorption spectroscopy of ultrafast polaron and triplet exciton formation in polythiophene films with different regioregularities. *J. Am. Chem. Soc.* 131: 16869–16880.

106 Tamai, Y., Ohkita, H., Benten, H., and Ito, S. (2013). Singlet fission in poly(9,9′-di-*n*-octylfluorene) films. *J. Phys. Chem. C* 117: 10277–10284.

107 Kasai, Y., Tamai, Y., Ohkita, H. et al. (2015). Ultrafast singlet fission in a push–pull low-bandgap polymer film. *J. Am. Chem. Soc.* 137: 15980–15983.

8

Material Transfer and Spontaneous Motion in Mesoscopic Scale with Molecular Technology

Yoshiyuki Kageyama[1], Yoshiko Takenaka[2], and Kenji Higashiguchi[3]

[1] *Hokkaido University, Department of Chemistry, Faculty of Science, North-10, West-8, Kita-ku, Sapporo, Hokkaido, 060-0810, Japan*
[2] *National Institute of Advanced Industrial Science and Technology (AIST), Research Institute for Sustainable Chemistry, 1-1-1, Higashi, Tsukuba, Ibaraki, 305-8565, Japan*
[3] *Kyoto University, Department of Synthetic Chemistry and Biological Chemistry, Graduate School of Engineering, Katsura, Nishikyo-ku, Kyoto, 615-8510, Japan*

8.1 Introduction

In this chapter, the motion of objects, especially movements of centroid positions, will be discussed. Objects can be classified into three types based on their size – those having a scale larger than a millimeter, a scale smaller than a micrometer, and the intermediate scale. For the scale larger than a millimeter, gravity plays an important role in the motion of objects; for example, we can make objects move using friction between objects and ground originated by gravity. For the scale smaller than a micrometer, the effects of viscosity and thermal fluctuation are dominant. At this scale, the effect of gravity on the motion of objects is negligible. At the intermediate, or mesoscale, which ranges between a micrometer and a millimeter, gravity has little effect on the motion of objects, and we can make objects move by using many kinds of chemical reactions, as shown below. The motion of objects at the mesoscale generated by molecular technology is interesting for applications and will be discussed in this chapter.

8.1.1 Introduction of Chemical Actuators

Robots have become more pervasive and are utilized in many fields. For instance, recent progress in machining technology has resulted in the development and production of small-size robots with high functionality, for example, drones and flying group robots [1, 2]. If robots become smaller, it may be possible to realize microrobots that deliver drugs and perform operations autonomously in our bodies. Toward such applications, smaller actuator systems are expected.

Traditional robots generally mount electric motors or heat engines for their work. For the operation of these motors, complex connections to the power-supplying unit, the electric cell or fuel tank, and the exhaust heat paths are required. It is almost impossible to build in complex wiring or piping in

Molecular Technology: Energy Innovation, Volume 1, First Edition.
Edited by Hisashi Yamamoto and Takashi Kato.
© 2018 Wiley-VCH Verlag GmbH & Co. KGaA. Published 2018 by Wiley-VCH Verlag GmbH & Co. KGaA.

smaller-sized robots; thus, such types of systems should be excluded. By this standpoint, chemical actuators, which are molecular-based devices that generate kinetic energy from chemical reactions, are expected.

Another important point in the design of smaller-sized robots is the feeble effect of gravity. As thermal fluctuation must be dominant, smaller-sized robots in such conditions do not move efficiently or autonomously. Another mechanism other than friction derived from gravity is necessary. The candidates for such mechanisms are attractive molecular interactions and the flows of media. From this perspective, chemical actuators are widely studied for locomotive work at the mesoscale.

Next, one of the chemical actuators working in our bodies is introduced. The well-known ultimate chemical actuator is muscle. In a sarcomere, the minimum unit of muscle, actin and myosin assemble to form thin filament and thick filament, respectively. Myosin is an ATPase motor protein having actin-binding sites and changes its structure periodically with its catalytic cycle. Along with the periodic structural changes of myosin, the interaction between myosin and actin changes in cycles. Repeating the change in the interaction induces the long-distance sliding motion of bundled myosin into the channel of sarcomere. Although the sliding distance generated by one sarcomere is only approximately 1 µm, muscle shows macroscopic motion (mm to m) and strong force (~kPa) since many sarcomeres make a sliding motion cooperatively. The energy efficiency of a muscle is estimated to be 50–75% depending on the type of species [3]. Heat engines cannot realize this high efficiency because they follow Carnot's principle; that is, the maximum energy efficiency of a heat engine that works at 37 °C (body temperature) under ambient temperature (25 °C) is 4%. This highly efficient movement can be realized by the direct conversion of chemical energy to kinetic energy (chemical actuation process) and by the directional motion generated by the molecular interactions.

8.1.2 Composition of This Chapter

The minimum unit for extracting kinetic energy from a chemical reaction is one molecule. On the other hand, the maximum size of molecular devices that shows motion triggered by a chemical reaction is the size where a large number of relevant molecules work cooperatively. This ranges from a nanometer to a centimeter (in many cases, it ranges from 10 nm to several hundred micrometers, which we refer to as the mesoscale). By combining such molecular devices together, larger devices can be expected to be created. Therefore, the mesoscale is an important scale for connecting the microscopic motion (i.e., molecular motion) with the movement that can be observed or recognized macroscopically.

In this chapter, we will discuss the mechanism and the system for creating macroscopic movements utilizing mobile molecules and their interactions. Although the energy efficiency of photochemical actuators is insufficient for our goal, we will mention them because these actuators lead the early stage of the study.

In Section 8.2, we present an overview of various strategies where the molecules generate mesoscopic motion. The generation of the environmental gradients or the flows, which are necessary for the translational motion of objects, will be discussed. In Section 8.3, the mechanical motion of molecules induced by the chemical/photochemical reactions is described. These molecules, known as molecular machines, show stimuli-responsive motion in the range between sub-angstroms and subnanometers. In Section 8.4, we focus on the recent results regarding the mesoscopic motion generated by the hybridization of molecular functions. In Section 8.5, we present a summary and review of related researches.

8.2 Mechanism to Originate Mesoscale Motion

In this section, we will discuss "molecular power," which realizes the translational motion of the order from submicrometers to centimeters. "Molecular power," as we introduce here, means the molecular function to generate mechanical power by converting supplied energy into kinetic energy (or its precedence energy), and the molecular ability to interact with other molecules or outer media to realize directional mesoscale motion. Each "molecular power" is not large enough to generate the mesoscopic motion. However, when a large number of molecules work cooperatively, translational motion in mesoscopic size (which is thousands of times larger than that of one molecule) can be produced.

8.2.1 Motion Generated by Molecular Power

There are many types of molecular power that function at the molecular level, for instance, molecular isomerization, rotary motion, recombination of molecular binding and photothermal conversion, with consumption of external energy – the energy of light, heat, electricity, magnetism, chemical substrates, and so on. The reactions induced by molecular power will be referred to as "primary reactions." Following the primary reaction, the phenomena that lead to the mesoscale motion occur. These phenomena are generated by the cooperation of many molecules. Several examples of translational motion induced by "molecular power" are summarized in Table 8.1.

Among these cases, a particle or a molecule moves by the gradient of the environment that is externally produced in (D) and (J). On the other hand, objects move by themselves in other cases. We can classify these into two types: the motions generated by the mechanical motion of objects (A, B, C, and K) and by the gradient produced by their own (E, F, G, H, and I) .

8.2.2 Gliding Motion of a Mesoscopic Object by the Gradient of Environmental Factors

For cases (E)–(J) in Table 8.1, particles flow, as if they are gliders, following the microscopic environmental gradient produced by their molecular power. Here, microscopic environmental gradient means the gradient of temperature, surface tension, or concentration. For instance, in case (E), a particle generates a local gradient of temperature around itself. Then, the particle moves due to the gradient

8 Material Transfer and Spontaneous Motion in Mesoscopic Scale

Table 8.1 Examples of translational motion induced by "molecular power."

Index	Input(external energy)	Primary reaction	Following phenomena	Output(mesoscopic motion)
A [4, 5]	Two-color light	Azobenzene photoisomerization	Deformation of a film [4] or a crystal [5]	Rotation of millimeter-sized wheels [4] or translational motion of the crystal [5]
B [6, 7]	Intermittent light	Azobenzene photoisomerization	Deformation of elastomers	Swimming of the elastomers with generation of flow of media due to deformation of the elastomers
C [8]	Continuous light	Azobenzene photoisomerization	Phase transition of a crystal and resultant feedback to the reaction	Swimming of the crystal with generation of flow of media by self-oscillatory flipping
D [9, 10]	Continuous light	Photothermal effect	Gradient of density of a liquid crystal media	Translational motion of a particle in the liquid crystal
E [11]	Continuous light	Photothermal effect	Gradient of surface tension	Translational motion of a Janus particle with generation of flow of media
F [12, 13]	Chemical substances	Hydrolysis	Gradient of surface tension	Translational motion of a micrometer-sized droplet
G [14]	Chemical substances	Dissolution and sublimation	Gradient of surface tension	Translational motion of a particle by flow of a media
H [15]	Chemical substances	Reaction-diffusive behavior	Gradient of surface tension	Translational motion of an oil droplet
I [16]	AC electric field	Polarization in a particle	Anisotropic flow of media	Translational motion of a Janus particle
J [17]	Heat	Change in chemical potential	Mass transfer (Soret effect)	Generation of gradient of concentration
K	Chemical substances (ATP)	Structural change	Cooperative change in molecular interactions	Motion of a muscle

of surface tension caused by the gradient of temperature (generally, the surface tension is high in the region of low temperature).

The entropy of a system with a gradient of some environmental factors is lower than that without a gradient of the factors. According to the second law of thermodynamics, the entropy of a system tends to increase. Therefore, to relax the gradient, solute and solvent flow, and particles glide, in mesoscopic or macroscopic scales. As the generation of the environmental gradient continues as long as there is supplying energy, continuous motion can be observed.

8.2.3 Mesoscopic Motion of an Object by Mechanical Motion of Molecules

In cases (A), (B), (C), and (K) of Table 8.1, objects advance themselves like a kayak by the mechanical motion of the materials composed of stimuli-responsive molecules. In these cases, objects move by themselves using the mechanical motion originated from the bundling of molecular power. How can objects advance themselves with molecular power? To create mesoscopic motion, organization of stimuli-responsive molecules into mesoscale objects is necessary.

When structurally changing molecules are assembled in a random manner, the assembly shows the three-dimensional (3D) transformation, that is, expansion or contraction, with a change in density. Fischer et al. reported movement of a rod in media by external control of expansion–contraction of the rod [7]. One- or two-dimensional motion can be created by restriction of the 3D motion by the boundary condition or the devised system. For instance, a 2D bending motion can be realized by laminating films showing different expansion coefficients. In the film made of photoresponsive molecules, they change their occupied volume by the irradiation of light. Although the increase in volume is a 3D change, the film shows a 2D bending motion because the photoisomerization conversion is larger on the irradiated surface of the film than on the opposite surface of the film [4, 18, 19]. Some previous studies reported that directionally controlled motion of objects can be created by utilizing molecular orientation in the objects [20–22]. By restriction of the 3D propagation wave of a Belousov–Zhabotinsky (BZ) reaction to a 1D gel, Yoshida et al. showed the creation of 1D propagative irregularities on the gel surface and realized 1D motion of a micro object [23].

8.2.4 Toward the Implementation of a One-Dimensional Actuator: Artificial Muscle

There is an actuator that realizes a one-dimensional mesoscopic motion by the mechanism at the molecular level. As we mentioned in Section 8.1, most animals have an actuator called muscle that moves in a one-dimensional way with high energy efficiency. The mechanism of a muscle is explained by the flashing ratchet. In the system of flashing ratchets, the translational motion of objects appears by turning on and off the thermal motion of objects moving on a substrate with an asymmetric potential. In this section, we will consider the bottom-up technology to fabricate a one-dimensional actuator using a flashing ratchet.

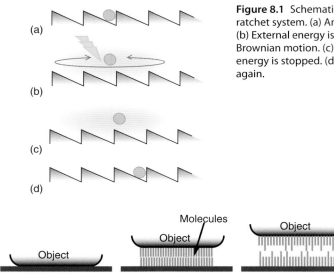

Figure 8.1 Schematic image of a flashing ratchet system. (a) An object is trapped. (b) External energy is injected, resulting in Brownian motion. (c) The injection of energy is stopped. (d) The object is trapped again.

Figure 8.2 Interaction between a substrate and an object. (a) Bare surfaces. (b) Modified surfaces with molecules. (c) Modified surfaces with molecules of various lengths.

Figure 8.1 shows the schematic image of a flashing ratchet. There is an asymmetric potential on the surface of a substrate. When the thermal energy of an object is not high enough to overcome this potential, the object is trapped by the potential (Figure 8.1a). Mesoscale objects are hard to move on a substrate because they are strongly attracted to the substrate due to van der Waals interaction (Figure 8.2a). The attractive force between a 1-μm object and the substrate is $\sim 10^6 k_B T$ [24]. To move the objects, it is necessary to tune the attractive and repulsive forces between the objects and the substrate.

To move the object, it is theoretically possible to appropriately control the interaction between the object and the substrate in the following ways: (i) decreasing the contact area, (ii) reducing the friction between the object and the substrate, and (iii) selecting the proper media. The concrete idea is drawn schematically in Figure 8.2. At first, by modifying the surfaces of the substrate and the object with molecules, the surfaces of the substrate and the object are prevented from closely sticking at the atomic scale (Figure 8.2b). By tuning the molecular density on the surfaces of the substrate and the object, and reducing the number of contact points by 10^{-3}, the attractive force will decrease by an order of 10^3. This modification corresponds to the implementation of the above idea (i). Next, the friction between the objects and the substrate will decrease by an order of 10 using molecules having various lengths (Figure 8.2c) [25]. This modification corresponds to the implementation of the above idea (ii). Lastly, the effect of van der Waals interaction between the object and the substrate in a solution, which is related to the Hamaker constant, will decrease by using a medium with a high dielectric constant, such as water [24]. When the dielectric constant of the medium becomes larger by an order of 10, the attractive force is reduced by an order of 10. This modification corresponds to the implementation of the above idea (iii).

From the above, the attractive force will be reduced by an order of 10^5. Then, the force of attraction between object and the substrate results in several tens of $k_B T$. As this energy is not large, it is possible to control desorption/adsorption of the object from the substrate by the structural change of stimuli-responsive molecules (against light, heat, chemical substances, and so on) modified on the object or by the on–off switching of thermal motion of the object.

Let us go back to the discussion of Figure 8.1. External energy is injected into the object, and the kinetic energy of the object is increased so as not to be affected by the potential. The object shows isotropic Brownian motion without the effect of the potential (Figure 8.1b). Then, the injection of external energy is stopped and the object is again trapped by the potential (Figure 8.1c). Although the Brownian motion is isotropic, in this case, the position of the trapped object tends to be shifted to the right, from the initial state, because of an asymmetric potential (Figure 8.1d). Thus, by turning on and off the injection of the external energy, the object moves to the right side stochastically [26].

A system that extracts mesoscopic motion to the outside using a flushing ratchet such as a muscle is not yet realized with artificial substances. If such artificial muscle is realized, it is expected to realize an actuator with high energy efficiency.

8.3 Generation of "Molecular Power" by a Stimuli-Responsive Molecule

"Molecular power" is generated by stimuli-responsive compounds, whose chemical structure and state are changed by external stimuli, such as temperature, pH, metal ion, voltage, and light. In particular, the structural change of elaborately designed molecules is supposed to develop a mechanical movement. This type of molecule is called a molecular machine. In this section, some examples focusing on the change of chemical structure of molecules are introduced. In addition, the fundamentals of photochromism, which is widely used for the motif of molecular machines, are explained.

8.3.1 Structural Changes of Molecules and Supramolecular Structures

Stimuli-responsive molecules change their shape by external stimuli. By conducting into a supramolecular structure, the change triggers programmed motion of supramolecules. For example, Sauvage et al. achieved the rotation of catenane [27], which has phenanthroline and terpyridine moiety, by the oxidization and reduction of copper, as shown in Figure 8.3a. In their system, electrochemical change in the oxidation number of a copper ion triggers the change in affinity between the ion and nitrogen-containing heterocycles, yielding molecular-level motion. Another example is reported by Stoddart et al. They achieved

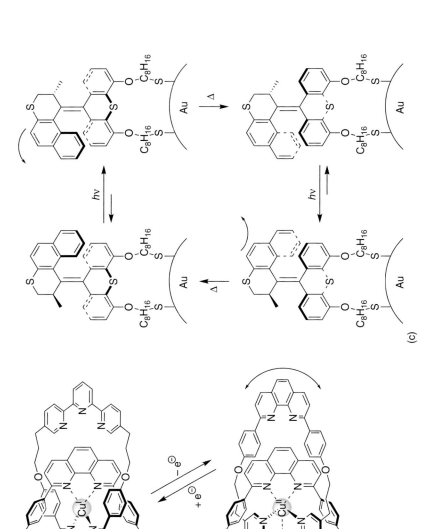

Figure 8.3 Representative molecular machines. (a) Pirouette [27]. (b) Shuttle [28]. (c) Rotor [29].

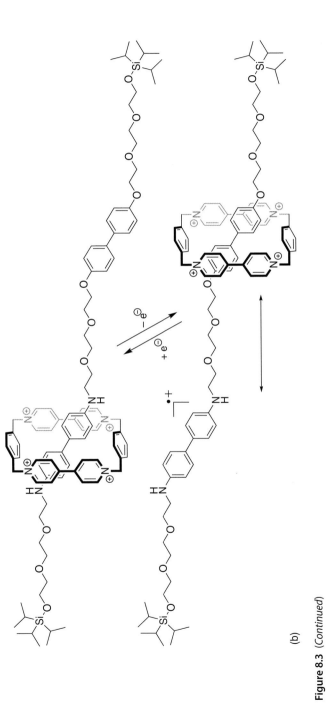

Figure 8.3 *(Continued)*

shuttle-slide of rotaxane [28], which is composed of a viologen macrocycle and a phenylene axial group, by the change of electrostatic interaction along with redox reaction of the viologen moiety, as shown in Figure 8.3b. In their system, a redox reaction of viologens (1,1′-disubstituted-4,4′-bipyridinium), which is known as the representative material showing reversible electrochromism between colorless dication and colored monoradical cation, is utilized. Further development is being carried out, and many types of directional mechanical motion have been achieved, such as molecular elevators [30] and molecular pumps [31].

Photoisomerization is a typical example of the mechanical motion in one molecule, for example, a swinging door motion that corresponds to *trans–cis* photoisomerization. A photoreactive molecule incorporating the ethene group shows repetitive isomerization upon continuous irradiation with light; however, the direction of the rotation is random. Feringa et al. achieved the molecular rotor [29], in which the rotation is restricted in one direction, as shown in Figure 8.3c. A key of the mechanism is the introduction of a ratchet-like structure; that is, there is a steric hindrance, but the potential barrier for rotation becomes lower after the photoisomerization.

8.3.2 Structural Changes of Photochromic Molecules

Photochromic compounds, which show color change between two or more isomers reversibly upon photoirradiation, show selectivity for forward and backward reactions via irradiation wavelength because of the shift in absorption spectra of each isomer. Although organic photochromic compounds are introduced in this section, many inorganic compounds [32] and complexes [33] also show photochromism.

Azobenzenes are the most representative compounds showing photochromism (Figure 8.4a). Although the average structure of the *trans*-isomer is planar, the *cis*-isomer is twisted due to steric hindrance of hydrogen atoms at the *ortho* position. Although the π-conjugation of these isomers is composed by the same atoms, the transition probability corresponding to the UV and visible range changes due to the difference of planarity. As a result, the observed color of each isomer is different. Furthermore, substitution of the aromatic group of azobenzenes leads to a shift of the absorption maximum wavelength and a change of the molar absorption coefficient [34].

Spiropyrans are also representative photochromic compounds (Figure 8.4b). The molecular structure changes from an orthogonally twisted two-plane (spiro isomer) to an expanded single plane (merocyanine isomer) due to bond formation and cleavage around the spiro carbon. A recent topic of spiropyrans related to mechanical response is the bond cleavage of spiro carbons by applied stress [35]. The accompanying coloration is applicable for sensing materials of stress.

Diarylethenes, which have thiophene derivatives as aryl groups, are known as a representative photochromic compound because of high fatigue resistance, thermal irreversibility, and tunability of photochemical properties by substitution (Figure 8.4c). Although diarylethenes undergo photoisomerization even in condensed matter [36], actuation of objects in mesoscopic regions is considered

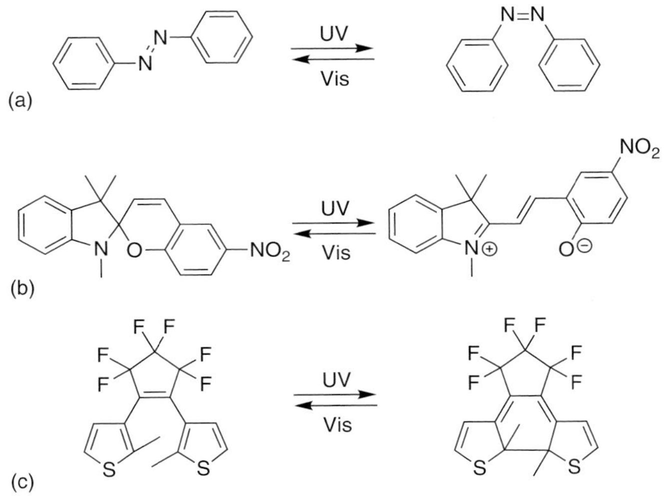

Figure 8.4 Isomerization of photochromic compounds. (a) Azobenzene. (b) Spiropyran. (c) Diarylethene.

to be relatively difficult because the change in molecular shape accompanied by isomerization is smaller than for azobenzenes and spiropyrans. However, a recent report [37] shows that drastic diameter changes of supramolecular assembly composed of diarylethenes are achieved using transitions in self-assembling due to the change in packing orientation between the aryl-rings.

8.3.3 Fundamentals of Kinetics of Photochromic Reaction

Two modes of mechanism for the energy conversion from light to kinetic energy are significant. One is heat mode, in which light is absorbed into materials and subsequently heat is generated due to vibrational relaxation. A representative application is photoactuation of a bead in media by thermal expansion [9–11]. The other mode is the photon mode, where absorbed photons cause excitation of molecules and subsequently the molecules show kinetic motion. A representative example is photochromism. It is considered that molecular machines showing photoinduced mechanical motion can be composed by photochromic compounds whose properties and shapes are elaborately controlled.

Photoreaction kinetics in solution is described as follows. A reversible photoisomerization reaction between reactant **a** and photoproduct **b** is expressed as follows, with negligible thermal isomerization:

$$\mathbf{a} \underset{\Phi_{\text{b-a}}}{\overset{\Phi_{\text{a-b}}}{\rightleftarrows}} \mathbf{b}$$

The differential equations between the concentrations, C, and some optical parameters are expressed as

$$\frac{dC_a}{dt} = -\Phi_{\text{a-b}}F_a + \Phi_{\text{b-a}}F_b, \tag{8.1a}$$

$$\frac{dC_b}{dt} = \Phi_{a\text{-}b}F_a - \Phi_{b\text{-}a}F_b, \tag{8.1b}$$

$$F_a = I(\lambda)\cdot(1 - 10^{-A(\lambda)})\cdot\left(\frac{\varepsilon_a(\lambda)C_a}{\varepsilon_a(\lambda)C_a + \varepsilon_b(\lambda)C_b}\right), \tag{8.2a}$$

$$F_b = I(\lambda)\cdot(1 - 10^{-A(\lambda)})\cdot\left(\frac{\varepsilon_b(\lambda)C_b}{\varepsilon_a(\lambda)C_a + \varepsilon_b(\lambda)C_b}\right), \tag{8.2b}$$

where Φ_{a-b} and Φ_{b-a} are the quantum yields of the corresponding reactions; $I(\lambda)$, $\varepsilon(\lambda)$, and $A(\lambda)$ are the irradiated light intensity, the molar absorption coefficients, and the absorbance of the solution at the irradiated wavelength (λ), respectively; and F are factors that indicate distribution of photons used for excitation of each chemical species. Conversion ratio to **b** at photostationary state (PSS), which is determined as $dC_a/dt = dC_b/dt = 0$ in (8.1), is expressed as

$$\frac{C_b}{C_a + C_b} = \frac{\varepsilon_a(\lambda)\Phi_{a-b}}{\varepsilon_a(\lambda)\Phi_{a-b} + \varepsilon_b(\lambda)\Phi_{b-a}}. \tag{8.3}$$

The conversion ratio depends on the molar absorption coefficient and quantum yield of each isomer. Usually, both isomers have absorption bands at the UV region; that is, a finite value for the molar absorption coefficient is expected not only for reactant **a** but also for photoproduct **b**. Under this condition, the conversion ratio does not become 100% unless the quantum yield of backward reaction is 0; therefore, a mixture of **a** and **b** is generally obtained. In order to obtain a conversion ratio as high as possible, an irradiation wavelength has to be selected having a large ratio of molar absorption coefficients ($\varepsilon_b/\varepsilon_a$) at the irradiated wavelength. Thus, the molecular design is also carried out for control of the absorption band and the molar absorption coefficient.

The conversion ratio is affected not only by molar absorption coefficients but also by quantum yields. When the ratio of quantum yields for forward/backward reactions (Φ_{a-b}/Φ_{b-a}) is large, high conversion is expected. Although quantum yields depend, to some extent, on irradiation wavelength, the change is smaller than molar absorption coefficients, as described above. The reason is that photoreactions generally proceed via the lowest excited state (S_1) because the rate of internal conversion from the higher excited state (S_n) to S_1 is much faster than the lifetime of state S_1. This principle is known as Kasha's rule [38]. However, quantum yields of azobenzenes have relatively high dependence on irradiated wavelength. The quantum yield of nonsubstituted azobenzene in ethyl acetate is reported [39] as follows: forward reaction (*trans–cis*), $\Phi_{360nm} = 0.09$, $\Phi_{440nm} = 0.25$; backward reaction (*cis–trans*) $\Phi_{360nm} = 0.14$, $\Phi_{440nm} = 0.48$.

On the other hand, the surrounding environment, where photoreactive molecules are located, strongly affects quantum yields. The isomerization reaction has to be completed within a few nanoseconds, which is the lifetime of the excited state; otherwise, deactivation of the excited state occurs by fluorescence or vibration relaxation. Therefore, properties of media, for example, polarity and dynamism, are significant for control of quantum yields. For example, photoisomerization of azobenzenes, which requires large structural change, is strongly suppressed in rigid media having small free volume, such as crystalline phase and polymers; therefore, the conversion ratio becomes a few percent due to the very small value of $\Phi_{a-b} \sim 0\%$.

8.3.4 Photoisomerization and Actuation

Functions of photological actuators are achieved by propagation of the structural changes of the photochromic molecules to changes of mesoscale morphologies. Generally, changes in molecular shape are averaged in solution due to the isotropicity of media and the random motion of the molecules. Although scale amplification of the changes of molecular shapes into the mesoscopic scale is expected in condensed anisotropic material, the photoisomerization is restricted, as described above. When condensed objects have large free volume, the changes of molecular shapes are absorbed into the free volume; therefore, the morphological change becomes small. On the other hand, when the change of molecular shape is small, a high conversion ratio is required for actuation. The conversion ratio in whole object is generally small because irradiated photons are almost absorbed near the surface of object, including a large number of photochromic molecules, due to the filter effect [40]. These factors work in a trade-off relationship between photoreactivity and morphological change. However, several recent examples showing large morphological change have been reported. Further explanation is provided in Section 8.4.

8.4 Mesoscale Motion Generated by Cooperation of "Molecular Power"

In this section, studies that demonstrate mesoscale movement using chemical methods are reviewed. The smallest limit of the objects introduced in this section is with a size that allows larger movement than thermal motion. Such movements are realized with sizes that can create heterogeneous fields or temporal asymmetries such as "flow," as described in Section 8.2.

8.4.1 Motion in Gradient Fields

The pioneering research for chemical actuated macroscopic locomotion of small objects is reported by Whitesides et al. [41]. They prepared poly(dimethylsiloxane) boats, to the bottom of which platinum-coated glasses are equipped, and floated on the surface of aqueous hydrogen peroxide. Due to the decomposition of hydrogen peroxide on the platinum catalyst, the boats move in the two-dimensional surface and self-assemble each other. The 9-mm diameter of the handmade boat is somewhat larger than the sizes of the targeting objects in this chapter. However, such asymmetrically platinum-modified objects with nanometer to micrometer sizes can also be created by nanofabrication techniques. Natan and Malluouk pioneered the preparation method of gold/platinum Janus rod [42], and Malluouk et al. realized the propelling motion of the micrometer-length rod in aqueous hydrogen peroxide (Figure 8.5a) [43].

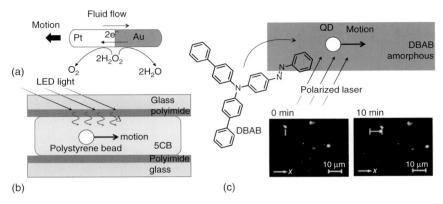

Figure 8.5 Illustrations of translational motions of a rod and beads. (a) Self-propelled motion of Au/Pt Janus rod with consumption of hydrogen peroxide. Source: From Paxton et al. [43]. Reproduced with permission of American Chemical Society. (b) Passive motion of polystyrene beads in liquid crystal by photothermal conversion at polyimide film. Source: From Takenaka and Yamamoto [10]. Reproduced with permission of Royal Society of Chemistry. (c) Passive motion of quantum dots by repetitive photoisomerization of DBAB. Source: From Nakano and Suzuki [46]. Reproduced with permission of Royal Society of Chemistry.

After their studies, numerous aggressive studies on the directional movement in the microscopic environmental gradient due to the reaction occurring on the surface of Janus particles have been reported [11, 16]. Many superior review articles have also been published [44, 45].

Spontaneous movements of oil/water or water/oil droplets composed of amphiphiles are also studied. The movements of droplets are realized by the self-induced Marangoni effect, which works due to the surface tension gradient induced by physical or chemical processes around the droplets. For example, Yoshikawa et al. reported a periodic two-dimensional motion of nitrobenzene droplets in water in the presence of stearyltrimethylammonium ion [15]. The difference in the kinetics of the reaction-diffusive behavior of stearyltrimethylammonium ion between the front and back of the droplet causes the spontaneous movement. Sugawara et al. reported movement of oleate/oleic anhydride droplets caused by hydrolysis of oleic anhydride [13]. The difference in the kinetics of the hydrolysis reaction between the front and back of the droplets causes its spontaneous movement.

A mesoscale object in a potential gradient moves in the direction where the force works due to the potential. In the movement of the Janus particle and droplets, spontaneous movements are realized by the self-induced potential gradient around themselves with consumption of externally supplied energy. It is also notable that motion of a particle can be realized in a gradient field generated externally. For example, Takenaka created a one-dimensional density gradient in a liquid crystal (LC) medium by photothermals of the polyimide film around the medium and realized 1D motion of beads in the medium (Figure 8.5b) [10]. Nakano et al. showed movements of quantum dots in an azobenzene-containing amorphous film (Figure 8.5c) [46]. In this study, oblique incident light may cause the asymmetry in the motions of azobenzene photoisomerization.

8.4.2 Movement Triggered by Mobile Molecules

Researches aiming to create movement of objects using mobile functions of "molecular machines," as described in Section 8.3, have been carried out, especially in the field of synthetic chemistry. Although the creation of mesoscopic movement using a single molecular machine is difficult, it becomes possible to realize the movements of objects by constructing a hierarchical structure.

Photoisomerization of azobenzene is one of the most popular sources of "molecular power" for producing mesoscopic movement. By arranging the scale changes in structure and polarity, the directional changes in molecular orientation and the transition moment due to the isomerization, mesoscopic motion of azobenzene-containing objects have been realized. For example, Ichimura et al. realized 2D movement of water droplets using the change in surface tension due to the photoisomerization of an azobenzene modified on a substrate [47]. Norikane et al. succeeded in the locomotion of some azobenzene crystals by irradiating ultraviolet light from one side of the crystal and visible light from the other side [5]. Photoisomerization-induced melting of the crystal on its surface is the key for this movement. The swimming motions of azobenzene elastomers are also demonstrated. Flows of water are generated by flipping of the elastomer in Camacho-Lopez's study [6] and by changing the thickness of the rod in Fischer's study (Figure 8.6a) [7]. Koshima et al. succeeded in producing reversible single-crystal shape transitions of a chiral azobenzene derivative [48]. Supplying light and thermal energy to the crystal, the crystal moves with a rotating motion. As described in Section 8.3, under a PSS, azobenzene repeats *trans–cis–trans* photoisomerization under continuous light irradiation. This feature causes the movements of quantum dots under polarized visible light, as mentioned above [46]. The repetitive photoisomerization under the light with interference fringes causes the formation of a concavo–convex pattern – a so-called surface relief structure – and this pattern has been applied to mass transportation [49]. Similarly to azobenzene, spiropyran derivatives change polarity and molecular structure by light irradiation [50] and have the ability to form surface relief structures [51]. Mass transportation using spiropyran may also be realized.

Light-induced motion of materials composed of diarylethene derivatives is also noteworthy. Irie et al. succeeded in demonstrating large deformation of a crystal [52] and a cocrystal [53], utilizing a slight change in molecular shape due to the photoisomerization. Using this characteristic, they showed the lifting of a metal ball and movement of a gear. The change in the π-electron system of the compound can also trigger the structural change of self-assemblies [37]. Together with these findings, Higashiguchi designed the structure of amphiphilic side chains to create a reversibly transforming system between a spherical shape and a nanofiber-amoeba shape at the mesoscale [54]. Moreover, by controlling the formation/deformation between sphere and nanofiber, collective transportation,

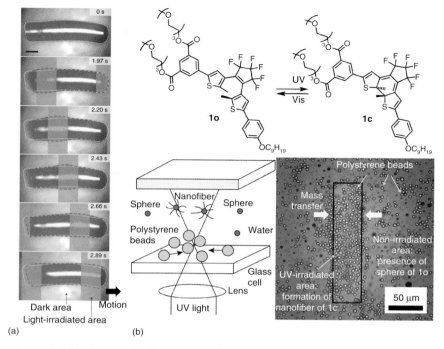

Figure 8.6 (a) Light-controlled locomotion of azobenzene elastomer. *Source*: From Palagi et al. [7]. Reproduced with permission of Nature Publishing Group. (b) Clustering of polystyrene beads by morphological change of self-assembly composed of **1c** and **1o**.

clustering, maintaining the cluster structure, and the releasing of polystyrene beads have also been realized (Figure 8.6b).

Researches have also been undertaken to create the intended mesoscale motion with highly designed mobile molecules, so-called molecular machines. Following Sauvage's study of the dual-rotaxane molecular muscle [55], Stoddart realized a muscular motion to bend a gold substrate utilizing the stimuli-responsive molecular motion of a dual-ring rotaxane molecule bound on the substrate [56]. These studies to create the desired motion with nanoscale molecular machinery are interesting and were awarded the Nobel Prize in Chemistry (2016). Feringa, who also won the Nobel Prize in Chemistry for his creation of directionally working molecular rotors with axial asymmetry, prepared a LC film composed of one of the molecular rotors and E7 (a commercially available mixture of liquid crystalline molecules) and demonstrated macroscopic rotation of a glass rod placed on the film under continuous UV light irradiation (Figure 8.7a) [57, 58]. It is noteworthy that in this research, the glass rod rotates continuously with dissipation of energy supplied steadily, considering the fact that to repeat the movements of molecular machine-actuated materials basically requires irregularities of stimuli, such as inclination of stimulus intensity and switching of stimuli. As dissipative characteristics are one of the keys for creation of autonomously working objects, we will describe systems driven by such far-from-equilibria in the next subsection.

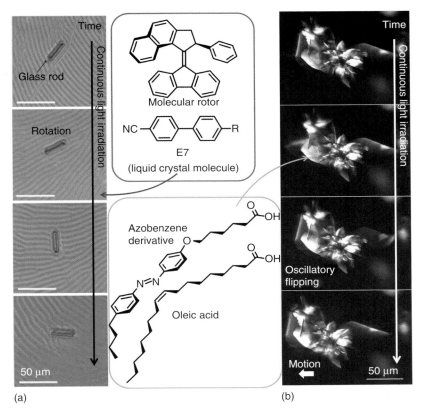

Figure 8.7 Sequential micrographs of autonomous mesoscopic motion under continuous light irradiation. (a) Rotation of a glass rod. *Source*: From Eelkema et al. [57]). Reproduced with permission of Nature Publishing Group. (b) Swimming of a self-assembly. *Source*: From Ikegami et al. [8]. Reproduced with permission of John Wiley and Sons.

8.4.3 Autonomous Motion with Self-Organization

The self-propulsions of Janus particles and droplets discussed above have the ability to move continuously with the consumption of supplied energy or chemical substrates under nonequilibrium conditions. Long-distance mass transfer using traveling waves, reported by Yoshida, is allowed by the far-from-equilibrium reaction known as a BZ reaction [23]. However, there are few studies that create continuous movements using molecular machines, despite the importance from the viewpoint of mesoscale movement.

Let us again consider the glass rod rotation reported by Feringa (Figure 8.7a). In a solution, the time scale of the swinging door motion of the photoisomerization of the molecular rotor is less than 300 ps, and the reaction time half-life for the rotation is 9.9 min [58]. On the other hand, the duration time for the revolution of the glass rod and the texture of LC is approximately 2 min. The differences in the time scales indicate that the collective revolution is a delayed motion of the LC molecules following the photoisomerization of the molecular rotor. The idea of synchronizing several dynamics having different time scales must be a primary strategy to construct an organized mesoscale motion with the molecular level machines, especially in the case aiming to create autonomous and sustainable movements. Remarkable works having this viewpoint are reported by Kageyama. After creation of spatially organized motion of mesoscale assemblies composed of azobenzene derivatives (photoisomerization lifetimes of which are assumed to be less than nanosecond order) and fatty acids (the self-orientation timespans of which are larger than microsecond order) [20, 59], he produced temporally organized motion of a cocrystal composed of same compounds [8]. The self-oscillatory flipping under PSS is realized by the combination of the repetitive photoisomerization under blue-light irradiation and time-delayed first-order phase transition of the co-crystal (Figure 8.7b). The assembly has the ability to swim in water with its continuous flipping to create water flow. The rotation of the glass rod by Feringa and the oscillation of assembly by Kageyama are autonomously continued motion, which is realized by molecular technology for chemical cooperation.

8.5 Summary and Outlook

The Nobel Prize in Chemistry (2016) was awarded for the design and synthesis of nanometer-sized artificial molecular machines. However, how to create mesoscale movement using such artificial molecular machines remains a subject of discussion. In this chapter, the authors reviewed previous studies that have created or discussed mesoscale movement with consideration of the future perspective of converting the mechanical motion of artificial molecular machines into mesoscale motion.

In addition, the creation of mesoscale movement using biomolecular machines is progressing. For example, it has been revealed that urease self-propels with consumption of urea, and the mechanism of the movement is assumed to be a self-electrophoresis due to the decomposition of urea [60]. This result shows that a macromolecule that is large enough to produce field heterogeneity can swim autonomously. Studies on transportation of artificial or biological substances using ATP molecular motors have also been actively researched. Developments in this field have revealed the mechanisms of biological movement. In a recent study, an artificial molecular motor walking on the actin filament was created by hybridizing an actin recognition domain with dynein–protein, which is a biological ATP motor walking on microtubules [61]. Such biotechnological approaches may produce bio-related motor devices beyond the function of biological motors.

Synthetic molecular systems that realize material transfer and spontaneous motion by the structural change of molecules have not yet been implemented in the real world. However, progress in research on molecular motors and their systems, comprising molecular technology, will lead to the realization of efficient mobile devices at the mesoscale and the creation of intelligent microbots in the near future.

References

1 Ma, K.Y., Chirarattananon, P., Fuller, S.B., and Wood, R.J. (2013). Controlled flight of a biologically inspired, insect-scale robot. *Science* 340: 603–607.
2 Saito, K., Maezumi, K., Naito, Y. et al. (2014). Neural networks integrated circuit for biomimetics MEMS microrobot. *Robotics* 3: 235.
3 Chapman, J.B. and Gibbs, C.L. (1972). An energetic model of muscle contraction. *Biophys. J.* 12: 227–236.
4 Yamada, M., Kondo, M., Mamiya, J.-I. et al. (2008). Photomobile polymer materials: towards light-driven plastic motors. *Angew. Chem. Int. Ed.* 47: 4986–4988.
5 Uchida, E., Azumi, R., and Norikane, Y. (2015). Light-induced crawling of crystals on a glass surface. *Nat. Commun.* 6: 7310.
6 Camacho-Lopez, M., Finkelmann, H., Palffy-Muhoray, P., and Shelley, M. (2004). Fast liquid-crystal elastomer swims into the dark. *Nat. Mater.* 3: 307–310.
7 Palagi, S., Mark, A.G., Reigh, S.Y. et al. (2016). Structured light enables biomimetic swimming and versatile locomotion of photoresponsive soft microrobots. *Nat. Mater.* 15: 647–653.
8 Ikegami, T., Kageyama, Y., Obara, K., and Takeda, S. (2016). Dissipative and autonomous square-wave self-oscillation of a macroscopic hybrid self-assembly under continuous light irradiation. *Angew. Chem. Int. Ed.* 55: 8239–8243.
9 Kim, Y.-K., Senyuk, B., and Lavrentovich, O.D. (2012). Molecular reorientation of a nematic liquid crystal by thermal expansion. *Nat. Commun.* 3: 1133.
10 Takenaka, Y. and Yamamoto, T. (2017). Light-induced displacement of a microbead through the thermal expansion of liquid crystals. *Soft Matter* 13: 1116–1119.
11 Jiang, H.-R., Yoshinaga, N., and Sano, M. (2010). Active motion of a Janus particle by self-thermophoresis in a defocused laser beam. *Phys. Rev. Lett.* 105: 268302.
12 Miura, S., Banno, T., Tonooka, T. et al. (2014). PH-induced motion control of self-propelled oil droplets using a hydrolyzable gemini cationic surfactant. *Langmuir* 30: 7977–7985.
13 Hanczyc, M.M., Toyota, T., Ikegami, T. et al. (2007). Fatty acid chemistry at the oil–water interface: self-propelled oil droplets. *J. Am. Chem. Soc.* 129: 9386–9391.
14 Ikura, Y.S., Heisler, E., Awazu, A. et al. (2013). Collective motion of symmetric camphor papers in an annular water channel. *Phys. Rev. E* 88: 012911.
15 Sumino, Y., Magome, N., Hamada, T., and Yoshikawa, K. (2005). Self-running droplet: emergence of regular motion from nonequilibrium noise. *Phys. Rev. Lett.* 94: 068301.
16 Gangwal, S., Cayre, O.J., Bazant, M.Z., and Velev, O.D. (2008). Induced-charge electrophoresis of metallodielectric particles. *Phys. Rev. Lett.* 100: 058302.

17 Maeda, Y.T., Buguin, A., and Libchaber, A. (2011). Thermal separation: interplay between the Soret effect and entropic force gradient. *Phys. Rev. Lett.* 107: 038301.
18 Iwaso, K., Takashima, Y., and Harada, A. (2016). Fast response dry-type artificial molecular muscles with [c2]daisy chains. *Nat. Chem.* 8: 625–632.
19 Kitagawa, D. and Kobatake, S. (2013). Crystal thickness dependence of photoinduced crystal bending of 1,2-bis(2-methyl-5-(4-(1-naphthoyloxymethyl)phenyl)-3-thienyl)perfluorocyclopentene. *J. Phys. Chem. C* 117: 20887–20892.
20 Kageyama, Y., Tanigake, N., Kurokome, Y. et al. (2013). Macroscopic motion of supramolecular assemblies actuated by photoisomerization of azobenzene derivatives. *Chem. Commun.* 49: 9386–9388.
21 Iamsaard, S., Asshoff, S.J., Matt, B. et al. (2014). Conversion of light into macroscopic helical motion. *Nat. Chem.* 6: 229–235.
22 Hosono, N., Kajitani, T., Fukushima, T. et al. (2010). Large-area three-dimensional molecular ordering of a polymer brush by one-step processing. *Science* 330: 808–811.
23 Maeda, S., Hara, Y., Yoshida, R., and Hashimoto, S. (2008). Peristaltic motion of polymer gels. *Angew. Chem. Int. Ed.* 47: 6690–6693.
24 Israelachvili, J.N. (1985). *Intermolecular and Surface Forces*. London: Academic Press.
25 Ohzono, T. and Fujihira, M. (2000). Molecular dynamics simulations of friction between an ordered organic monolayer and a rigid slider with an atomic-scale protuberance. *Phys. Rev. B* 62: 17055–17071.
26 Takenaka, Y. (2015). Development of a sliding nano-actuator (written in Japanese). In: *Public Engagement with Nano-based Emerging Technologies*, vol. 5, 103–105.
27 Cárdenas, D.J., Livoreil, A., and Sauvage, J.-P. (1996). Redox control of the ring-gliding motion in a Cu-complexed catenane: a process involving three distinct geometries. *J. Am. Chem. Soc.* 118: 11980–11981.
28 Bissell, R.A., Cordova, E., Kaifer, A.E., and Stoddart, J.F. (1994). A chemically and electrochemically switchable molecular shuttle. *Nature* 369: 133–137.
29 Van Delden, R.A., Ter Wiel, M.K., Pollard, M.M. et al. (2005). Unidirectional molecular motor on a gold surface. *Nature* 437: 1337–1340.
30 Badjic, J.D., Balzani, V., Credi, A. et al. (2004). A molecular elevator. *Science* 303: 1845–1849.
31 Cheng, C., Mcgonigal, P.R., Schneebeli, S.T. et al. (2015). An artificial molecular pump. *Nat. Nanotechnol.* 10: 547–553.
32 Lee, O.I. (1936). A new property of matter: reversible photosensitivity in Hackmanite from Bancroft, Ontario. *Am. Mineral.* 21: 764–776.
33 Rack, J.J. (2009). Electron transfer triggered sulfoxide isomerization in ruthenium and osmium complexes. *Coord. Chem. Rev.* 253: 78–85.
34 Siewertsen, R., Neumann, H., Buchheim-Stehn, B. et al. (2009). Highly efficient reversible Z-E photoisomerization of a bridged azobenzene with visible light through resolved $S_1(n\pi^*)$ absorption bands. *J. Am. Chem. Soc.* 131: 15594–15595.

35 Davis, D.A., Hamilton, A., Yang, J. et al. (2009). Force-induced activation of covalent bonds in mechanoresponsive polymeric materials. *Nature* 459: 68–72.

36 Irie, M., Fukaminato, T., Matsuda, K., and Kobatake, S. (2014). Photochromism of diarylethene molecules and crystals: memories, switches, and actuators. *Chem. Rev.* 114: 12174–12277.

37 Hirose, T., Matsuda, K., and Irie, M. (2006). Self-assembly of photochromic diarylethenes with amphiphilic side chains: reversible thermal and photochemical control. *J. Org. Chem.* 71: 7499–7508.

38 Kasha, M. (1950). Characterization of electronic transitions in complex molecules. *Discuss. Faraday Soc.* 9: 14–19.

39 Mita, I., Horie, K., and Hirao, K. (1989). Photochemistry in polymer solids. 9. Photoisomerization of azobenzene in a polycarbonate film. *Macromolecules* 22: 558–563.

40 Shibata, K., Muto, K., Kobatake, S., and Irie, M. (2002). Photocyclization/cycloreversion quantum yields of diarylethenes in single crystals. *J. Phys. Chem. A* 106: 209–214.

41 Ismagilov, R.F., Schwartz, A., Bowden, N., and Whitesides, G.M. (2002). Autonomous movement and self-assembly. *Angew. Chem. Int. Ed.* 41: 652–654.

42 Martin, B.R., Dermody, D.J., Reiss, B.D. et al. (1999). Orthogonal self-assembly on colloidal gold-platinum nanorods. *Adv. Mater.* 11: 1021–1025.

43 Paxton, W.F., Kistler, K.C., Olmeda, C.C. et al. (2004). Catalytic nanomotors: autonomous movement of striped nanorods. *J. Am. Chem. Soc.* 126: 13424–13431.

44 Sánchez, S., Soler, L., and Katuri, J. (2015). Chemically powered micro- and nanomotors. *Angew. Chem. Int. Ed.* 54: 1414–1444.

45 Teo, W.Z. and Pumera, M. (2016). Motion control of micro−/nanomotors. *Chem. Eur. J.* 22: 14796–14804.

46 Nakano, H. and Suzuki, M. (2012). Photoinduced mass flow of photochromic molecular materials. *J. Mater. Chem.* 22: 3702–3704.

47 Ichimura, K., Oh, S.-K., and Nakagawa, M. (2000). Light-driven motion of liquids on a photoresponsive surface. *Science* 288: 1624–1626.

48 Taniguchi, T., Fujisawa, J., Shiro, M. et al. (2016). Mechanical motion of chiral azobenzene crystals with twisting upon photoirradiation. *Chem. Eur. J.* 22: 7950–7958.

49 Viswanathan, N.K., Yu Kim, D., Bian, S. et al. (1999). Surface relief structures on azo polymer films. *J. Mater. Chem.* 9: 1941–1955.

50 Rosario, R., Gust, D., Garcia, A.A. et al. (2004). Lotus effect amplifies light-induced contact angle switching. *J. Phys. Chem. B* 108: 12640–12642.

51 Ubukata, T., Takahashi, K., and Yokoyama, Y. (2007). Photoinduced surface relief structures formed on polymer films doped with photochromic spiropyrans. *J. Phys. Org. Chem.* 20: 981–984.

52 Kobatake, S., Takami, S., Muto, H. et al. (2007). Rapid and reversible shape changes of molecular crystals on photoirradiation. *Nature* 446: 778–781.

53 Terao, F., Morimoto, M., and Irie, M. (2012). Light-driven molecular-crystal actuators: rapid and reversible bending of rodlike mixed crystals of diarylethene derivatives. *Angew. Chem. Int. Ed.* 51: 901–904.

54 Higashiguchi, K., Taira, G., Kitai, J.-I. et al. (2015). Photoinduced macroscopic morphological transformation of an amphiphilic diarylethene assembly: reversible dynamic motion. *J. Am. Chem. Soc.* 137: 2722–2729.

55 Jiménez, M.C., Dietrich-Buchecker, C., and Sauvage, J.-P. (2000). Towards synthetic molecular muscles: contraction and stretching of a linear rotaxane dimer. *Angew. Chem. Int. Ed.* 39: 3284–3287.

56 Liu, Y., Flood, A.H., Bonvallet, P.A. et al. (2005). Linear artificial molecular muscles. *J. Am. Chem. Soc.* 127: 9745–9759.

57 Eelkema, R., Pollard, M.M., Vicario, J. et al. (2006). Molecular machines: nanomotor rotates microscale objects. *Nature* 440: 163–163.

58 Eelkema, R., Pollard, M.M., Katsonis, N. et al. (2006). Rotational reorganization of doped cholesteric liquid crystalline films. *J. Am. Chem. Soc.* 128: 14397–14407.

59 Kageyama, Y., Ikegami, T., Kurokome, Y., and Takeda, S. (2016). Mechanism of macroscopic motion of oleate helical assemblies: cooperative deprotonation of carboxyl groups triggered by photoisomerization of azobenzene derivatives. *Chem. Eur. J.* 22: 8669–8675.

60 Muddana, H.S., Sengupta, S., Mallouk, T.E. et al. (2010). Substrate catalysis enhances single-enzyme diffusion. *J. Am. Chem. Soc.* 132: 2110–2111.

61 Furuta, A., Amino, M., Yoshio, M. et al. (2017). Creating biomolecular motors based on dynein and actin-binding proteins. *Nat. Nanotechnol.* 12: 233–237.

9

Molecular Technologies for Photocatalytic CO_2 Reduction

Yusuke Tamaki, Hiroyuki Takeda, and Osamu Ishitani

Tokyo Institute of Technology, School of Science, Department of Chemistry, O-okayama 2-12-1-NE-1, Meguro, Tokyo 152-8550, Japan

9.1 Introduction

'Artificial photosynthesis', which has been recently also called 'solar fuels', has been one of the dreaminess technologies for not only scientists but also the human race for long time. It became full-scale research field in the 1980s because many countries first had serious experiences on shortage of fossil resources, so-called oil shock, which started in 1973. We saw serious panic especially in developed nations, and many people were afraid of shortage of not only energy but also materials made from oil because in Japan, for example, about 20% of consumed oil has been used as most important raw materials in the chemical industries (Figure 9.1) [1]. Notably, the shortage of fossil resources directly means shortage of chemical resources as well. As the oil shock was caused by political and economic reasons, we could fortunately overcome it. However, it is an undeniable fact that the fossil resources are limited; therefore, we need to develop new technologies for getting both energy and chemical feedstocks.

We are also facing another serious problem related to fossil resources, that is, global warming. The main reason of this problem is increase of CO_2 concentration in the atmosphere, which is dominantly caused by consuming a huge amount of the fossil resources. Because of serious high risk of this issue, we have to develop sources of energy that do not increase atmospheric CO_2. Nocera and Lewis summarised the present situation and a future perspective of energy in their famous paper published in 2006 as follows [2]. Total energy consumed by human race will increase from 12.8 TW in 2000 to 28 TW in 2050 even we will be able to achieve 0.8% increase in efficient use of energy every year. Although some technologies create energy and energy sources, that is, biomass energy, wind power generation and geothermal power generation, they should not be able to produce enough energy for human race owing to competition of food production and/or limitation of installation location. However, we can find another hopeful candidate, that is, utilisation of solar energy, which can supply enough energy for us; energy of sunlight shining onto the earth for 1 h is an enough amount for covering total energy that the human race has consumed in 1 year.

Molecular Technology: Energy Innovation, Volume 1, First Edition.
Edited by Hisashi Yamamoto and Takashi Kato.
© 2018 Wiley-VCH Verlag GmbH & Co. KGaA. Published 2018 by Wiley-VCH Verlag GmbH & Co. KGaA.

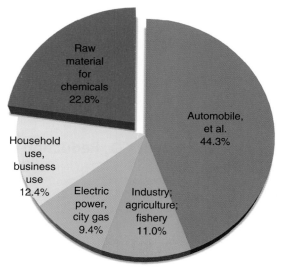

Figure 9.1 Use of oil in Japan (2014) [1].

Solar batteries have already been a practical technology for utilisation of solar energy. Although this technology is very useful and the generated output has recently increased in global scale, it should not be able to create all of required energy for all countries. In the case of Japan, for example, about the area of 38 000 km^2 should be covered by solar batteries for creating necessary energy (annual energy production 150 kWh m^{-2}, energy consumption in 2015 (5800 TWh)) [3]. This is about 10% of the land in Japan. If we would set up such huge amounts of solar batteries in Japan, ground that is suitable for vegetation must be used. This should induce rivalry between creating energy and vegetation including food production. In order to avoid this situation, most facilities for solar energy conversion should be set up in arid regions such as deserts that are not suitable for vegetation. Therefore, transportability of created energy is an important factor for many countries because the arid regions are localised in some parts of the earth. From this viewpoint, liquid-carbon-based high-energy materials such as hydrocarbons have a big advantage over electricity and H_2 because of their high-energy densities.

If we could construct practical solar energy conversion facilities to convert CO_2 to CO and water to H_2, we might use the Fischer–Tropsch process for making liquid hydrocarbons (Eq. (9.1)) and constructing an artificial carbon cycle system like the biosphere (Figure 9.2). Another way might be making formic acid from CO_2 by using solar energy because formic acid is a liquid and can be efficiently converted to H_2 and CO_2 (Eq. (9.2)). In other words, formic acid is a precursor of H_2 with high transportability [4].

$$CO + H_2 \xrightarrow{\text{catalyst}} \text{hydrocarbons} \tag{9.1}$$

$$HCOOH \xrightarrow{\text{catalyst}} H_2 + CO_2 \tag{9.2}$$

Artificial photosynthesis should include many functions (Figure 9.3): light harvesting because solar light is dilute, visible-light utilisation because the

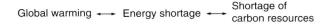

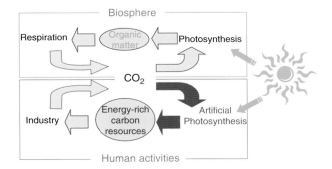

Figure 9.2 Purpose of artificial photosynthesis.

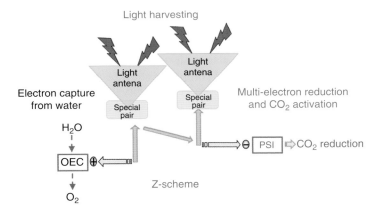

Figure 9.3 Requirements for constructing artificial photosynthesis.

content of UV light is low (a few per cent), an efficient photochemical charge separation system, a water oxidation catalyst, and a catalyst for CO_2 reduction.

In this chapter, we focus on molecular technologies for metal-complex photocatalysts for reducing CO_2. As CO_2 is the most oxidised carbon compound, high energy is generally required for its activation. For example, one-electron reduction of CO_2 requires a high reduction potential, i.e. $E^0 = -1.9\,\text{V}$ vs normal hydrogen electrode (NHE) (Eq. (9.3)). This one-electron reduction process has another problem with respect to practical use; the unstable product $CO_2^{\bullet-}$ is so active that unexpected products are produced under the reaction conditions. Multi-electron reduction of CO_2 coupled with a subsequent chemical reaction such as protonation is a useful method for solving these problems. Equations (9.4)–(9.8) show various redox reactions affording stable reduction products of CO_2, which require much lower reduction potentials for CO_2 reduction [5, 6]. Therefore, molecular technologies for constructing photocatalysts that can induce multi-electron redox reactions are necessary for useful CO_2 reduction systems.

		E^0 / V vs NHE	
$CO_2 + e^- \longrightarrow CO_2^-$		~ −2	(9.3)
$CO_2 + 2e^- + 2H^+ \longrightarrow HCO_2H$		−0.61	(9.4)
$CO_2 + 2e^- + 2H^+ \longrightarrow CO + H_2O$		−0.52	(9.5)
$CO_2 + 4e^- + 4H^+ \longrightarrow HCHO + H_2O$		−0.48	(9.6)
$CO_2 + 6e^- + 6H^+ \longrightarrow CH_3OH + H_2O$		−0.38	(9.7)
$CO_2 + 8e^- + 8H^+ \longrightarrow CH_4 + 2H_2O$		−0.24	(9.8)

The photochemical electron transfer phenomenon has been applied to various redox reactions. Although this fundamental process should also be applicable to CO_2 reduction systems, we might face a difficult situation because one-photon excitation can only induce one-electron transfer, as shown in Figure 9.4. We need an efficient system to convert photochemical one-electron transfer to multi-electron reduction of CO_2. Two-component systems have been developed for this purpose; one component is a 'redox photosensitiser' (PS) that can photochemically mediate one-electron transfer from one molecule to another and the other is a catalyst (CAT) that can accept electrons and reduce CO_2.

Photocatalytic systems consisting of metal complexes show prominent performances with respect to their efficiency and product selectivity for CO_2 reduction. Some metal complexes, typically $[Ru(N^\wedge N)_3]^{2+}$ ($N^\wedge N$ = diimine ligand), have suitable PS properties, and other complexes such as *fac*-$[Re(N^\wedge N)(CO)_3Cl]$ and *cis*-$[Ru(N^\wedge N)_2(CO)_2]^{2+}$ can function as efficient CATs for CO_2 reduction (see the following section). There are two types of mechanisms in redox photosensitised reactions, which initiate the photocatalytic reactions using the above two components, as shown in Figure 9.5. In the case of Figure 9.5a, the excited state of the PS (PS*) accepts an electron from the sacrificial reductant (SD, Process 3) to form the corresponding one-electron-reduced species (PS-OERS). If this PS-OERS has strong reduction power, subsequent electron transfer can proceed to the CAT and the PS-OERS returns to its original state (Process 4).

In the case shown in Figure 9.5b, oxidative quenching of the PS* (Process 5) proceeded as an initial process. If the reaction between the PS* and electron acceptor (EA) is favourable, the EA can accept an electron from the PS*. The produced one-electron-oxidised species of the PS (PS-OEOS) returns to its original state by receiving one electron supplied from the SD (Process 6). As described above, such a photochemical electron transfer generally resulted only

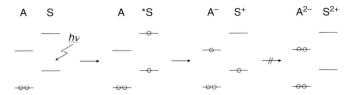

Figure 9.4 Photochemical electron transfer.

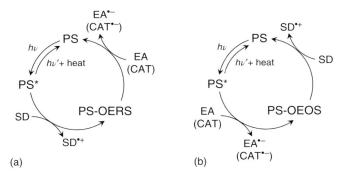

Figure 9.5 Two types of initiation mechanism in photocatalytic reactions.

in single electron transfer. Ideally, the CAT should accept multiple electrons from these redox photosensitisation processes and convert photochemical one-electron transfer to multi-electron reduction of CO_2. The reduced CAT releases a ligand and accepts a CO_2 molecule onto the central metal. As, in photochemical systems, the PS-OERS can supply only one electron to the CAT, a second electron must be supplied by another source, such as another PS-OERS, the deprotonation product of the oxidised SD (which is usually a strong reductant) or another reduced CAT (disproportionation).

In this chapter, we focus on molecular technologies for producing efficient and durable photocatalytic systems for CO_2 reduction that consist of metal complexes as the main players in the photocatalytic reactions. Two different types of metal-complex photocatalysts are described in the first section: (i) two-component systems consisting of PS and CAT molecules and (ii) supramolecular photocatalysts where the PS and CAT units are connected to each other with a bridging ligand. In the latter part of this chapter, hybrid photocatalytic systems consisting of a metal-complex photocatalyst and semiconductor particles are introduced, in which oxidation power is strengthened because of step-by-step excitation of both the PS unit and the semiconductor, the so-called Z-scheme.

As strongly related processes, photoelectrochemical systems for CO_2 reduction consisting of a metal-complex and semiconductor electrodes are described in Chapter 10. Readers are strongly encouraged to refer to this to obtain information on further applications of metal-complex photocatalysts.

9.2 Photocatalytic Systems Consisting of Mononuclear Metal Complexes

9.2.1 Rhenium(I) Complexes

The rhenium complex, *fac*-Re(bpy)(CO)$_3$Cl (**1**), was first found by Lehn and co-workers to function as a 'photocatalyst' that selectively converted CO_2 to CO with a quantum yield (Φ_{CO}) of 0.14 using triethanolamine (TEOA) as SD [7, 8]. This predominant feature of the rhenium(I) complex originates from both its

Table 9.1 Photocatalytic CO_2 reduction using rhenium(I) bipyridine complexes.

	Rhenium complex					Φ_{CO}[b)]	References
Entry	X	L	L'	n	$E_{1/2}^{red\ a)}$(V)		
(1) *fac*-[Re(4,4'-X$_2$-bpy)(CO)$_3$L]$^{n+}$							
1	H	Cl$^-$	CO	0	−1.67	0.14	[8, 10]
2	H	NCS$^-$	CO	0	−1.61	0.30	[10]
3	H	CN$^-$	CO	0	−1.67	0	[10]
4a	H	P(OEt)$_3$	CO	1	−1.43	0.38	[11, 12]
4b	MeO	P(OEt)$_3$	CO	1	−1.67	0.33	[10]
4c	CF$_3$	P(OEt)$_3$	CO	1	−1.03	0.005	[12]
5	H	PPh$_3$	CO	1	−1.40	0.05	[12, 13]
6	H	py	CO	1		0.03	[14]
7	H	OCHO$^-$	CO	0	−1.72[c)]	0.05	[13]
MS	MeO	P(OEt)$_3$	CO	1	−1.67	0.59	[10]
	H	CH$_3$CN	CO	1	−1.49[c)]		
8	H	CH$_3$CN	CO	1	−1.49[c)]	0.04	[10]
(2) *cis,trans*-[Re(4,4'-X$_2$-bpy)(CO)$_2$LL']$^{n+}$							
9a	H	P(OEt)$_3$	P(OEt)$_3$	1	−1.69	∼0	[15]
9b	Me	P(*p*-FPh)$_3$	P(OMe)$_3$	1	−1.74	0.09	[16, 17]
9c	Me	P(*p*-FPh)$_3$	P(*p*-FPh)$_3$	1	−1.73	0.20	[16, 17]
9d	Me	P(O^{i}Pr)$_3$	P(O^{i}Pr)$_3$	1	−1.86	0.02	[16, 17]

a) Reduction potentials of the complexes vs Ag/AgNO$_3$ measured in CH$_3$CN.
b) Quantum yields of CO formation.
c) Reduction potential converted from Ref. [18] with $E_{1/2}^{red}$ vs Fc/Fc$^+$ + 0.09 V.

long excitation lifetimes even in solution (a few tens of ns ∼ 1 μs) and its function as an electrochemical catalyst for CO_2 reduction [9]. Thus, the Re(I) complex functions as both a PS and a CAT.

As shown in Table 9.1 and Scheme 9.1, the effect of ligand substitution on the photocatalytic performance has been investigated. Regarding the monodentate ligand on the Re centre, *fac*-Re(bpy)(CO)$_3$(NCS) (**2**) shows better photocatalytic ability than **1**, whereas *fac*-Re(bpy)(CO)$_3$(CN) (**3**) has no photocatalytic ability [10, 19]. The most efficient photocatalyst is *fac*-[Re(bpy)(CO)$_3${P(OEt)$_3$}]$^+$ (**4a**) (Φ_{CO} = 0.38) [11]. However, the photocatalytic ability of *fac*-[Re(bpy)(CO)$_3$(PPh$_3$)]$^+$ (**5**) is low (Φ_{CO} = 0.05) [13]; also, neither *fac*-[Re(bpy)(CO)$_3$(py)]$^+$ (**6**) nor *fac*-[Re(bpy)(CO)$_3$(OCHO)] (**7**) function effectively as photocatalysts [14].

For **5** and **6**, the monodentate ligand dissociates via a chain reaction from the one-electron-reduced (OER) species [Re(bpy$^{•-}$)(CO)$_3$(L)] (L = PPh$_3$ or py), which is produced by reductive quenching of the excited state of the rhenium complexes by TEOA as the SD (Eq. (9.9)). Because the generated OER species

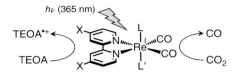

Scheme 9.1 Photocatalytic CO$_2$ reduction using [Re(4,4'-X$_2$-bpy)(CO)$_2$LL']$^{n+}$.

of [Re(bpy)(CO)$_3$(S)] (**10**, S = DMF, TEOA) has a more negative $E_{1/2}^{red}$ than that of the original complexes, the added electron on **10**$^{•-}$ transfers to the original complexes (Eq. (9.10)). The quantum yields of this ligand substitution are 16.9 for **5** and 50.3 for **6** [13, 14]. The generated solvent complex (**10**) is converted to a formato complex (**7**), which is known to be generated through CO$_2$ insertion into a rhenium hydride complex [8], and has only low photocatalytic ability. This is the reason why **5** and **6** show low photocatalytic ability. On the contrary, the reason why **3** does not function as a photocatalyst is quite different, and this will be discussed later.

$$[Re(bpy^{•-})(CO)_3(L)] + S \rightarrow [Re(bpy^{•-})(CO)_3(S)] + L \ (L = PPh_3 \text{ or } py) \quad (9.9)$$

$$[Re(bpy)(CO)_3(PPh_3)]^+ + [Re(bpy^{•-})(CO)_3(S)]$$
$$\rightarrow [Re(bpy^{•-})(CO)_3(PPh_3)] + [Re(bpy)(CO)_3(S)]^+ \quad (9.10)$$

The substituents on the bipyridine ligand also affect the photocatalytic ability [12]. The electron-withdrawing CF$_3$ groups on the 4,4'-position of the bipyridine ligand lower the CO$_2$ reducing ability (**4c**, $\Phi_{CO} = 0.005$), whereas the electron-donating MeO groups show almost no effect on the photocatalytic ability (**4b**, $\Phi_{CO} = 0.33$). Because an electronically reduced state nucleophilically activates CO$_2$, a correlation between the reduction potential and the photocatalytic efficiency is found. The $E_{1/2}$ value of **4c**, which has lower photocatalytic ability, is more positive than those of **4a** and **4b** [12].

The biscarbonyl complexes, *cis,trans*-[Re(4,4'-X$_2$-2,2'-bpy)(CO)$_2$(PR$_3$)(PR'$_3$)]$^+$, have also been reported as photochemical CO$_2$ reduction catalysts. However, simple substitution of one CO ligand with P(OEt)$_3$ (**9a**) has a negative effect on the photocatalytic ability, resulting in almost no photocatalytic performance [15]. This is also the case for **9d**, which has two P(O^{i}Pr)$_3$ ligands. The reason for this is the weaker ligand field of PR$_3$ compared with that of CO. The weaker π back-donation increases the energy of the dπ (t$_{2g}$) orbitals of the Re centre. Because the lowest excited state of these types of Re(I) complexes is the ^{3}MLCT (triplet metal(dπ)-to-ligand(bpy) charge transfer) state, their redox potentials, especially the reduction potential, in the excited state are lowered. Thus, in the case of **9a**, the excited state cannot be reduced by TEOA.

This situation can be improved by introducing π–π interactions between the aryl groups of the phosphine ligands and the bpy ligand. As can be seen in Table 9.1, **9c** shows a good photocatalytic performance with a quantum yield of 0.20, and **9b** also has better photocatalytic ability than **9d**. This is caused by the π–π interaction originating from the phenyl groups on the PR$_3$ ligands [16, 17], which improves the oxidation power of the excited state. This phenomenon is understood as the restriction of structural changes in the excited state, which also improves the excited-state lifetimes of these Re(I) complexes.

9.2.2 Reaction Mechanism

The initial step of these photocatalytic reactions is the same as reductive quenching, as shown in Figure 9.5a [20–23]. The lowest ^{3}MLCT excited state generated by the photoirradiation is reductively quenched in the presence of a suitable SD, which generates a one-electron-reduced species (OERS) (Eq. (9.11)).

$$[Re^I(LL)(CO)_3X]^{n+} + D + h\nu$$
$$\rightarrow [Re^I(LL^{\cdot-})(CO)_3X]^{(n-1)+} + D^{\cdot+} \quad (D = TEOA, TEA) \quad (9.11)$$

Because amines such as TEOA ($E_0 = +0.80$ V vs SCE [22]) or triethylamine (TEA, $E_{1/2}^{ox} = +1.15$ V vs SCE [24]) are suitable reductants, the Re(I) complexes in the excited state can be reduced. For example, the reduction potential of the excited state of **1** becomes more positive ($^*E_{1/2}^{red} = +1.15$ V vs SCE) than that of the ground state ($E_{1/2}^{red} = -1.35$ V vs SCE [25]) by accepting 2.50 eV of excitation energy [26] from the irradiated light. Therefore, the reaction represented by Eq. (9.11) can proceed to form an OERS. This OERS is an important intermediate in the photocatalytic CO_2 reduction.

OERS of *fac*-Re(bpy)(CO)$_3$L (L = Cl$^-$, Br$^-$), which are attributable to bpy anion radical complexes [27], were detected in transient absorption spectra by Kutal et al. [20, 21] and Kalyanasundaram [22] as species with lifetimes below 1 s. On the contrary, OERS of *fac*-[Re(LL)(CO)$_3$(PR$_3$)]$^+$-type complexes are relatively stable and accumulate even during photocatalytic CO_2 reduction [12]. Owing to the stability of the OERS, its dark reaction behaviour can be investigated by turning off the continuum light source after a short period of photolysis of the reaction solution. Thus, decay of the OERS of *fac*-[Re(dmb)(CO)$_3$\{P(OEt)$_3$\}]$^+$ could be analysed as pseudo-first-order kinetics under CO_2 ($k_1 = 5.6 \times 10^{-4}$ M^{-1} s^{-1} at [CO_2] = 0.139 M) [12]. In Table 9.2, the k_1 values of the OERS of *fac*-[Re(LL)(CO)$_3$(PR$_3$)]$^+$ and Φ_{CO} are summarised. A correlation can be found between k_1 and Φ_{CO}, indicating that the reactivity of the OERS and CO_2 determines the CO_2 reduction efficiency as a rate-determining step. Such efficient reactivity is found in complexes with $E_{1/2}^{red} < -1.4$ V.

A comparison of the three rhenium complexes with anionic ligands, *fac*-[Re(bpy)(CO)$_3$(L)] (**2**: L = NCS$^-$, **3**: CN$^-$, **1**: Cl$^-$), shows very different photocatalytic abilities in spite of their similar photophysical and electrochemical characteristics (Table 9.3). Although **2** is a better photocatalyst than **1**, **3** shows no photocatalytic behaviour. These results indicate that the reactivity of the OERS determines the efficiency of the photocatalytic CO_2 reduction.

It is known that the Cl$^-$ ligand dissociates from the OERS of **1** (Eq. (9.12)) [18, 27–30] due to partial charge penetration of the bpy anion radical into the d_{z^2} orbital through hyperconjugation between the π^* and d_{z^2} orbitals, which destabilises the Re—Cl bond in the OER state. On the other hand, CN$^-$ and NCS$^-$ have stronger ligand fields than that of Cl$^-$ and thus the energy of their d_{z^2} orbitals increased, which reduced the charge population on the d_{z^2} orbitals. Therefore, ligand dissociation should be restrained. In the case of **3**, CN$^-$ does not dissociate

Table 9.2 Reaction rates of OERS of fac-[Re(4,4'-X_2bpy)(CO)$_3$(PR'$_3$)]$^+$ and quantum yields of the photocatalytic CO_2 reduction.[a]

[Re(4,4'-X_2-bpy)(CO)$_3$(PR'$_3$)]$^+$			k_1[b]/10^{-5}	Φ_{CO}[c]	$E^{red}_{1/2}$[d](V)
Entry	X	PR'$_3$	M^{-1}s^{-1}		
4d	H	P(n-Bu)$_3$	10.7	0.013	−1.39
4e	H	PEt$_3$	3.5	0.024	−1.39
4f	H	P(O-i-Pr)$_3$	94.2	0.20	−1.44
4a	H	P(OEt)$_3$	56.0	0.16	−1.43
4g	H	P(OMe)$_3$	60.0	0.17	−1.41
4i	Me	P(OEt)$_3$	186	0.18	−1.55
4c	CF$_3$	P(OEt)$_3$	5.2	0.005	−1.03

a) A DMF–TEOA (5:1 v/v) solution containing a complex (2.6 mM) was irradiated at 365 nm (light intensity, 1.27×10^{-8} einstein s^{-1}).
b) Second-order reaction rate constants of the OERS and CO_2 (0.139 M).
c) Quantum yield of CO formation.
d) Reduction potentials of the complexes vs Ag/AgNO$_3$.

Table 9.3 Photocatalytic CO_2 reduction using fac-[Re(bpy)(CO)$_3$(L)] with an anionic ligand L (L = Cl$^-$, NCS$^-$, CN$^-$) and their photophysical and electrochemical properties.

Complex	Φ_{CO}[a]	TON$_{CO}$[b]	τ[c](ns)	$E^{red}_{1/2}$[d](V)
1	0.14	15	25[e]	−1.67
2	0.30	30	30	−1.61
3	0	0	87	−1.67

a) Quantum yields of CO formation (365 nm light irradiation with 7.5×10^{-9} einstein s^{-1}).
b) Turnover numbers of CO formation.
c) Excited-state lifetimes in DMF at room temperature under an Ar atmosphere.
d) Reduction potentials of the complexes vs Ag/AgNO$_3$.
e) Value in MeCN from Ref. [22].

from the OERS of **3**, which is the reason why **3** did not function as a photocatalyst despite of its photosensitising ability.

$$[Re^I(bpy^{\cdot-})(CO)_3Cl]^- \rightarrow [Re^I(bpy^{\cdot-})(CO)_3] + Cl^-$$
$$\leftrightarrow [Re^0(bpy)(CO)_3] + Cl^-. \quad (9.12)$$

In the dark reaction of the OERS of **1**, a 17-electron species forms after Cl$^-$ elimination (Eq. (9.12)) and reacts with CO_2, generating a 'CO_2 adduct' (Eq. (9.13)). As the source of the second electron for the two-electron reduction of CO_2, another molecule of the OERS is proposed (Eq. (9.14)) due to its strong

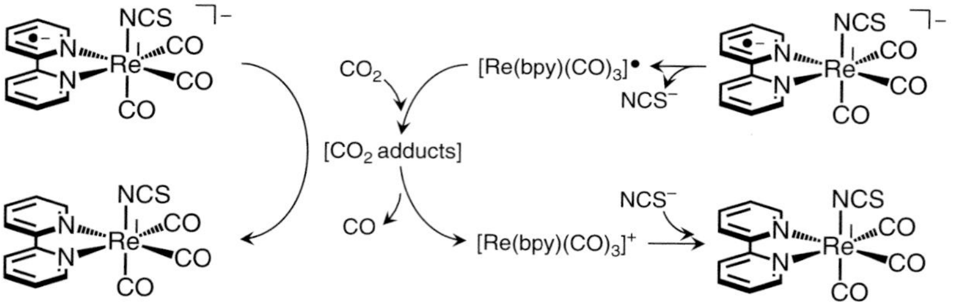

Scheme 9.2 Reaction mechanism of the OERS of **2** and CO_2 in the dark.

reducing power.

$$[Re(bpy)(CO)_3] + CO_2 \rightarrow \text{'CO}_2 \text{ adduct'}, \tag{9.13}$$

$$\text{'CO}_2 \text{ adduct'} + [Re^I(bpy^{\cdot-})(CO)_3Cl]^- + Cl^-$$
$$\rightarrow CO + 2[Re^I(bpy)(CO)_3Cl]. \tag{9.14}$$

Similarly, **2** also acts as a photocatalyst with the same mechanism as **1** because NCS^- can dissociate in the OERS state due to the moderate ligand field between Cl^- and CN^-. Thus, in this case, the ligand dissociation is slower than that of **1** as discussed above, showing that the OERS accumulated to some extent in the solution during the photocatalytic reaction. Therefore, the OERS of **2** can also function efficiently as a PS (Scheme 9.2) [10].

As candidates for the 'CO$_2$ adducts', a CO_2-bridged binuclear complex, $(CO)_3(dmb)Re-CO_2-Re(dmb)(CO)_3$ (**11**, dmb = 4,4'-dimethyl-2,2'-bipyridine) and a rhenium carboxylate complex, $Re^I(bpy)(CO)_3(COOH)$ (**12**) were synthesised. Fujita et al. [31–33] reported that **11** is produced by the reaction of $[Re(bpy)(CO)_3]^-$ generated from the rhenium dimer complex, $(CO)_3(dmb)Re-Re(dmb)(CO)_3$, under photoirradiation with CO_2 in THF. Quantitative CO formation was confirmed by photoreaction of $(CO)_3(dmb)Re-CO_2-Re(dmb)(CO)_3$ with CO_2 accompanying CO_3^{2-} generation. Compound **12** can be synthesised by nucleophilic reaction of OH^- with a carbon atom of a CO ligand in $Re(bpy)(CO)_4^+$. Because this reaction is reversible, $Re(bpy)(CO)_4^+$ can be regenerated from **12**. CO generation was also confirmed by light irradiation of $Re(bpy)(CO)_4^+$, accompanying fac-$[Re(bpy)(CO)_3S]^+$ (S = solvent) generation [34–36]. Another type of 'CO$_2$ adduct' is fac-$[Re^{II}(bpy)(CO)_3(COO^-)]^+$. This was observed by FT-IR and ESI-MS during the photocatalytic reaction [37].

9.2.3 Multicomponent Systems

Thus, the OERS of the rhenium complex has two roles in photocatalytic CO_2 reduction: (i) as a CO_2 activator similar to a CAT, which adds CO_2 as an adduct on the vacant site of the Re centre, and (ii) as a simple one-electron donor similar to a simple PS. A highly efficient photocatalytic system has been achieved by assigning each function to two types of rhenium complexes. As PS, fac-$[Re\{4,4'-(MeO)_2bpy\}(CO)_3\{P(OEt)_3\}]^+$ (**4b**) was selected because it generates the OERS efficiently by photoirradiation in the presence of TEOA ($\Phi = 1.6$), and

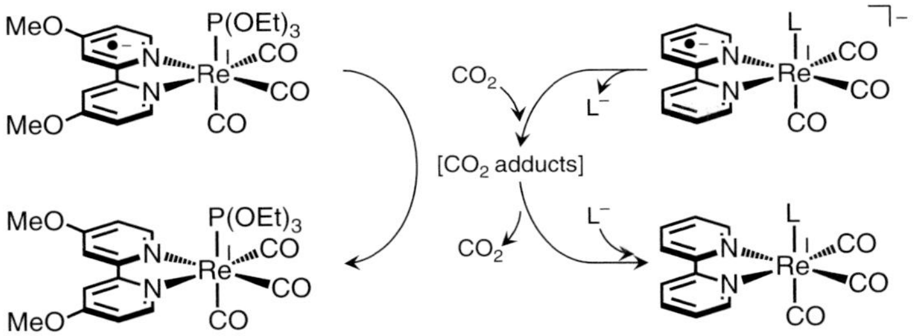

Scheme 9.3 Efficient multicomponent system with **4b** as PS and **8** as CAT.

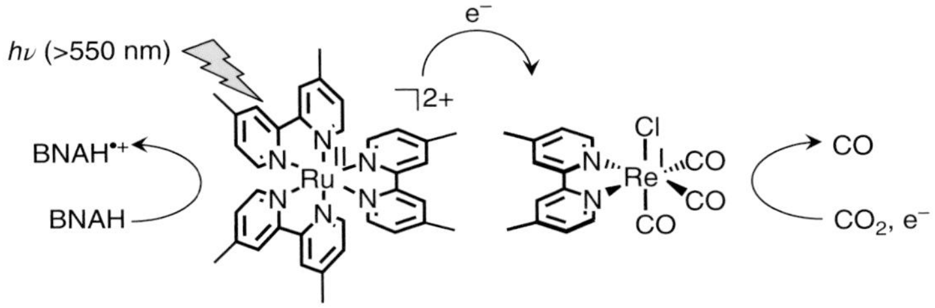

Scheme 9.4 Mixed system of [Ru(dmb)$_3$]$^{2+}$ and **13** functions under visible-light irradiation.

the OERS has strong reducing power ($E_{1/2}^{red} = -1.67$ V vs Ag/AgNO$_3$). As CAT, *fac*-[Re(bpy)(CO)$_3$(CN$_3$CN)]$^+$ (**8**) was used because it rapidly dissociates the solvent ligand for the CO$_2$ activation site. Thus, the mixed system (MS) of **8** and **4b** is a highly efficient system for photocatalytic CO$_2$ reduction ($\Phi_{CO} = 0.59$, Scheme 9.3) [10].

By changing the PS to one that absorbs visible light, the system can be improved because *fac*-[Re(bpy)(CO)$_3$L]$^{n+}$ shows only weak absorption in the visible-light region. Thus, the MS with [Ru^{II}(dmb)$_3$]$^{2+}$ (dmb = 4,4′-dimethyl-2,2′-bipyridine) as PS and *fac*-Re(dmb)(CO)$_3$Cl (**13**) as CAT shows a good photocatalytic performance by converting CO$_2$ to CO selectively with $\Phi_{CO} = 0.062$ under light irradiation at >500 nm in the presence of 1-benzyl-1,4-dihydronicotinamide (BNAH) as SD (Scheme 9.4) [38].

The ring-shaped Re(I) multinuclear complexes have excellent PS properties; these include strong absorption in the visible region, a long excited-state lifetime, strong oxidation power in the excited state, a high quantum yield of photochemical formation of the OERS and strong reduction power of the OERS [39]. Therefore, some of these have been used as PSs for photocatalytic CO$_2$ reduction. As a typical example, a DMF–TEOA (5 : 1 v/v) mixed solution containing the ring-shaped trinuclear complex (**Re-ring**, Scheme 9.5), *fac*-[Re(bpy)(CO)$_3$(CH$_3$CN)]$^+$ (**8**), as CAT and 1,3-dimethyl-2-phenyl-2,3-dihydro-1*H*-benzo[*d*] imidazole (BIH, Scheme 9.5) as a reductant was irradiated at $\lambda_{ex} = 436$ nm under a CO$_2$ atmosphere, resulting in selective formation of CO with high efficiency [40]. The Φ_{CO} was 0.82, which indicates that this is the most efficient photocatalytic system for CO$_2$ reduction so far. We also successfully

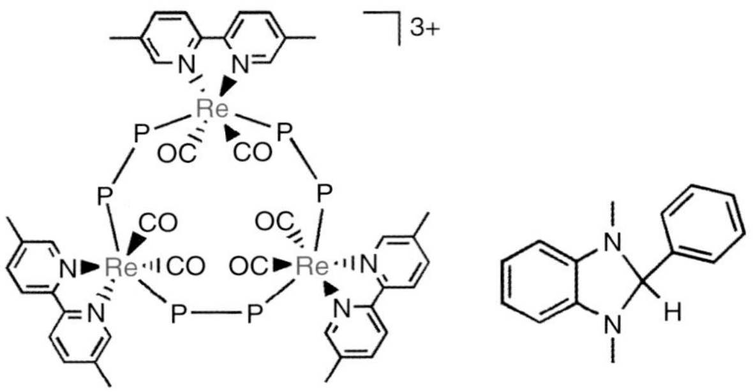

Scheme 9.5 Structures of **Re-ring** and BIH.

applied the ring-shaped Re(I) multinuclear complex photosensitiser to the photocatalytic reduction of CO_2 with *cis,trans*-[Ru{(*t*-Bu)$_2$bpy}(CO)$_2$(Cl)$_2$] and *fac*-[Mn{(*t*-Bu)$_2$bpy}(CO)$_3$(MeCN)]$^+$ as catalysts. These photocatalytic reactions selectively produced HCOOH with the highest quantum yields among the reported systems using each catalyst and TEOA as a reductant without BIH (Φ_{HCOOH} = 58%, 48%) [41].

9.2.4 Photocatalytic CO_2 Reduction Using Earth-Abundant Elements as the Central Metal of Metal Complexes

An important development in photocatalytic CO_2 reduction is the expansion of the limits of the metal elements contained in the metal complexes. In particular, the first transition elements such as Mn, Fe, Co, Ni and Cu are attracting much attention as earth-abundant metals [42], whereas Ru, Re, Os and Ir complexes have been known to function as both good PSs and/or CATs. However, this type of reaction is challenging because these complexes have low stability in reaction solutions with the polar coordinating solvents that are typically required in photocatalytic CO_2 reduction.

To overcome this disadvantage, multidentate ligands have been utilised for constructing CATs. In earlier studies in 1980–2000, Co^{II} and Ni^{II} cyclams and Fe and Co porphyrins and their derivatives were reported to function as CATs for reducing CO_2 to produce CO [43]. The metal cyclams formed both CO and HCOO$^-$ as the main products of CO_2 reduction in the presence of the PS and SD. Interestingly, the metal porphyrins could function as photocatalysts to reduce CO_2 to CO even in the absence of PS because the porphyrins themselves functioned not only as CATs but also as PSs [44]. However, the reaction efficiency of these systems depended largely on an additional PS. Thus, photocatalytic CO_2 reduction using *p*-terphenyl as PS proceeded efficiently (Scheme 9.6) [44].

Another important factor in CO_2 reduction using the first transition elements as CAT is proton conjugation. In electrochemical studies on an Fe porphyrin (Scheme 9.7) as CAT for CO_2 reduction, protons were found to facilitate the electrochemical reduction of CO_2 [45]. Based on this study, the Fe porphyrins with phenol moieties around the central Fe enhanced the electrochemical CO_2

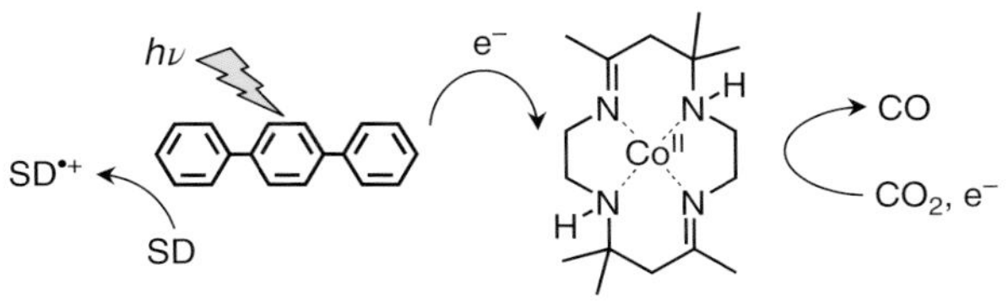

Scheme 9.6 Photocatalytic CO_2 reduction using a dye PS and a Co macrocyclam.

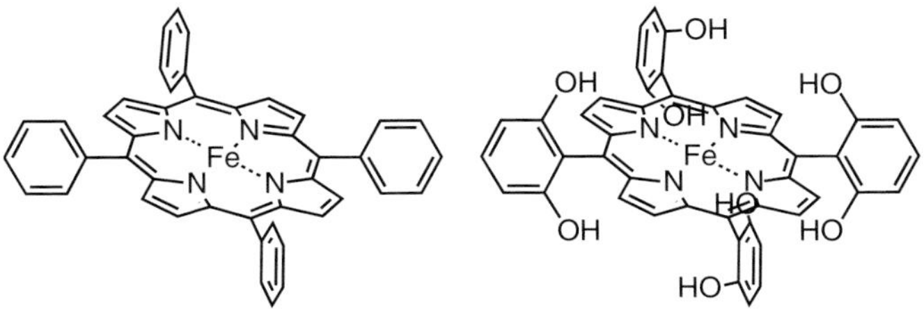

Scheme 9.7 Structure of Fe porphyrins as CATs for CO_2 reduction.

reduction [46]. In the presence of organic dyes (Scheme 9.7, right) as PS, photocatalytic CO_2 reduction using the Fe porphyrin with attached phenols proceeds more efficiently and selectively compared with the Fe porphyrin with phenyl moieties instead of phenols [47].

As interesting cases, simple α-diimine complexes also function as CAT in photocatalytic CO_2 reduction. Lehn and co-workers reported photocatalytic CO_2 reduction using $Co^{II}Cl_2 \cdot 6H_2O$ as CAT in combination with $Ru(bpy)_3^{2+}$ as PS and TEA or TEOA as SD in organic solvents such as CH_3CN and N,N-dimethylformamide (DMF). This reaction was improved by adding bpy, phen or their derivatives. However, the product selectivity against H_2 evolution is low [48]. Fe ions and bpy complexes are also active species for CO_2 reduction as CATs for producing CO in organic solvents such as CH_3CN or N-methyl-2-pyrrolidone (NMP) in the presence of a PS such as p-terphenyl or $Ru(bpy)_3^{2+}$ and TEA or TEOA as SD [49]. These systems also illustrate the importance of bpy with respect to reaction efficiency. However, a similar amount of H_2 was also produced along with the photocatalytic CO_2 reduction [49b].

A successful reaction is photocatalytic CO_2 reduction using $Fe^{II}(dmp)_2(NCS)_2$ (**14**, dmp = 2,9-dimethyl-1,10-phenanthroline) as CAT and emissive Cu^I complexes as PS [50]. Emissive Cu^I complexes have attracted attention as PS in photocatalytic systems for H_2 evolution [51] and organic synthesis [52]. Especially, heteroleptic Cu^I complexes, $[Cu(dmp)(P)_2]^{2+}$ (dmp = 2,9-dimethyl-1,10-phenanthroline, P = phosphine ligand: Scheme 9.8), were used as PS because their oxidation power in the excited state is stronger than that of the corresponding homoleptic complexes, $Cu(dmp)_2^+$ [53]. In the presence of BIH as the SD [54], visible-light (436 nm) irradiation of a solution containing a Cu PS (**15**, Scheme 9.8) and an Fe CAT generates CO with a maximum selectivity of 78%, with H_2 as the minor product. When the Cu dimer **15** was used as the PS, the efficiency and durability of CO formation were the

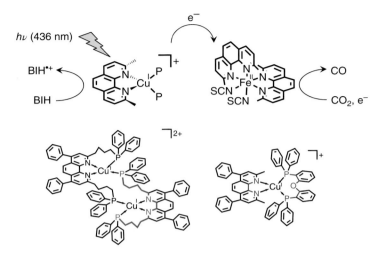

Scheme 9.8 Cu^I complexes used as PSs in the photocatalytic reduction of CO_2.

highest, i.e. $\Phi_{CO} = 6.7\%$ and $TON_{CO} > 270$ based on the quantity of **14** used ($TON_{CO} > 54$ based on the quantity of **14** used). This reaction was initiated by reductive quenching of the excited state of the Cu complex by BIH.

Stabilisation of the complex using multidentate ligands also results in an effective PS. Although Cu^I complexes such as **16** show relatively low thermal and photostabilities, a dimer **15** containing tetradentate ligands that link a dmp moiety and two phosphine moieties improves the stability, especially in the OERS state. Thus, the photocatalytic activity for CO_2 reduction of the system using **15** as the PS was much higher than that using **16** with the bidentate phosphine ligand ($\Phi_{CO} = 2.6\%$).

The Mn(I) complex *fac*-Mn(bpy)(CO)$_3$Br (**17**, Scheme 9.9) has functions as a CAT for CO_2 reduction, its molecular structure is similar to that of photocatalyst **1** and the Mn centre is the same group and the same valence as the Re centre [55, 56]. However, the same reaction conditions using Ru(dmb)$_3^{2+}$ as PS and BNAH as SD as with the rhenium complex afford HCOOH selectively and not CO in the photocatalytic CO_2 reduction, although the reaction product is CO in the electrochemical CO_2 reduction using the manganese complex. The values of Φ_{HCOOH} and TON_{HCOOH} are 5.3% and 149, respectively [57], which represent almost the same efficiency as that of the rhenium complex [38]. In this reaction, a Mn dimer ([Mn(bpy)(CO)$_3$]$_2$) with a Mn—Mn bond was produced via one-electron reduction of the original Mn complex. During this period, CO is generated as the main product; however, HCOOH is generated efficiently after this period.

This photocatalytic reaction starts with reductive quenching of the ^{3}MLCT excited state of [Ru(dmb)$_3$]$^{2+}$ to give the corresponding OERS. This has enough reduction power ($E^{red}_{1/2} = -1.82$ V vs Ag/AgNO$_3$) to transfer one electron to **17**

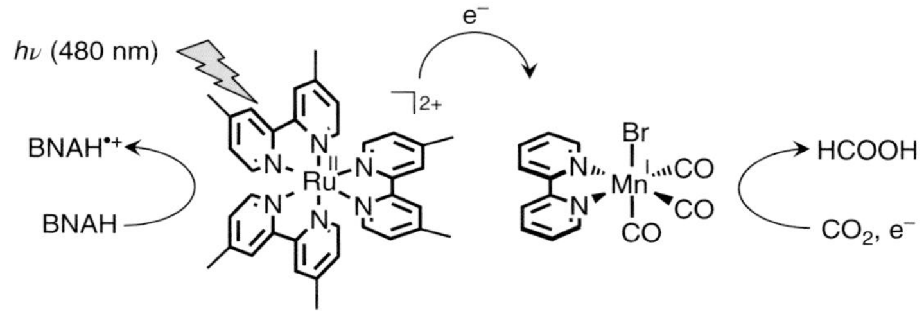

Scheme 9.9 Mixed system of [Ru(dmb)$_3$]$^{2+}$ and **17** functions under visible-light irradiation.

($E_p = -1.65$ V), forming the Mn dimer as described above. Although the subsequent processes are not clear, some mechanistic investigations have been performed. As the reduction potential of the Mn dimer is comparable ($E_p = -1.84$ V) with that of [Ru(dmb)$_3$]$^{2+}$, reduction of the Mn dimer might proceed to give [Mn(bpy)(CO)$_3$]$^-$, but it would be a slow process. The use of [Ru(bpy)$_3$]$^{2+}$, which has a more positive reduction potential ($E_{1/2}^{red} = -1.72$ V), instead of [Ru(dmb)$_3$]$^{2+}$, did not affect the photocatalytic activity of the system (TON$_{HCOOH}$ = 157 after 12-h irradiation). Thus, [Mn(bpy)(CO)$_3$]$^-$ does not perform as a dominant intermediate in the photocatalytic CO$_2$ reduction [57].

9.3 Supramolecular Photocatalysts: Multinuclear Complexes

The two functional metal complexes described above, i.e. a **PS** and a **CAT**, are combined via a bridging ligand giving the so-called supramolecular photocatalysts, as shown in Scheme 9.10 [58].

Electron transfer between two units of the supramolecular photocatalysts can be accelerated to improve the photocatalytic ability by multinucleation. This also increases the durability of the photosensitiser unit because the unstable excited and/or reduced states of the photosensitiser units can be consumed more rapidly, accelerating the photocatalytic reaction.

Especially, these advantages should be strengthened in hybrid systems of supramolecular photocatalysts and photofunctional solid materials. When complex photocatalysts are attached to the surfaces of heterogeneous materials by anchoring groups, in the case of supramolecular photocatalysts, intramolecular

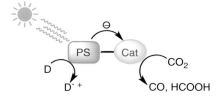

Scheme 9.10 Conceptual image of supramolecular photocatalysts for the reduction of CO$_2$.

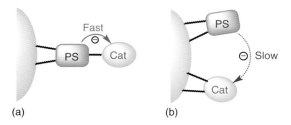

Scheme 9.11 Conceptual images of the hybrid systems of (a) the supramolecular photocatalyst and (b) the mixed system of mononuclear complexes on the surfaces of heterogeneous materials.

electron transfer should proceed rapidly from the photosensitiser unit to the catalyst unit even on the surface (Scheme 9.11a). On the other hand, this strongly depends on the densities of metal complexes on the surface and becomes much slower or does not proceed if the photosensitiser and the catalyst are separately attached to the surface with a low density (Scheme 9.11b). Such hybrid systems of supramolecular photocatalysts and the photofunctional solid materials developed in our group are summarised in Section 9.5 and Chapter 10.

9.3.1 Ru(II)—Re(I) Systems

The structures and abbreviations of Ru(II)—Re(I) supramolecular photocatalysts are shown in Chart 9.1 and their photocatalytic properties are summarised in Table 9.4. We reported the first successful supramolecular photocatalyst for CO_2 reduction in 2005 [38]. The supramolecular complex **Ru—Re1** consisted of a $[Ru(N^{\wedge}N)_3]^{2+}$ ($N^{\wedge}N$ = diimine ligand) photosensitiser unit and a *fac*-Re($N^{\wedge}N$)(CO)$_3$Cl catalyst unit, which are connected with a —CH$_2$CH(OH)CH$_2$— chain between two 4-methyl-bpy moieties. **Ru—Re1** photocatalysed the reduction of CO_2 to CO under visible-light irradiation ($\lambda_{ex} > 500$ nm) in the presence of BNAH as a sacrificial electron donor. CO was produced with high selectivity, efficiency and durability (entry 10: $\Phi_{CO} = 0.12$, $TON_{CO} = 170$), which were much higher than those of the MS of the corresponding mononuclear model complexes, i.e. $[Ru(4dmb)_3]^{2+}$ (4dmb = 4,4′-dimethyl-2,2′-bipyridine) and *fac*-Re(4dmb)(CO)$_3$Cl (entry 11: $\Phi_{CO} = 0.062$, $TON_{CO} = 101$). The following reaction mechanism was clarified as shown in Scheme 9.12: [Process 1] selective light absorption by the Ru photosensitiser unit giving the triplet metal-to-ligand-charge-transfer (^{3}MLCT) excited state via intersystem crossing from the ^{1}MLCT excited state, [Process 2] the reductive quenching of the excited Ru unit by BNAH giving the OERS of the Ru unit, [Process 3] intramolecular electron transfer from the OERS of the Ru photosensitiser unit to the Re unit and [Process 4] CO_2 reduction on the Re unit.

On the other hand, **Ru—Re2** (entry 12: $TON_{CO} = 50$) and **Ru—Re3** (entry 13: $TON_{CO} = 3$) with bpy or 4,4′-bis(trifluoromethyl)-2,2′-bipyridine ((CF_3)$_2$bpy) as peripheral ligands instead of 4dmb on the Ru units exhibited much lower photocatalytic abilities in comparison with both **Ru—Re1** and the MS (entry 11: $[Ru(4dmb)_3]^{2+}$ + *fac*-Re(4dmb)(CO)$_3$Cl) [38]. This difference was induced

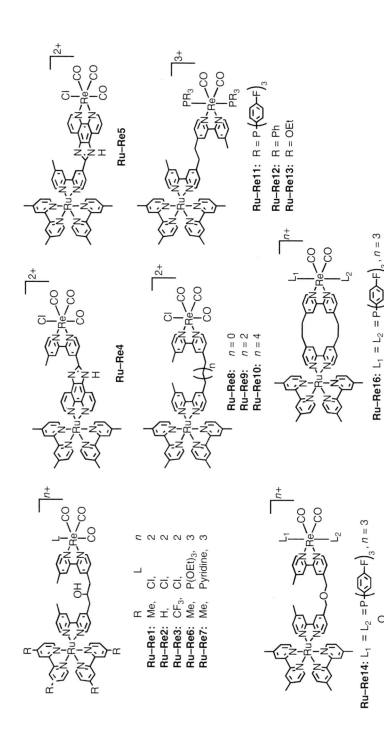

Chart 9.1 Structures and abbreviations of Ru(II)—Re(I) complexes.

Table 9.4 Photocatalytic properties of Ru(II)—Re(I) systems.

Entry	Photocatalyst	Donor[a]	Product	Γ.[b] (%)	$\Phi_{product}$	TON	TOF (min^{-1})	References
10	Ru—Re1	BNAH	CO	—	0.12	170	—	[38]
11	[Ru(4dmb)$_3$]$^{2+}$ + *fac*-Re(4dmb)(CO)$_3$Cl	BNAH	CO	—	0.062	101	—	[38]
12	Ru—Re2	BNAH	CO	—	—	50	—	[38]
13	Ru—Re3	BNAH	CO	—	—	3	—	[38]
14	Ru—Re4	BNAH	CO	—	—	14	—	[38]
15	Ru—Re5	BNAH	CO	—	—	28	—	[38]
16	Ru—Re6	BNAH	CO	—	0.16	232	—	[59]
17	Ru—Re7	BNAH	CO	—	—	97	—	[59]
18	Ru—Re8	BNAH	CO	—	0.13	180	—	[60]
19	Ru—Re9	BNAH	CO	—	0.11	120	—	[60]
20	Ru—Re10	BNAH	CO	—	0.11	120	—	[60]
21	Ru—Re11	BNAH	CO	94	0.15	207	4.7	[61]
22		BIH	CO	>99	0.45	3029	35.7	[54]
23	Ru—Re12	BNAH	CO	91	0.10	144	—	[61]
24	Ru—Re13	BNAH	CO	73	0.10	27	—	[61]
25	Ru—Re14	BNAH	CO	95	0.18	253	—	[62]
26	Ru—Re15	BNAH	CO	—	0.12	—	—	[62]
27		BIH	CO	>99	0.50	>1000	—	[63]
28	Ru—Re16	BNAH	CO	—	0.09	123	—	[64]
29	Ru—Re17	BNAH	CO	—	0.16	204	—	[64]

a) BNAH: 1-benzyl-1,4-dihydronicotinamide; BIH: 1,3-dimethyl-2-phenyl-2,3-dihydro-1*H*-benzo[*d*]imidazole.
b) Selectivity of the product calculated as [target product (mol)]/[reduced compounds (mol)].

by unfavourable intramolecular electron transfer from the OERS of the Ru unit to the Re unit because the unpaired electron should be mainly localised on the peripheral diimine ligands with a lower energy level of the LUMO, so that the intramolecular electron transfer becomes endergonic (in the case of **Ru—Re3**, for example, $E_{1/2}^{red}(Ru) = -1.23$ V; $E_{1/2}^{red}(Re) = -1.76$ V vs Ag/AgNO$_3$). In the case of **Ru—Re1**, one-electron reduction of the Ru and Re units occurred at almost equal potentials ($E_{1/2}^{red} = -1.77$ V). These results clearly indicate that the photosensitiser unit should be reduced at an equal or more negative potential compared with the catalyst unit in order to construct an efficient supramolecular photocatalyst for the reduction of CO_2.

The bridging ligand connecting the photosensitiser and catalyst units also strongly affected the performance of the supramolecular photocatalysts. Both **Ru—Re4** (entry 14: TON$_{CO}$ = 14) and **Ru—Re5** (entry 15: TON$_{CO}$ = 28), in which bridging ligands are conjugated, were not effective as photocatalysts [38]. In the

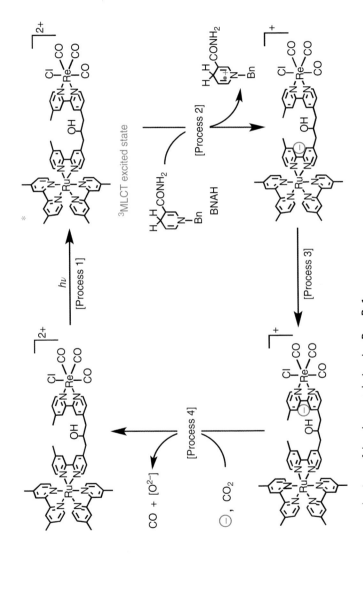

Scheme 9.12 Reaction mechanisms of the photocatalysis using **Ru—Re1**.

Table 9.5 The relationship between the photocatalyses of fac-[Re(N^N)(CO)$_3$(PR$_3$)]$^+$ and their first reduction potentials.[a]

fac-[Re(N^N)(CO)$_3$(PR$_3$)]$^+$		$E^{red}_{1/2}$	Φ_{CO}	TON$_{CO}$
N^N	PR$_3$			
4dmb	P(OEt)$_3$	−1.55	0.18	4.1
bpy	P(O-i-Pr)$_3$	−1.44	0.20	6.2
bpy	P(OEt)$_3$	−1.43	0.16	5.9
bpy	P(OMe)$_3$	−1.41	0.17	5.5
bpy	PEt$_3$	−1.39	0.024	0.83
bpy	P(n-Bu)$_3$	−1.39	0.013	0.65
(CF$_3$)$_2$bpy	P(OEt)$_3$	−1.03	0.005	0.10

a) A 4-ml DMF solution containing the complex (2.6 mM) and TEOA (1.26 M) as a sacrificial electron donor was irradiated at 365 nm under a CO$_2$ atmosphere. The light intensity was 1.27×10^{-8} einstein s^{-1}.

case of **Ru—Re4**, the electron obtained photochemically should be localised mainly on the Ru site of the bridging ligand (inefficient Process 3) because the π* level of the phenanthroline–imidazolyl motif of the bridging ligand is lower than that of the bpy motif in the Re site. In the case of **Ru—Re5**, although intramolecular electron transfer should proceed rapidly, the photocatalytic ability remained poor. This is because the conjugated bridging ligand lowers the reduction power of the OERS of the Re unit ($E^{red}_{1/2} = -1.10$ V), which induces slower catalysis of the Re unit. Table 9.5 shows the relationship between the photocatalytic activities of the mononuclear Re complexes, i.e. fac-[Re(N^N)(CO)$_3$(PR$_3$)]$^+$ and their first reduction potentials. [12] These data indicate that the threshold of the photocatalysis exists at $E^{red}_{1/2}$(N^N/N^N$^{·-}$) = −1.4 V.

From these results and discussions, the architecture for constructing efficient supramolecular photocatalysts for the reduction of CO$_2$ can be proposed as follows: (i) The electron captured photochemically by the photosensitiser unit should be mainly localised on the bridging ligand rather than on the peripheral ligands to induce efficient intramolecular electron transfer to the catalyst unit (Process 3). (ii) A non-conjugated linker should be used in the bridging ligand between the photosensitiser and catalyst units because the introduction of a conjugated linker drastically lowers the reducing power of the catalyst unit (Process 4). Based on this architecture, we have successfully developed various efficient and durable supramolecular photocatalysts for the reduction of CO$_2$ as described below.

The photocatalytic abilities of the supramolecular photocatalysts were affected by the length of the alkyl chain between the diimine moieties in the bridging ligand [60]. **Ru—Re8** bridged by a —C$_2$H$_4$— chain exhibited a higher photocatalytic activity (entry 18: $\Phi_{CO} = 0.13$, TON$_{CO} = 180$) in comparison with the supramolecular photocatalysts with a —C$_4$H$_8$— or —C$_6$H$_{12}$— chain (**Ru—Re9**,

Ru—Re10), whereas the photocatalytic abilities of **Ru—Re9** (entry 19) and **Ru—Re10** (entry 20) were almost the same. A weak but definite electronic interaction between the photosensitiser and catalyst units was observed only through the —C_2H_4— chain in the excited **Ru—Re8**. This induced more efficient quenching of the ^{3}MLCT excited state of **Ru—Re8** by the sacrificial electron donor BNAH than in the other systems (Process 2) and therefore should be the main reason for the differences in the photocatalyses. The interaction increased when two —C_2H_4— chains were introduced between the Ru and Re units (**Ru—Re17**), improving the photocatalytic ability (entry 29: $\Phi_{CO} = 0.16$, $TON_{CO} = 204$) [64].

The peripheral ligand of the Re unit also affected the photocatalyses of supramolecular photocatalysts [59]. When triethylphosphite was employed as a peripheral ligand (**Ru—Re6**) instead of Cl$^-$, the photocatalytic ability was improved (entry 16: $\Phi_{CO} = 0.16$, $TON_{CO} = 232$), whereas the introduction of a pyridine ligand (**Ru—Re7**) induced lower photocatalytic activity (entry 17: $TON_{CO} = 97$). We found that, in the initial stage of the photocatalytic reaction, **Ru—Re6** was converted rapidly into an aminoethylcarbonato complex with an $^-OC(O)OC_2H_4N(C_2H_4OH)_2$ ligand, which was produced via CO_2 capture by the Re centre with deprotonated TEOA (Eq. (9.15)) [59, 63, 65]. The formation of this CO_2 adduct and its use in the photocatalytic reduction of a low concentration of CO_2 are described in detail below [63]. This was an actual catalytic species in the photocatalytic reduction of CO_2. Even though similar rapid ligand substitution occurred when using **Ru—Re7**, the free pyridine accelerated the decomposition of the photocatalyst.

$$(9.15)$$

Re dicarbonyl complexes, *cis,trans*-[Re(N^N)(CO)$_2$(PR$_3$)$_2$]$^+$ (R = *p*-F-C$_6$H$_4$, Ph, OEt) [16, 17], also functioned as catalyst units for CO_2 reduction instead of Re tricarbonyl complexes [54, 61]. These Ru(II)—Re(I) binuclear complexes (**Ru—Re11** ~ **Ru—Re13**) photocatalysed the reduction of CO_2 to CO. A supramolecular photocatalyst with P(*p*-F-C$_6$H$_4$)$_3$ ligands on the Re unit, **Ru—Re11**, was the best photocatalyst in this series (entry 21: $\Phi_{CO} = 0.15$, $TON_{CO} = 207$, $TOF_{CO} = 4.7$ min^{-1}) [61].

The introduction of a —CH_2OCH_2— chain in the bridging ligand further increased the photocatalytic activity (**Ru—Re14**, entry 25: $\Phi_{CO} = 0.18$, $TON_{CO} = 253$) [62]. Owing to the electron-withdrawing property of the ether group, the stronger oxidation power of the excited photosensitiser unit caused Process 2 to be more efficient. The reaction mechanism following intramolecular electron transfer from the OERS of the photosensitiser unit to the catalyst unit was partially clarified as illustrated in Scheme 9.13. In the initial stage of the photocatalysis, a subset of the biscarbonyl Re unit was converted to the tricarbonyl Re species with an aminoethyl carbonate ligand,

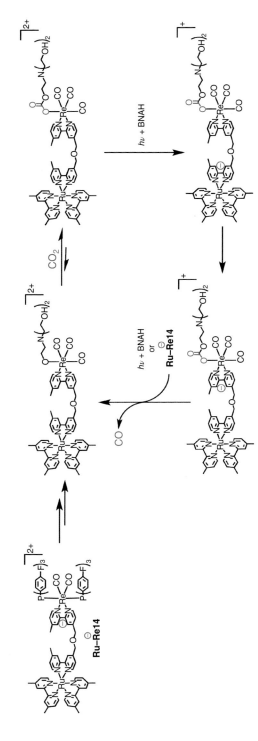

Scheme 9.13 Reaction mechanisms using **Ru—Re14**.

[Re(N^N)(CO)$_3${OC(O)OC$_2$H$_4$N(C$_2$H$_4$OH)$_2$}]$^{2+}$, i.e. **Ru—Re14** was partially converted into **Ru—Re15** in the reaction solution, where a CO$_2$ molecule was captured by the Re unit with the aid of deprotonated TEOA. In a photostationary state with the **Ru—Re14** system, a 3 : 1 mixture of **Ru—Re14** and **Ru—Re15** was observed. Even though **Ru—Re15** itself photocatalysed the reduction of CO$_2$ to CO with $\Phi_{CO} = 0.12$ (entry 26), the mixture of **Ru—Re14** and **Ru—Re15** exhibited better photocatalytic activity ($\Phi_{CO} = 0.19$) because **Ru—Re15** functioned as a photocatalyst and **Ru—Re14** assisted the photocatalysis as an external photosensitiser. It is noteworthy that the oxidising power of **Ru—Re14** in the excited state is stronger than that of **Ru—Re15** ($\eta_q = 0.58$), inducing a larger quenching fraction by BNAH as $\eta_q = 0.82$.

For the first time in the field of photocatalytic CO$_2$ reduction using metal-complex photocatalysts, the oxidised compounds of a sacrificial electron donor were quantitatively analysed in a photocatalytic system using **Ru—Re11** and BNAH [61]. BNAH functioned as a one-electron and a one-proton donor and finally dimerised, giving BNA dimers. We have successfully determined the electron and material balance of the photocatalysis as shown in Eq. (9.16).

$$CO_2 + 2\,BNAH \xrightarrow{\text{Ru–Re11}/h\nu,\ \text{DMF–TEOA}} CO + 4,4'\text{-BNA}_2 + [O^{2-}] + 2H^+ \quad (9.16)$$

The efficiency and durability of the Ru—Re supramolecular photocatalysts can be drastically improved by using 1,3-dimethyl-2-phenyl-2,3-dihydro-1H-benzo[d]imidazole (BIH) as a sacrificial electron donor instead of BNAH (**Ru—Re11**, entry 22: $\Phi_{CO} = 0.45$, TON$_{CO} = 3029$, TOF$_{CO} = 35.7$ min^{-1}) [54]. The reasons for this fascinating effect are (i) the stronger reducing power of BIH and (ii) the lack of inhibition of the photocatalysis by the oxidised compound of BIH. The electron and material balance of the photocatalysis using BIH were determined according to Eq. (9.17).

$$2CO_2 + \text{BIH} \xrightarrow{\text{Ru–Re11}/h\nu,\ \text{DMF–TEOA}} CO + \text{BI}^+ + HCO_3^- \quad (9.17)$$

In the photocatalytic CO$_2$ reduction using the MS of a [Ru(4dmb)$_3$]$^{2+}$ photosensitiser and Re(I) tricarbonyl complex catalysts, we found that one of the deactivation processes is the decomposition of the [Ru(4dmb)$_3$]$^{2+}$ photosensitiser, which should be induced by the photochemical reaction of the OERS

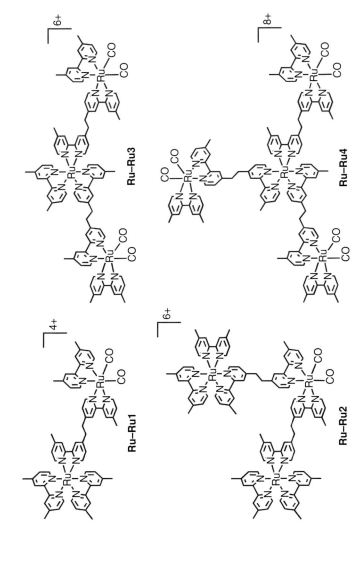

Chart 9.2 Structures and abbreviations of Ru(II)—Ru(II) complexes.

of $[Ru(4dmb)_3]^{2+}$ to give an Ru solvento complex, $[Ru(4dmb)_2(solvent)_2]^{2+}$, although the Re(I) catalysts still maintained their tricarbonyl structure [66].

9.3.2 Ru(II)—Ru(II) Systems

The structures and abbreviations of Ru(II)—Ru(II) supramolecular photocatalysts are shown in Chart 9.2 and their photocatalytic properties are summarised in Table 9.6.

Based on the molecular architecture obtained from the Ru(II)—Re(I) supramolecular photocatalysts, other types of multinuclear complexes with an Ru(II) carbonyl complex as a catalyst unit instead of the Re unit have been developed [67, 68]. $[Ru(N^\wedge N)_2(CO)_2]^{2+}$-type complexes are well-known electrochemical catalysts for CO_2 reduction to HCOOH with high selectivity under basic conditions; these were originally reported by Tanaka and co-workers [69–71]. The $[Ru(4dmb)_{3-m}(BL)_m]^{2+}$ ($m = 1$, 2, 3; BL = 1,2-bis(4′-methyl-[2,2′]bipyridin-4-yl)-ethane) photosensitiser(s) were connected to cis-$[Ru(4dmb)_{2-n}(BL)_n(CO)_2]^{2+}$ ($n = 1$, 2) catalyst(s) via a non-conjugated bridging ligand(s), where the ratio of the photosensitiser to the catalyst units was tuned from 1 : 3 to 2 : 1 (**Ru—Ru1**–**Ru—Ru4**). All of them photocatalytically reduced CO_2 to HCOOH in the presence of BNAH under irradiation at $\lambda_{ex} > 500$ nm [67]. The photocatalytic activities depended strongly on the ratio between the photosensitiser and the catalyst units, and a higher ratio of the photosensitiser to the catalyst led to a higher yield of HCOOH; **Ru—Ru2** (entry 31: selectivity of HCOOH production $\Gamma_{HCOOH} = 91\%$) and **Ru—Ru1** (entry 30: $\Gamma_{HCOOH} = 90\%$) exhibited high selectivity in photocatalytic HCOOH production, whereas **Ru—Ru3** (entry 33: $\Gamma_{HCOOH} = 77\%$) and **Ru—Ru4** (entry 34: $\Gamma_{HCOOH} = 70\%$) were deactivated rapidly during the photocatalyses. The deactivation of the photocatalytic reaction was caused by conversion of the catalyst units. Electrochemical reduction of $[Ru(bpy)_2(CO)_2]^{2+}$ has been reported to induce dissociation of a bpy ligand, giving a black polymer with a Ru—Ru bond, that is, $[Ru(bpy)(CO)_2]_n$ (Eq. (9.18)) [72].

Table 9.6 Photocatalytic properties of Ru(II)—Ru(II) systems.

Entry	Photocatalyst	Donor[a]	Product(s)	Γ(%)	$\Phi_{product}$	TON	TOF (min^{-1})	References
30	Ru—Ru1	BNAH	HCOOH	90	0.038	315	—	[67]
31	Ru—Ru2	BNAH	HCOOH	91	0.041	562	7.8	[67]
32		BI(OH)H	HCOOH	87	0.46	2766	44.9	[68]
33	Ru—Ru3	BNAH	HCOOH	77	0.030	353	—	[67]
34	Ru—Ru4	BNAH	HCOOH	70	0.017	234	—	[67]

a) BNAH: 1-benzyl-1,4-dihydronicotinamide; BI(OH)H: 1,3-dimethyl-2-(o-hydroxyphenyl)-2,3-dihydro-1H-benzo[d]imidazole.

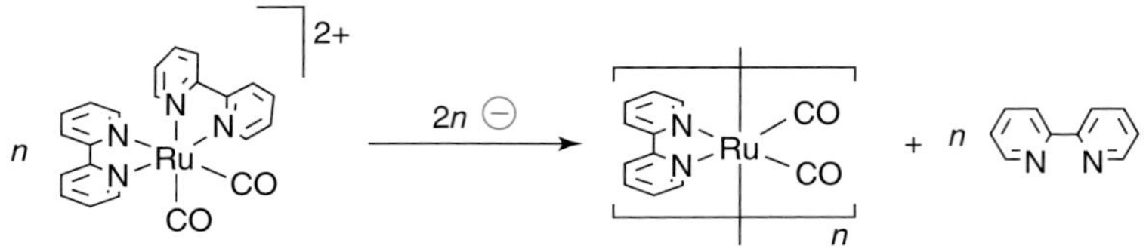

$$\text{(9.18)}$$

When **Ru—Ru3** and **Ru—Ru4** were used, the oligomerisation/polymerisation of the catalyst units proceeded, whereas **Ru—Ru1** and **Ru—Ru2** maintained their structure for a long time. HCOOH production from CO_2 is a two-electron reduction process. In the **Ru—Ru3** and **Ru—Ru4** systems, a larger number of catalyst units in one photocatalyst should reduce the chance for each catalyst unit to accept a second electron from the OERS of the photosensitiser unit. This should trigger the loss of the diimine ligand and formation of the Ru—Ru bond. On the other hand, in **Ru—Ru1** and **Ru—Ru2** systems, the higher ratio of the photosensitiser unit increases the opportunity for a second electron injection into the catalyst unit and promotes CO_2 reduction over the deactivation process.

Also in the photocatalysis using **Ru—Ru2**, the use of 1,3-dimethyl-2-(o-hydroxyphenyl)-2,3-dihydro-1H-benzo[d]imidazole (BI(OH)H) instead of BNAH (entry 31: $\Phi_{HCOOH} = 0.041$, $TON_{CO} = 562$, $TOF_{CO} = 7.8\,min^{-1}$) as a sacrificial electron donor substantially improved the photocatalytic activities (entry 32: $\Phi_{HCOOH} = 0.46$, $TON_{CO} = 2766$, $TOF_{CO} = 44.9\,min^{-1}$) [68].

9.3.3 Ir(III)—Re(I) and Os(II)—Re(I) Systems

The structures and abbreviations of Ir(III)—Re(I) and Os(II)—Re(I) supramolecular photocatalysts are shown in Chart 9.3 and their photocatalytic properties are summarised in Table 9.7.

As described above, the decomposition of the $[Ru(N^{\wedge}N)_3]^{2+}$-type photosensitiser unit in the Ru—Re supramolecular photocatalysts induced both the deactivation of the photocatalyses and the catalysis of HCOOH formation by $[Ru(N^{\wedge}N)_2(solvent)_2]^{2+}$ species, which can catalyse CO_2 reduction to HCOOH under basic conditions. This makes evaluation of the photocatalysis difficult, especially with regard to the product distribution. Therefore, the

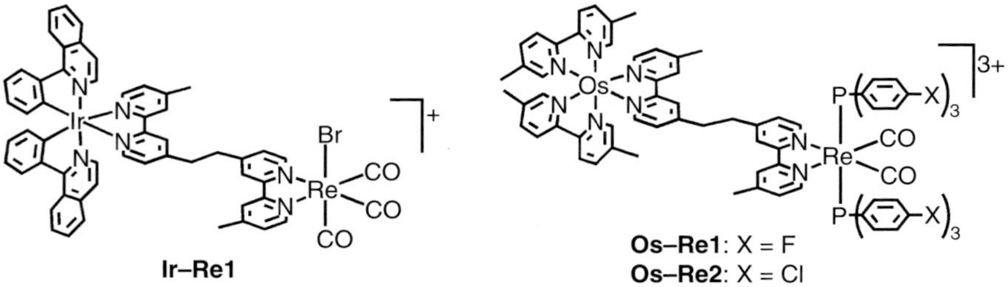

Os–Re1: X = F
Os–Re2: X = Cl

Ir–Re1

Chart 9.3 Structures and abbreviations of Os(II)—Re(I) and Ir(III)—Re(I) complexes.

Table 9.7 Photocatalytic properties of Ir(III)—Re(I) and Os(II)—Re(I) systems.

Entry	Photocatalyst	Donor[a]	Product(s)	Γ(%)	$\Phi_{product}$	TON	TOF (min^{-1})	References
35	Ir—Re1	BNAH	CO	>99	0.21	130	—	[73]
36		BIH	CO	>99	0.41	1700	—	[73]
37	Os—Re1	BIH	CO	>99	0.10	762	1.6	[74]
38	Os—Re2	BIH	CO	>99	0.12	1138	3.3	[74]

a) BNAH: 1-benzyl-1,4-dihydronicotinamide; BIH: 1,3-dimethyl-2-phenyl-2,3-dihydro-1H-benzo[d]imidazole.

development of alternative photosensitisers to [Ru(N^N)$_3$]$^{2+}$-type complexes, where the photochemically decomposed products do not catalyse CO$_2$ reduction, should be important with respect to evaluating the catalysis of various metal-complex catalysts for the reduction of CO$_2$. An Ir(III) complex with two 1-phenyl-isoquinoline (piq) ligands exhibited favourable properties as a photosensitiser, such as visible-light absorption (λ_{abs} < 560 nm) and an excited state (τ_{em} = 2.8 μs) with a long lifetime. A supramolecular photocatalyst of the Ir(III) complex connected to a Re(I) tricarbonyl complex (**Ir—Re1**) photocatalysed the reduction of CO$_2$ to CO with high selectivity, efficiency and durability (entry 35: Γ_{CO} > 99%, Φ_{CO} = 0.21, TON$_{CO}$ = 130) under visible-light irradiation (λ_{ex} > 500 nm) in the presence of BNAH as a sacrificial electron donor [73]. Formic acid was not detected even after 15 h irradiation, whereas the corresponding supramolecular photocatalyst with the Ru(II) photosensitiser (**Ru—Re8**) produced a significant amount of HCOOH (TON$_{CO}$ = 190, TON$_{HCOOH}$ = 55), resulting in lower selectivity for CO formation (Γ_{CO} = 76%). Note that, in the initial stage of photocatalysis with **Ru—Re8**, CO produced mainly (97% after 1 h irradiation). The efficiency and durability were significantly increased when using BIH instead of BNAH (entry 36: Φ_{CO} = 0.41, TON$_{CO}$ = 1700). Even though a longer irradiation time induced gradual deactivation of **Ir—Re1**, ligand-substituted Ir(III) species were not detected but the isoquinoline moieties in the Ir(III) unit were hydrogenated.

Visible-light absorption of the Ru(II) photosensitiser units in efficient supramolecular photocatalysts is limited to λ_{abs} < 560 nm. For the use of visible light over a longer wavelength region, an Os(II) complex was employed as a photosensitiser unit because the [Os(N^N)$_3$]$^{2+}$-type complexes absorb light at much longer wavelengths (λ_{abs} < 730 nm) owing to a relatively strong singlet-to-triplet (S-T) forbidden absorption band. A supramolecular photocatalyst consisting of a [(5dmb)$_2$Os(N^N)]$^{2+}$ (5dmb = 5,5'-dimethyl-2,2'-bipyridine) photosensitiser and a cis,trans-[Re(N^N)(CO)$_2${P(p-X-C$_6$H$_4$)$_3$}$_2$]$^+$ (X = F, Cl) catalyst (**Os—Re1, Os—Re2**) absorbed visible light over a much wider wavelength range (λ_{abs} < 730 nm) in comparison with the corresponding Ru analogue (**Ru—Re11**) and photocatalysed the reduction of CO$_2$ to CO in the presence of BIH under red-light irradiation at λ_{ex} > 620 nm [74]. **Ru—Re11** did not function as a photocatalyst under the same conditions as there was no absorption at $\lambda_{ex} > 560$ nm. The reaction proceeded via a similar mechanism to that of Ru(II)—Re(I) photocatalysts, i.e. (i) selective red-light absorption by the Os unit forming its ^{3}MLCT excited state; (ii) reductive quenching of the excited Os unit by BIH to produce OERS; (iii) intramolecular electron transfer from the OERS of the Os unit to the Re unit; and (iv) reduction of CO_2 at the reduced Re unit. The photocatalytic activities depended on the phosphine ligands in the Re unit, and **Os—Re2** with Cl substituents in the phenyl groups exhibited better photocatalysis (entry 38: $\Phi_{CO} = 0.12$, $TON_{CO} = 1138$, $TOF_{CO} = 3.3$ min^{-1}) in comparison with that of **Os—Re1** with F substituents (entry 37: $\Phi_{CO} = 0.10$, $TON_{CO} = 762$, $TOF_{CO} = 1.6$ min^{-1}).

As described above, we can now choose a photosensitiser unit from three options, i.e. Ru(II), Ir(III) and Os(II) complexes, to develop efficient supramolecular photocatalysts. [Ru(N^N)$_3$]$^{2+}$-type complexes combine properties that are appropriate for a redox photosensitiser, such as (i) absorption of visible light ($\lambda_{abs} < 560$ nm), (ii) a long lifetime of their excited state ($\tau_{em} \sim 1$ μs), (iii) an OERS with a reducing power strong enough to transfer an electron to the catalyst unit for the reduction of CO_2 and (iv) a relatively stable OERS. However, this type of complex decomposes to [Ru(N^N)$_2$(solvent)$_2$]$^{2+}$ species, which cause confusion with respect to the evaluation of the product distribution [75, 76]. On the other hand, both Ir(III) and Os(II) complexes are free from such defects because of their higher stability in comparison with the Ru(II) photosensitiser and the inability of the decomposed species to catalyse CO_2 reduction. The [Ir(C^N)$_2$(N^N)]$^+$-type (C^N = cyclometalated ligand) photosensitiser has another advantage with respect to its anisotropy, which might induce a rectification effect when immobilised on a semiconducting surface [73]. The Os(II) photosensitiser exhibits S-T absorption, which makes it possible to utilise visible light over a much wider wavelength region ($\lambda_{abs} < 730$ nm), although the high toxicity of Os^{VIII}O$_4$ inhibits its wider application [74].

9.4 Photocatalytic Reduction of Low Concentration of CO_2

Exhaust gas that discharges from heavy industries such as heat power plants usually includes relatively low concentrations of CO_2 (less than 20%). Although there are some methods to condensate CO_2, e.g. collection by using amines and membrane filtration, have been developed, they require consumption of a lot of energy for these processes. Therefore, direct reduction of CO_2 is one of the important technologies for practical usages of the photocatalytic systems. However, most reported photocatalytic systems on reduction of CO_2 had been investigated using pure CO_2 [42, 77].

We found that rhenium(I) complexes having a Re—O bond with deprotonated triethanolamine as a monodentate ligand efficiently captures CO_2 to be converted to a CO_2 adduct [65]. The complex with a labile MeCN ligand *fac*-[Re^I(bpy)(CO)$_3$(CH$_3$CN)]$^+$ (**Re—MeCN**) was dissolved in pure DMF giving a DMF complex *fac*-[Re^I(bpy)(CO)$_3$(DMF)]$^+$ (**Re—DMF**) (Eq. (9.19)). Addition

Figure 9.6 IR spectra of a DMF–TEOA (5 : 1 v/v) solution containing the equilibrated mixture of **Re—DMF** and **Re—OC$_2$H$_4$NR$_2$** before (blue) and after (red) CO$_2$ bubbling.

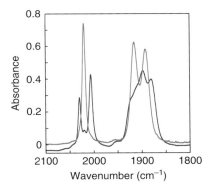

of TEOA into this solution (DMF:TEOA = 5 : 1 v/v) yielded a 1 : 2 mixture of **Re—DMF** and fac-[Re^I(bpy)(CO)$_3$(OCH$_2$CH$_2$NR$_2$)] (**Re—OC$_2$H$_4$NR$_2$**, R = CH$_2$CH$_2$OH) [21], (an equilibrium constant between **Re—DMF** and **Re—OC$_2$H$_4$NR$_2$**: $K_{LS} = 19$) (Eq. (9.20)). The solution containing **Re—DMF** and **Re—OC$_2$H$_4$NR$_2$** was bubbled with CO$_2$ efficiently afforded the CO$_2$ adduct, i.e. fac-[Re^I(bpy)(CO)$_3$(R$_2$N—CH$_2$CH$_2$O—COO$^-$)] (**Re—OC(O)OC$_2$H$_4$NR$_2$**). This CO$_2$ insertion process can be followed by IR spectrum of solution as shown in Figure 9.6; the CO$_2$ adduct **Re—OC(O)OC$_2$H$_4$NR$_2$** displayed ν_{CO} peaks at 2020, 1915 and 1892 cm^{-1}, where the blue line shows the IR spectrum before the bubbling of CO$_2$, i.e. the mixture of **Re—DMF** and **Re—OC$_2$H$_4$NR$_2$**. The ^{13}C NMR spectrum of **Re—OC(O)OC$_2$H$_4$NR$_2$** without ^{1}H decoupling displayed an intense triplet signal at 158.4 ppm with a coupling constant of 3.6 Hz (Figure 9.7), which is a typical signal of carbonate coordinated to a metal ion and the coupling constant is comparable with that for the three-bond coupling between ^{13}C and ^{1}H. As bubbling of the solution containing **Re—OC(O)OC$_2$H$_4$NR$_2$** with Ar caused disappearance of **Re—OC(O)OC$_2$H$_4$NR$_2$** and concomitant reproduction of a mixture of **Re—DMF** and **Re—OC$_2$H$_4$NR$_2$**, capture and release of CO$_2$ is an equilibrium reaction, and its equilibrium constant between **Re—OC$_2$H$_4$NR$_2$** and **Re—OC(O)OC$_2$H$_4$NR$_2$** was $K_{CO_2} = 1.7 \times 10^3$ M^{-1}. The value suggests that

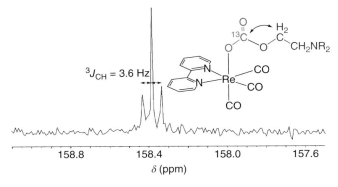

Figure 9.7 ^{13}C NMR spectrum of **Re—OC(O)OC$_2$H$_4$NR$_2$** without ^{1}H decoupling under a ^{13}CO$_2$ atmosphere.

even a very low concentration of CO_2 is enough to produce the CO_2 adduct **Re—OC(O)OC$_2$H$_4$NR$_2$**.

Both abilities of the CO_2 accumulation and reduction of CO_2 can be harmonised in one system, a novel photocatalytic system for reduction of low concentration CO_2 could be developed. Such a system with a heteronuclear bimetal complex comprising a Ru(II) redox photosensitizer unit and a Re(I) catalyst unit with high capability of CO_2 capture is introduced in the following part.

$$\text{Re–DMF} \xrightleftharpoons[+\text{DMF}]{+\text{TEOA–H}^+} \text{Re–OC}_2\text{H}_4\text{NR}_2 \quad K_{LS} = 19 \tag{9.19}$$

$$\text{Re–OC}_2\text{H}_4\text{NR}_2 \xrightleftharpoons[-CO_2]{+CO_2} \text{Ru–Re–OC(O)OC}_2\text{H}_4\text{NR}_2 \quad K_{CO_2} = 1.7 \times 10^3 \text{ M}^{-1} \tag{9.20}$$

The structure dinuclear complex composed of a photosensitiser [Ru^{II}(N^N)$_3$]$^{2+}$ unit (N^N = diimine ligand) and a catalyst [Re^I(N^N)(CO)$_3$(DMF)]$^{n+}$ unit connected with a dimethylene ether (—CH$_2$OCH$_2$—) linkage (**Ru—Re—DMF** in Figure 9.8a) worked as the efficient photocatalyst for CO_2 reduction under a pure CO_2 atmosphere (Table 9.4) [62]. The CO_2-capturing ability of the Re unit of this complex was examined in the presence of triethanolamine (TEOA) [65]. Similar equilibrium reactions proceeded among the DMF complex (**Ru—Re—DMF**), a TEOA complex (**Ru—Re—OC$_2$H$_4$NR$_2$**), and the CO_2 adduct (**Ru—Re—OC(O)OC$_2$H$_4$NR$_2$**) in this supramolecular photocatalyst as shown in Figure 9.8. An equilibrium constant for ligand substitution between DMF and the deprotonated TEOA (Process 1) was $K_{LS} = 60$, which is larger than that in the case of the mononuclear Re complex, and an equilibrium constant for the CO_2 capture reaction (Process 2) was $K_{CO_2} = 1.5 \times 10^3$ M^{-1}, which was comparable with that of the mononuclear complex. According to this large equilibrium constant for CO_2 capture, under a 10% CO_2 gas atmosphere, 94% of the complex was converted to **Ru—Re—OC(O)OC$_2$H$_4$NR$_2$**; moreover, the use of 1% and 0.5% CO_2 gave 65% and 48% formation yield of **Ru—Re—OC(O)OC$_2$H$_4$NR$_2$**, respectively (Figure 9.8b).

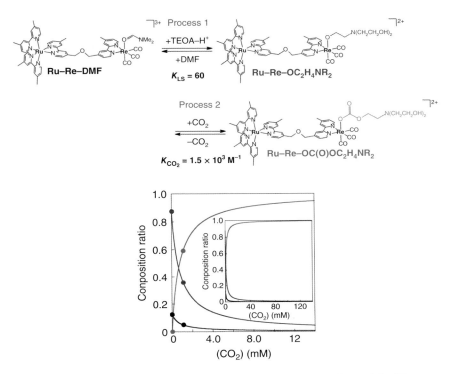

Figure 9.8 Ligand substitution of the dinuclear **Ru—Re—DMF** complex and the CO_2 capture behaviour in a mixed solvent system of DMF and TEOA (5 : 1 v/v). Three complexes, **Ru—Re—DMF**, **Ru—Re—OC$_2$H$_4$NR$_2$** and **Ru—Re—OC(O)OC$_2$H$_4$NR$_2$**, observed in the mixed solvent system under a CO_2 atmosphere. Composition ratios of the complexes depending on CO_2 concentration in a solution.

Figure 9.9 shows the photocatalytic ability of the Ru(II)—Re(I) complex for CO_2 reduction, which was investigated under two gas atmospheres, i.e. pure CO_2 and Ar-based gas containing 10% CO_2, in a DMF–TEOA solution (5 : 1 v/v) containing **Ru—Re—DMF** and **Ru—Re—OC$_2$H$_4$NR$_2$** (total concentration was 0.05 mM) as well as BIH (0.1 M) as a sacrificial reductant (λ_{ex} = 480 nm). Selective CO formation was observed in both cases, where the most important point is that this time course of CO formation using 10% CO_2 was similar to that of 100% CO_2. The turnover number for CO formation (TON_{CO}) after irradiation for 26 h reached 1020 even in the case with 10% CO_2. The quantum yield of CO formation was determined as Φ_{CO} = 0.50 under a 100% CO_2 atmosphere, and a similar value was obtained in cases with 10% CO_2.

Figure 9.10 shows the initial stage of the photocatalytic reactions with more variety of concentrations of CO_2 from 0.5% CO_2 to 100% CO_2. The time courses of CO formation were similar among the cases with 100%, 50% and 10% CO_2. Even in cases using gas with a much lower concentration of CO_2, i.e. 1% or 0.5%, CO was produced with about 80% and 60% efficiencies, respectively, compared with using 100% CO_2 for 1-h irradiation. In addition, the turnover numbers for CO formation were 215 and 205 after 19-h irradiation, respectively. Therefore,

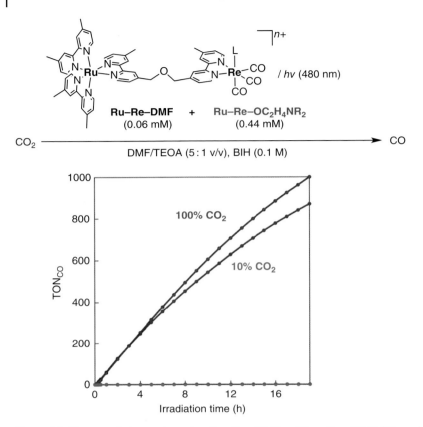

Figure 9.9 Photocatalytic reaction using **Ru—Re—DMF** and **Ru—Re—OC$_2$H$_4$NR$_2$** under a 100% or 10% CO$_2$ atmosphere, where blue, green, red and orange dots denote the turnover numbers of CO formation under 100% CO$_2$, CO formation under 10% CO$_2$, H$_2$ formation under 100% and H$_2$ formation under 10%, respectively.

this photocatalytic system can supply a good method for reducing CO$_2$ even with gas containing low concentrations of CO$_2$, without the use of other methods for CO$_2$ condensation.

The details of the reaction mechanism of the photocatalytic CO$_2$ reduction using the CO$_2$ adducts of the Re complexes have not been clear yet. However, the CO$_2$ adduct should work as a main player because of the following evidences. Figure 9.11a shows comparisons of the photocatalytic CO$_2$ reduction using the Ru—Re dinuclear complex between two solvent systems, i.e. with TEOA and without TEOA. In the case using only DMF as the solvent, CO could not be detected under 1% CO$_2$ atmosphere during irradiation although continuous CO formation was observed under 100% CO$_2$. In addition, there is a good correlation between the concentration of **Ru—Re—OC(O)OC$_2$H$_4$NR$_2$** in the reaction solution before irradiation and the formation rate of CO in the initial stages of the photocatalytic reaction (Figure 9.11b). These results strongly indicate that **Ru—Re—OC$_2$H$_4$NR$_2$** can play the role of a collector of CO$_2$ from gas containing

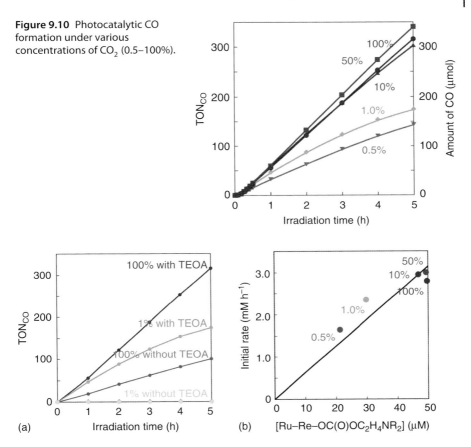

Figure 9.10 Photocatalytic CO formation under various concentrations of CO_2 (0.5–100%).

Figure 9.11 Profound influence on photocatalytic CO_2 reduction by triethanolamine: (a) photocatalytic reactions under 100% and 1% CO_2 atmosphere with and without TEOA. (b) Linear relationship between the initial rate of CO formation and the concentration of **Ru—Re—OC(O)OC$_2$H$_4$NR$_2$**.

low concentrations of CO_2 and the CO_2 adduct **Ru—Re—OC(O)OC$_2$H$_4$NR$_2$** can efficiently reduce CO_2; this molecular technology makes the photocatalytic reduction of low concentration of CO_2 possible.

9.5 Hybrid Systems Consisting of the Supramolecular Photocatalyst and Semiconductor Photocatalysts

As described earlier, the metal-complex photocatalysts that can drive efficient photocatalytic CO_2 reduction have been developed. However, these systems consisting only of metal complexes as the photosensitiser require an electron donor with strong reduction power such as BNAH, amines and BIH because oxidation powers of metal-complex photosensitisers are relatively weak. New molecular technologies for adding stronger oxidation power are necessary for applying the photocatalytic systems to 'practical' artificial photosynthesis.

For solving this, we have developed two systems that drive electron transfer vis step-by-step excitation with two photons: (i) hybrids of the metal-complex photocatalysts with visible-light-driven semiconductor particles and (ii) photoelectrochemical cells consisting of a supramolecular–photocatalyst photocathode and an n-type semiconductor photoanode. As the latter is interpreted in detail in Chapter 10, here, we focus on the hybrids with the supramolecular photocatalyst and semiconductor photocatalysts.

Although some semiconductor photocatalysts have stronger oxidation power than the metal-complex photocatalysts, selectivity of CO_2 reduction is much lower because of competition with H_2 production [78, 79]. If we can emergently hybridise a metal-complex photocatalyst for CO_2 reduction and a semiconductor photocatalyst with strong oxidation power, a novel type of photocatalytic systems for CO_2 reduction with both high selectivity of CO_2 reduction and strong oxidation power might be developed. In these systems, the supramolecular photocatalysts have structural advantages (Scheme 9.11). If both the redox photosensitiser and catalyst in the two-component systems are separately fixed on the surface of semiconductor, collision between the one-electron-reduced photosensitiser, which should be produced by interfacial electron transfer from the semiconductor, and the catalyst should be a very slow or impossible process. Therefore, the electron transfer between them should be much slower and less efficient than that in a homogeneous solution. We do not need to worry about this problem of molecular photocatalytic systems on surface when the supramolecular photocatalyst is used because the two components are always connected to each other. In addition, if anchor groups, which can attach to the surface, are introduced into the photosensitiser unit, the rapid intramolecular electron transfer might suppress the problematic back-electron transfer from the OERS of the photosensitiser unit to the semiconductor because of the longer distance between the catalyst and the semiconductor.

In the systems consisting of the supramolecular photocatalyst and semiconductor, a photochemically produced electron in the conduction band of the semiconductor transfers to the excited state of the photosensitiser unit of the supramolecular photocatalyst (Scheme 9.14). This step-by-step excitation of both the supramolecular photocatalyst and the semiconductor (the so-called Z-scheme process) can achieve both an added electron into the catalyst unit of the supramolecular photocatalyst, which has strong reduction power, and a hole in the valence band of the semiconductor, which has strong oxidation power.

We have reported several such hybrid systems. The first successful system is a supramolecular photocatalyst with a *cis,trans*-Ru(N^N)(CO)$_2$Cl$_2$ moiety as the catalyst for CO_2 reduction and a [Ru(4,4'-Me$_2$bpy)(N^N){4,4'-(H$_2$O$_3$PCH$_2$)$_2$bpy}]$^{2+}$ photosensitiser having methylphosphonate groups as anchoring groups (**(P)—Ru—Ru(CO)$_2$Cl$_2$**) were adsorbed on tantalum oxynitride powders (the bandgap is about 2.5 eV) [80] with silver nanoparticles on the surface (**Ag/TaON**) [81, 82]. This hybrid **(P)—Ru—Ru(CO)$_2$Cl$_2$—Ag/TaON** was dispersed in methanol and irradiated by visible light ($\lambda_{ex} > 400$ nm) under a CO_2 atmosphere, giving HCOOH as a main reduction product with H_2 and CO

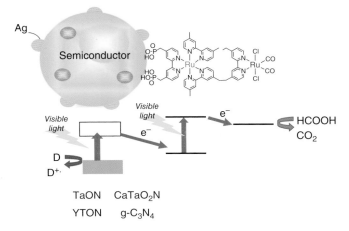

Scheme 9.14 Hybrid photocatalysts consisting of a supramolecular photocatalyst and semiconductors.

as minor products. HCHO was also detected, of which the produced amount was similar to the sum of the reductive products. Isotope experiments using $^{13}CH_3OH$ as the solvent or under a $^{13}CO_2$ atmosphere clearly indicated that the carbon sources of HCOOH and HCHO were CO_2 and methanol, respectively. The photocatalytic reduction of CO_2 did not proceed only using the supramolecular photocatalyst because the excited state of the photosensitiser unit could not be reductively quenched by methanol (Figure 9.12). Although irradiation to a suspension of **Ag/TaON** without the supramolecular photocatalyst under the CO_2 atmosphere catalytically produced H_2, CO_2 reduction did not proceed. In the case of using **Ag/TaON** adsorbed with only the mononuclear Ru complexes as the catalyst model, i.e. *cis,trans*-Ru{4,4′-($H_2O_3PCH_2$)$_2$bpy}(CO)$_2$Cl$_2$, instead of **(P)—Ru—Ru(CO)$_2$Cl$_2$—Ag/TaON**, very small amounts of HCOOH and CO were produced with a considerable amount of H_2. Even using **(P)—Ru—Ru(CO)$_2$Cl$_2$—Ag/TaON**, neither the CO_2 reduction products nor H_2 was produced without irradiation. From these results, we can conclude that the photocatalytic reduction of CO_2 proceeded via step-by-step absorption of two photons by both **(P)—Ru—Ru(CO)$_2$Cl$_2$** and **Ag/TaON**, so-called artificial Z-scheme mechanism. The Ag nanoparticles on **TaON** had important roles for the photocatalysis of the hybrid, probably assistance of the photochemical charge separation in **TaON** and roles as an electron pool located closely to the supramolecular photocatalysts and as an energy source via plasmon excitation. The photocatalytic reduction of CO_2 with methanol as the reductant is an endergonic reaction by $\Delta G° = +83.0\,kJ\,mol^{-1}$ (Eq. (9.21)), which indicates that visible-light energy is converted into chemical energy using this hybrid photocatalyst.

$$CO_2 + CH_3OH \rightarrow HCOOH + HCHO \quad \Delta G° = +83.0\ kJ\ mol^{-1} \qquad (9.21)$$

Several kinds of semiconductors can be used instead of **TaON**, i.e. perovskite **CaTaO$_2$N** (the bandgap: 2.5 eV) [83], yttrium–tantalum oxynitride

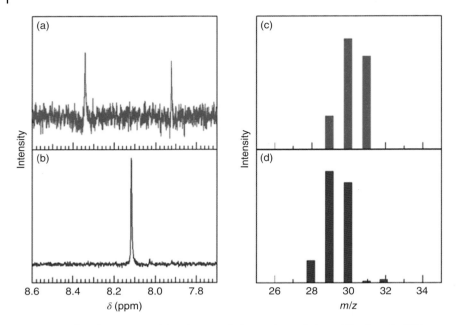

Figure 9.12 ^{1}H NMR spectra of the photocatalytic reaction solutions: **(P)—Ru—Ru(CO)$_2$Cl$_2$—Ag/TaON** was irradiated at $\lambda_{ex} > 400$ nm for 15 h in (a) CH$_3$OH under ^{13}CO$_2$ and (b) CH$_3$OH saturated with unlabelled CO$_2$. Mass spectra of formaldehyde peaks in the GC/MS analysis of the photocatalytic reaction solutions: **(P)—Ru—Ru(CO)$_2$Cl$_2$—Ag/TaON** (1 mg) was irradiated at $\lambda_{ex} > 400$ nm for 24 h in (c) unlabelled CO$_2$-saturated ^{13}CH$_3$OH and (d) unlabelled CH$_3$OH. Adapted with permission from ref. [36]. Copyright 2012 American Chemical Society.

(**YTON**, 2.1 eV) [84], and mesoporous graphitic carbon nitride (**g-C$_3$N$_4$**, 2.8 eV) [85, 86]. High selectivity of HCOOH (>99%) was achieved by using either **(P)—Ru—Ru(CO)$_2$Cl$_2$—Ag/CaTaO$_2$N** or **(P)—Ru—Ru(CO)$_2$Cl$_2$—Ag/YTON** as the Z-scheme CO$_2$ reduction system operable under visible light. Especially, the hybrid with the organic semiconductor **(P)—Ru—Ru(CO)$_2$Cl$_2$—Ag/g-C$_3$N$_4$** showed very high performance for CO$_2$ reduction under visible-light irradiation, a very high turnover number (>33 000 based on the adsorbed amount of **(P)—Ru—Ru(CO)$_2$Cl$_2$**) with relatively high selectivity for HCOOH production (87–99%) [85]. Enhancement of the photocatalysis with the silver nanoparticles was drastically observed in this system as well. The results of photocatalytic reactions, emission decay measurements and time-resolved infrared spectroscopy suggested that the Ag nanoparticles on the semiconductor accepted electrons with lifetimes of several milliseconds from the conduction band of **g-C$_3$N$_4$**, which were transferred to the excited photosensitiser unit of **(P)—Ru—Ru(CO)$_2$Cl$_2$**, promoting the high photocatalysis. The similar hybrid using carbon nitride nanosheets **(P)—Ru—Ru(CO)$_2$Cl$_2$/Ag/NS—C$_3$N$_4$** efficiently photocatalysed CO$_2$ reduction even in water containing a proper electron donor [86]. The selectivity of CO$_2$ reduction to formate was approximately 98%. Another supramolecular photocatalyst **(P)—Ru—Re** was hybridised with **g-C$_3$N$_4$**. This hybrid **(P)—Ru—Re—g-C$_3$N$_4$** photocatalysed CO$_2$ reduction giving CO with high selectivity of over 90%. The photocatalysis was much improved by modification of **(P)—Ru—Re—g-C$_3$N$_4$** with highly dispersed silica because this modification improved adsorptivity of the supramolecular photocatalyst **(P)—Ru—Re** on the surface.

9.6 Conclusion

In this chapter, we have described the molecular technologies for replacing efficient, durable and multifunctional metal-complex photocatalysts for CO$_2$ reduction. There are two types of metal-complex photocatalysts: (i) two-component systems consisting of the redox photosensitiser, which initiates photochemical electron transfer from a reductant to the catalyst, and the catalyst that totally accepts two electrons and reduces CO$_2$, and (ii) supramolecular photocatalysts where the photosensitiser unit(s) and the catalyst unit are connected to each other with the bridging ligand(s). The highest efficiency was obtained by using the former system, i.e. combination of the ring-shaped Re(I) trimer as the photosensitiser and the Re mononuclear complex as the catalyst. From the viewpoint of durability, the supramolecular photocatalysts have advantages because of more rapid intramolecular electron transfer from the reduced photosensitiser unit to the catalyst unit; one of the decomposition processes of the photocatalytic systems is the photochemical ligand substitution of the one-electron-reduced photosensitiser. The supramolecular photocatalysts can be applied to heterogeneous systems such as hybrids with semiconductor particles and electrodes. The former hybrids are introduced in this chapter, and the latter cases are described in Chapter 10. The readers are strongly recommended to read Chapter 10 as well. Low concentrations of CO$_2$ can be directly reduced by using the Re complexes with both functions of the CO$_2$ absorber and the catalyst. As described in the introduction section of this chapter, various functions are required for constructing 'practical' artificial photosynthesis. New molecular technologies should be developed for this purpose.

Acknowledgements

We acknowledge Japan Science and Technology Agency for financial support (CREST program, Grant Number: JPMJCR13L1).

References

1 Home page: Petroleum Industry in Japan (2016). http://www.paj.gr.jp/statis/data/.
2 Lewis, N.S. and Nocera, D.G. (2007). *Proc. Natl. Acad. Sci. U. S. A.* 104: 20142–20142.
3 Primary energy in Japan (2015): 20,928 PJ (Agency for Natural Resources and Energy, METI, Japan).

4 Johnson, T.C., Morris, D.J., and Wills, M. (2010). *Chem. Soc. Rev.* 39: 81–88.
5 Sahami, S. and Weaver, M.J. (1981). *Electroanal. Chem.* 122: 155–170.
6 Sahami, S. and Weaver, M.J. (1981). *Electroanal. Chem.* 122: 171–181.
7 Hawecker, J., Lehn, J.-M., and Ziessel, R. (1983). *J. Chem. Soc. Chem. Commun.* 536–538.
8 Hawecker, J., Lehn, J.-M., and Ziessel, R. (1986). *Helv. Chim. Acta* 69: 1990–2012.
9 Grice, K.A. and Kubiak, C.P. (2014). *Adv. Inorg. Chem.* 66: 163–188.
10 Takeda, H., Koike, K., Inoue, H., and Ishitani, O. (2008). *J. Am. Chem. Soc.* 130: 2023–2031.
11 Hori, H., Johnson, F.P.A., Koike, K. et al. (1996). *J. Photochem. Photobiol. A Chem.* 96: 171–174.
12 Koike, K., Hori, H., Ishizuka, M. et al. (1997). *Organometallics* 16: 5724–5729.
13 Hori, H., Johnson, F.P.A., Koike, K. et al. (1997). *J. Chem. Soc. Dalton Trans.* 1019–1023.
14 Hori, H., Ishihara, J., Koike, K. et al. (1999). *J. Photochem. Photobiol. A Chem.* 120: 119–124.
15 Ishitani, O., George, M.W., Ibusuki, T. et al. (1994). *Inorg. Chem.* 33: 4712–4717.
16 Tsubaki, H., Sekine, A., Ohashi, Y. et al. (2005). *J. Am. Chem. Soc.* 127: 15544–15555.
17 Tsubaki, H., Sugawara, A., Takeda, H. et al. (2007). *Res. Chem. Intermed.* 33: 37–48.
18 Johnson, F.P.A., George, M.W., Hartl, F., and Turner, J.J. (1996). *Organometallics* 15: 3374–3387.
19 Kurz, P., Probst, B., Spingler, B., and Alberto, R. (2006). *Eur. J. Inorg. Chem.* 2966–2974.
20 Kutal, C., Corbin, A.J., and Ferraudi, G. (1987). *Organometallics* 6: 553–557.
21 Kutal, C., Weber, M.A., Ferraudi, G., and Geiger, D. (1985). *Organometallics* 4: 2161–2166.
22 Kalyanasundaram, K. (1986). *J. Chem. Soc. Faraday Trans.* 82: 2401–2415.
23 George, M.W., Johnson, F.P.A., Westwell, J.R. et al. (1993). *J. Chem. Soc. Dalton Trans.* 2977–2979.
24 Loveland, J.W. and Dimeler, G.R. (1961). *Anal. Chem.* 33: 1196–1201.
25 Sullivan, B.P., Bolinger, C.M., Conrad, D. et al. (1985). *J. Chem. Soc. Chem. Commun.* 1414–1416.
26 Fujita, E. and Brunschwig, B.S. (2001). *Catalysis, Heterogeneous Systems, Gas Phase Systems*, vol. 4 (ed. V. Balzani), 88–126. Weinheim: Wiley-VCH.
27 Klein, A., Vogler, C., and Kaim, W. (1996). *Organometallics* 15: 236–244.
28 Christensen, P., Hamnett, A., Muir, A.V.G., and Timney, J.A. (1992). *J. Chem. Soc. Dalton Trans.* 1455–1463.
29 Stor, G.J., Hartl, F., van Outersterp, J.W.M., and Stufkens, D.J. (1995). *Organometallics* 14: 1115–1131.
30 Scheiring, T., Klein, A., and Kaim, W. (1997). *J. Chem. Soc. Perkin Trans.* 2: 2569–2571.
31 Shinozaki, K., Hayashi, Y., Brunschwig, B.S., and Fujita, E. (2007). *Res. Chem. Intermed.* 33: 27–36.

32 Hayashi, Y., Kita, S., Brunschwig, B.S., and Fujita, E. (2003). *J. Am. Chem. Soc.* 125: 11976–11987.
33 Fujita, E. and Muckerman, J.T. (2004). *Inorg. Chem.* 43: 7636–7647.
34 Gibson, D.H. and Yin, X. (1998). *J. Am. Chem. Soc.* 120: 11200–11201.
35 Gibson, D.H. and Yin, X. (1999). *Chem. Commun.* 1411–1412.
36 Gibson, D.H., Yin, X., He, H., and Mashuta, M.S. (2003). *Organometallics* 22: 337–346.
37 Kou, Y., Nabetani, Y., Masui, D. et al. (2014). *J. Am. Chem. Soc.* 136: 6021–6030.
38 Gholamkhass, B., Mametsuka, H., Koike, K. et al. (2005). *Inorg. Chem.* 44: 2326–2336.
39 (a) Rohacova, J., Sekine, A., Kawano, T. et al. (2015). *Inorg. Chem.* 54: 8769–8777. (b) Asatani, T., Nakagawa, Y., Funada, Y. et al. (2014). *Inorg. Chem.* 53: 7170–7180.
40 Morimoto, T., Nishiura, C., Tanaka, M. et al. (2013). *J. Am. Chem. Soc.* 135: 13266–13269.
41 Rohacova, J. and Ishitani, O. (2016). *Chem. Sci.* 7: 6728–6739.
42 Takeda, H., Cometto, C., Ishitani, O., and Robert, M. (2017). *ACS Catal.* 7: 70–88.
43 (a) Tinnemans, A.H.A., Koster, T.P.M., Thewissen, D.H.M.W., and Mackor, A. (1984). *Recl. Trav. Chim. Pays-Bas* 103: 288–295. (b) Grant, J.L., Goswami, K., Spreer, L.O. et al. (1987). *J. Chem. Soc. Dalton Trans.* 2105–2109. (c) Craig, C.A., Spreer, L.O., Otvos, J.W., and Calvin, M. (1990). *J. Phys. Chem.* 94: 7957–7960. (d) Matsuoka, S., Yamamoto, K., Pac, C., and Yanagida, S. (1991). *Chem. Lett.* 20: 2099–2100. (e) Matsuoka, S., Yamamoto, K., Ogata, T. et al. (1993). *J. Am. Chem. Soc.* 115: 601–609. (f) Ogata, T., Yamamoto, Y., Wada, Y. et al. (1995). *J. Phys. Chem.* 99: 11916–11922.
44 (a) Grodkowski, J., Behar, D., Neta, P., and Hambright, P. (1997). *J. Phys. Chem. A* 101: 248–254. (b) Dhanasekaran, T., Grodkowski, J., Neta, P. et al. (1999). *J. Phys. Chem. A* 103: 7742–7748.
45 Bhugun, I., Lexa, D., and Savéant, J.-M. (1996). *J. Am. Chem. Soc.* 118: 1769–1776.
46 Costentin, C., Drouet, S., Robert, M., and Savéant, J.-M. (2012). *Science* 338: 90–94.
47 (a) Bonin, J., Chaussemier, M., Robert, M., and Routier, M. (2014). *ChemCatChem* 6: 3200–3207. (b) Bonin, J., Robert, M., and Routier, M. (2014). *J. Am. Chem. Soc.* 136: 16768–16771.
48 Ziessel, R., Hawecker, J., and Lehn, J.-M. (1986). *Helv. Chim. Acta* 69: 1065–1084.
49 (a) Grodkowski, J. and Neta, P. (2000). *J. Phys. Chem. A* 104: 4475–4479. (b) Alsabeh, P.G., Rosas-Hernández, A., Barsch, E. et al. (2016). *Cat. Sci. Technol.* 6: 3623–3630.
50 Takeda, H., Ohashi, K., Sekine, A., and Ishitani, O. (2016). *J. Am. Chem. Soc.* 138: 4354–4357.
51 (a) Mejía, E., Luo, S.-P., Karnahl, M. et al. (2013). *Chem.-Eur. J.* 19: 15972–15978. (b) Karnahl, M., Mejía, E., Rockstroh, N. et al. (2014). *ChemCatChem* 6: 82–86. (c) Luo, S., Mejía, E., Friedrich, A. et al. (2013). *Angew.*

Chem. Int. Ed. 52: 419–423. (d) Tschierlei, S., Karnahl, M., Rockstroh, N. et al. (2014). *ChemPhysChem* 15: 3709–3713. (e) Khnayzer, R.S., McCusker, C.E., Olaiya, B.S., and Castellano, F.N. (2013). *J. Am. Chem. Soc.* 135: 14068–14070. (f) Lazorski, M.S. and Castellano, F.N. (2014). *Polyhedron* 82: 57–70.

52 Knorn, M., Rawner, T., Czerwieniec, R., and Reiser, O. (2015). *ACS Catal.* 5: 5186–5193.

53 Sakaki, S., Mizutani, H., Kase, Y. et al. (1996). *J. Chem. Soc. Dalton Trans.* 1909–1914.

54 Tamaki, Y., Koike, K., Morimoto, T., and Ishitani, O. (2013). *J. Catal.* 304: 22–28.

55 (a) Bourrez, M., Molton, F., Chardon-Noblat, S., and Deronzier, A. (2011). *Angew. Chem. Int. Ed.* 50: 9903–9906. (b) Bourrez, M., Orio, M., Molton, F. et al. (2014). *Angew. Chem. Int. Ed.* 53: 240–243.

56 (a) Smieja, J.M., Sampson, M.D., Grice, K.A. et al. (2013). *Inorg. Chem.* 52: 2484–2491. (b) Sampson, M.D., Nguyen, A.D., Grice, K.A. et al. (2014). *J. Am. Chem. Soc.* 136: 5460–5471. (c) Zeng, Q., Tory, J., and Hartl, F. (2014). *Organometallics* 33: 5002–5008.

57 Takeda, H., Koizumi, H., Okamoto, K., and Ishitani, O. (2014). *Chem. Commun.* 50: 1491–1493.

58 Tamaki, Y. and Ishitani, O. (2017). *ACS Catal.* 3394–3409.

59 Sato, S., Koike, K., Inoue, H., and Ishitani, O. (2007). *Photochem. Photobiol. Sci.* 6: 454–461.

60 Koike, K., Naito, S., Sato, S. et al. (2009). *J. Photochem. Photobiol. A Chem.* 207: 109–114.

61 Tamaki, Y., Watanabe, K., Koike, K. et al. (2012). *Faraday Discuss.* 155: 115–127.

62 Kato, E., Takeda, H., Koike, K. et al. (2015). *Chem. Sci.* 6: 3003–3012.

63 Nakajima, T., Tamaki, Y., Ueno, K. et al. (2016). *J. Am. Chem. Soc.* 138: 13818–13821.

64 Ohkubo, K., Yamazaki, Y., Nakashima, T. et al. (2016). *J. Catal.* 343: 278–289.

65 Morimoto, T., Nakajima, T., Sawa, S. et al. (2013). *J. Am. Chem. Soc.* 135: 16825–16828.

66 Tamaki, Y., Imori, D., Morimoto, T. et al. (2016). *Dalton Trans.* 45: 14668–14677.

67 Tamaki, Y., Morimoto, T., Koike, K., and Ishitani, O. (2012). *Proc. Natl. Acad. Sci. U. S. A.* 109: 15673–15678.

68 Tamaki, Y., Koike, K., and Ishitani, O. (2015). *Chem. Sci.* 6: 7213–7221.

69 Ishida, H., Tanaka, K., and Tanaka, T. (1985). *Chem. Lett.* 405–406.

70 Ishida, H., Tanaka, K., and Tanaka, T. (1987). *Organometallics* 6: 181–186.

71 Ishida, H., Fujiki, K., Ohba, T. et al. (1990). *J. Chem. Soc. Dalton Trans.* 2155–2160.

72 Chardon-Noblat, S., Collomb-Dunand-Sauthier, M.-N., Deronzier, A. et al. (1994). *Inorg. Chem.* 33: 4410–4412.

73 Kuramochi, Y. and Ishitani, O. (2016). *Inorg. Chem.* 55: 5702–5709.

74 Tamaki, Y., Koike, K., Morimoto, T. et al. (2013). *Inorg. Chem.* 52: 11902–11909.

75 Van Houten, J. and Watts, R.J. (1978). *Inorg. Chem.* 17: 3381–3385.
76 Lehn, J.-M. and Ziessel, R. (1990). *J. Organomet. Chem.* 382: 157–173.
77 Yamazaki, Y., Takeda, H., and Ishitani, O. (2015). *J Photochem Photobiol C: Photochem Rev* 25: 106–137.
78 Iizuka, K., Wato, T., Miseki, Y. et al. (2011). *J. Am. Chem. Soc.* 133: 20863–20868.
79 Wang, Z., Teramura, K., Huang, Z. et al. (2016). *Catal. Sci. Technol.* 6: 1025–1032.
80 Chun, W.J., Ishikawa, A., Fujisawa, H. et al. (2003). *J. Phys. Chem. B* 107: 1798–1803.
81 Sekizawa, K., Maeda, K., Domen, K. et al. (2013). *J. Am. Chem. Soc.* 135: 4596–4599.
82 Nakada, A., Nakashima, T., Sekizawa, K. et al. (2016). *Chem. Sci.* 7: 4364–4371.
83 Yoshitomi, F., Sekizawa, K., Maeda, K., and Ishitani, O. (2015). *ACS Appl. Mater. Interfaces* 7: 13092–13097.
84 Muraoka, K., Kumagai, H., Eguchi, M. et al. (2016). *Chem. Commun.* 52: 7886–7889.
85 Kuriki, R., Matsunaga, H., Nakashima, T. et al. (2016). *J. Am. Chem. Soc.* 138: 5159–5170.
86 Kuriki, R., Yamamoto, M., Higuchi, K. et al. (2017). *Angew. Chem. Int. Ed.* 56: 4867–4871.

10

Molecular Design of Photocathode Materials for Hydrogen Evolution and Carbon Dioxide Reduction

Christopher D. Windle[1], Soundarrajan Chandrasekaran[1], Hiromu Kumagai[2], Go Sahara[2], Keiji Nagai[3], Toshiyuki Abe[4], Murielle Chavarot-Kerlidou[1], Osamu Ishitani[2], and Vincent Artero[1]

[1] Univ Grenoble Alpes, CNRS, Commissariat à l'Energie Atomique (CEA), Laboratoire de Chimie et Biologie des Métaux, 17 rue des Martyrs, 38000 Grenoble, France
[2] Tokyo Institute of Technology, School of Science, Department of Chemistry, O-okayama 2-12-1-NE-1, Meguro-ku, Tokyo 152-8550, Japan
[3] Tokyo Institute of Technology, Institute of Innovative Research (IIR), Laboratory for Chemistry and Life Science, R1-26, Nagatsuda 4259, Midori-ku Yokohama, Kanagawa 226-8503, Japan
[4] Hirosaki University, Graduate School of Science and Technology, Department of Frontier Materials Chemistry, 3 Bunkyo-cho, Hirosaki 036-8561, Japan

10.1 Introduction

Solar energy research is the spearhead of non-carbon-based energy technologies because only solar energy can provide the additional >14 TW required to fulfil our societal needs by 2050 [1, 2]. This huge energy input, however, must be stored in a durable way because of the mismatch between solar energy availability and economic or domestic demand. Towards this objective, producing fuels from available resources such as water and carbon dioxide is an attractive solution. Such an endeavour is often referred to as artificial photosynthesis because it mimics the properties of plants, algae and some bacteria to harvest solar energy and to use it to produce the chemicals that fuel their metabolism. Most photosynthetic organisms convert carbon dioxide into organic compounds, but some micro-algae and cyanobacteria are able, under certain conditions, to produce molecular hydrogen from sunlight in water. Producing hydrogen through water splitting or combining CO_2 reduction with water oxidation thus appear as two attractive solutions to store the abundant flow of sunlight falling on Earth. A close-to-market way to convert solar energy into hydrogen is to interface photovoltaic technologies with electrolysis technologies. However, natural systems exploit molecular photosensitisers (such as chlorophylls, for example) and molecular catalytic active sites in enzymes to achieve this process. A closer look at the organisation of the photosynthetic chain (Figure 10.1) shows that (i) each enzymatic site (the oxygen-evolving centre (OEC) in photosystem II or FNR-producing NAD(P)H-/hydrogenases-producing H_2) is coupled to a photosystem, (ii) oxidative and reductive photocatalytic systems are coupled

Molecular Technology: Energy Innovation, Volume 1, First Edition.
Edited by Hisashi Yamamoto and Takashi Kato.
© 2018 Wiley-VCH Verlag GmbH & Co. KGaA. Published 2018 by Wiley-VCH Verlag GmbH & Co. KGaA.

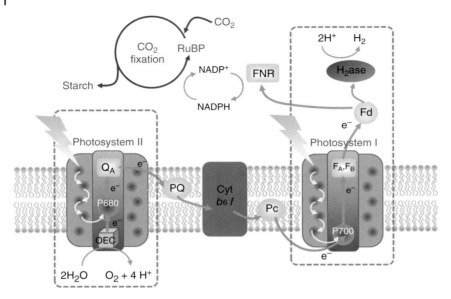

Figure 10.1 Schematic representation of the photosynthetic chain.

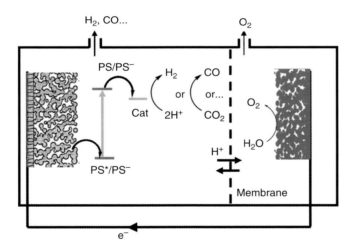

Figure 10.2 Schematic representation of photoelectrochemical cells (PECs) for water splitting or CO_2 reduction coupled with water oxidation.

through an electron transfer chain and (iii) the whole system is immobilised onto a membrane. A possible way to reproducing such a highly complex organisation in an artificial system is through the construction of so-called photoelectrochemical cells (Figure 10.2) that combine a photoanode achieving water oxidation and a photocathode able to produce hydrogen [3–5] or reduce CO_2 [6, 7]. A number of experimental demonstrations have already been reported, most of them relying on the use of solid-state semiconductor and catalysts for water oxidation, hydrogen evolution or CO_2 reduction [8]. In terms of efficiency, limitation currently arises from the photocathode performances and their design will constitute the topic of this chapter, with an emphasis on molecular design, that is, the integration of either a molecular-based photoactive material or a molecular catalyst. The role of a fuel-forming photocathode can be split into three parts. It must (i) absorb and convert sunlight energy into excitons, (ii) separate the charges and (iii) reduce H^+ to H_2 or reduce CO_2. In this chapter, we focus on photocathodes where at least one of those roles is performed by a molecular component.

10.2 Photocathode Materials for H_2 Evolution

We have broken this first part into three sections separated by the nature of the component performing the absorption and conversion of sunlight. We begin with systems where the light absorber is a solid-state semiconducting material and typically an inorganic extended solid. In most cases, this material is p-type and is also responsible for charge separation. The molecular component is the H^+ reduction catalyst. The second part discusses dye-sensitised systems where a molecular dye is responsible for solar energy conversion. Charge separation is provided by a solid-state semiconductor that does not efficiently absorb solar light. H^+ reduction is mediated by a molecular catalyst. The third part is concerned with systems in which an organic semiconducting material converts solar energy. These materials are blended to create a p–n junction that separates the charges and a catalyst is added for H^+ reduction.

10.2.1 Molecular Photocathodes for H_2 Evolution Based on Low Bandgap Semiconductors

This part is focused on p-type inorganic semiconductors grafted with molecular catalysts [9] as photocathodes for H_2 evolution. The inorganic semiconductors currently in the limelight are silicon (Si, 1.1 eV bandgap), gallium phosphide (GaP, 2.2 eV bandgap), indium phosphide (InP, 1.3 eV bandgap), cuprous oxide (Cu_2O, 2.1 eV bandgap) and gallium indium phosphide ($GaInP_2$, 1.8 eV bandgap). They all possess a low bandgap responsible for visible-light absorption [10, 11].

In the following, we will present the different photoelectrodes reported so far with a specific focus on the various methodologies used to construct such hybrid systems.

10.2.1.1 Molecular Catalysts Physisorbed on a Semiconductor Surface

The simplest method that has been used is to simply physisorb a molecular catalyst onto the semiconductor surface. This can be done by drop-casting or soaking method. Nann et al. [12] used an iron sulfur carbonyl catalyst [$Fe_2S_2(CO)_6$], which can be considered as a mimic of the catalytic subsite of [FeFe] hydrogenase enzyme and therefore capable for catalytic proton reduction, to fabricate an InP-based photocathode (Figure 10.3). The photocathode fabrication involves the layer-by-layer coating (10 layers) of InP quantum dots (QDs) onto gold support using thiol chemistry. The modified photocathode was immersed in

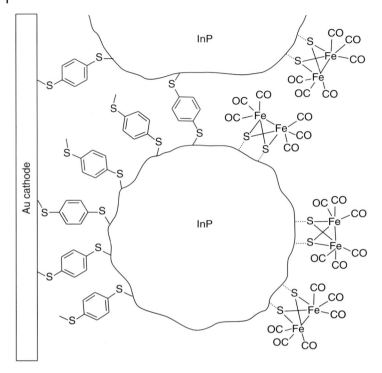

Figure 10.3 Structure of the InP-based electrode with physisorbed diiron catalyst. *Source:* From Nann et al. [12]. Reproduced with permission of John Wiley & Sons.

a toluene solution of the synthetic mimic overnight, to allow spontaneous attachment of the diiron mimic onto InP QDs. Photoelectrolysis at −0.4 V vs Ag/AgCl in sodium tetrafluoroborate aqueous electrolyte in a three-electrode system produced ∼250 nA cm^{-2} current density under illumination with a 395 nm LED array. The current density was stable for 1 h.

The same [Fe$_2$S$_2$(CO)$_6$] catalyst has also been physisorbed on planar silicon [13] and nanostructured silicon such as porous silicon [14, 15], porous silicon nanoparticles [16], silicon nanowires [13] and diatom-derived silicon [17] to fabricate photoelectrodes for H$_2$ production. Photoelectrolysis was carried out in 0.5 M sulfuric acid (∼pH 0.3) as an electrolyte in a three-electrode system under illumination with an Abet solar simulator (1 sun irradiation). A comparison of the studies using [Fe$_2$S$_2$(CO)$_6$] catalyst based on the approximate current density and faradaic efficiency is shown in Table 10.1. Stable photocurrents were observed for porous silicon and silicon nanowire photocathodes coated with [Fe$_2$S$_2$(CO)$_6$] catalyst for a reasonable amount of time. Although planar silicon coated with [Fe$_2$S$_2$(CO)$_6$] catalyst showed higher photocurrents than porous silicon with [Fe$_2$S$_2$(CO)$_6$] catalyst, stability was a major concern with planar silicon. In the case of particle systems coated with [Fe$_2$S$_2$(CO)$_6$] catalyst such as InP QDs, porous silicon nanoparticles and diatom-derived silicon, lower photocurrent values and modest stability were observed when compared with

Table 10.1 Comparison of the studies involving molecular photocathodes coated with $[Fe_2S_2(CO)_6]$ catalyst at pH = 0.3 under illumination with an Abet solar simulator.

Photocathodes coated with $[Fe_2S_2(CO)_6]$	Applied potential (V vs RHE)	Current density (A cm^{-2})	Faradaic efficiency (%)	References
Gold electrode coated with InP QDs	−0.14	0.25×10^{-6}	60	[12]
Planar silicon	−0.24	5.00×10^{-3}	77	[13]
Porous silicon	−0.24	2.80×10^{-3}	90	[14, 15]
Porous silicon nanoparticles	−0.13	2.20×10^{-6}	14	[16]
Diatom-derived silicon	−0.24	6.00×10^{-6}	12	[17]
Silicon nanowire	−0.24	17.00×10^{-3}	99	[13]

robust porous silicon and silicon nanowire systems, respectively. The reason perhaps the low electronic conduction between particles and varied amount of particles attached on the electrode surface. However, silicon-based molecular photocathodes improved the current density and, therefore, H_2 production, by several orders of magnitude when compared with previous work by Nann et al. [12].

Chorkendorff and coworkers used the same strategy to decorate naked H-terminated p-Si(100) photocathodes with various sulfide clusters of molybdenum and tungsten (Figure 10.4) [18, 19]. Excellent photocurrent densities (12–14 mA cm^{-2} at saturation) were observed under red light irradiation. Clearly, molybdenum-based clusters (8 mA cm^{-2} at 0 V vs. reversible hydrogen electrode (RHE)) are more efficient than tungsten-based clusters (8 mA cm^{-2} at 0.3–0.35 V

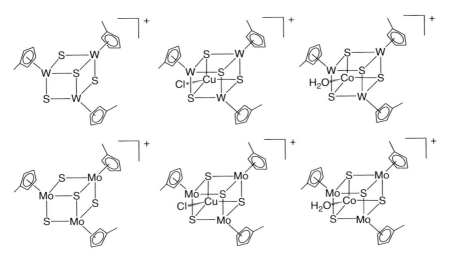

Figure 10.4 Structure of the bioinspired sulfide clusters of molybdenum and tungsten used as hydrogen-evolving catalysts by Chorkendorff and coworkers [18] to decorate silicon-based photocathodes.

vs RHE) to promote catalysis at low applied potentials. The presence of Cu in the tetranuclear clusters does not influence the results, whereas Co-based clusters yield unstable performances. Slightly better performances are obtained when using silicon nanopillars [19]. Except for electrodes based on Co/W and Co/Mo clusters, stable performance over 1 h was observed with less than 5% loss of the photocurrent density over the first 30 min.

10.2.1.2 Covalent Attachment of the Catalyst to the Surface of the Semiconductor

To better control the stability of the catalyst linkage, molecular technologies based on semiconductor surface chemistry have been developed. Two approaches have been used. The first one consists in directly functionalising the surface of the naked semiconductor, whereas the second relies on the coating of the semiconductor with a layer of transparent metal oxide onto which the catalyst is covalently grafted.

The surface chemistry of silicon and other II–V semiconductors is rich and allows for the creation of stable covalent E—C (E = Si, Ga, …) bonds. For example, UV light can be used to induce the coupling of alkene to p-Si(111) and p-GaP(100) surfaces. This allowed Moore and Sharp to decorate such surfaces by protected pendant amino groups [20]. Further deprotection of the amino groups followed by amide coupling with a phthalimide-functionalised nickel bis(diphosphine) catalyst, initially developed for a covalent attachment onto carbon nanotubes (CNTs) [21, 22], gave semiconductor substrates bearing the nickel complex at the surface as shown by grazing angle-attenuated total reflectance (GATR) FTIR and XPS spectroscopies. GATR-FTIR spectra of the functionalised substrates also display a vibrational band characteristic of surface oxide layers likely formed during surface functionalisation. No photoelectrochemical activity is reported for this construct, likely due to the presence of this passivating layer. In another study, the same methodology was used to graft a cobalt or iron porphyrins at the surface of p-GaP(100) substrate (Figure 10.5) [24]. These electrodes proved active for light-driven hydrogen evolution with photocurrents up to 1.3 mA cm^{-2} at 0 V vs RHE in 0.1 M phosphate buffer (pH 7) and 1 mA cm^{-2} at 0.35 V vs RHE for the cobalt porphyrin system. The current density was found to be stable for >5 min.

Rose and coworkers were successful in immobilising a nickel bis(diphosphine) complex (Figure 10.5) starting for a chlorinated p-Si(111) wafer by a stepwise surface modification procedure exploiting palladium-catalysed coupling chemistry [23]. Unreactive sites are blocked with methyl residues, which provide a good stability upon cycling in CH_3CN in the presence of trifluoroacetic acid (TFA). A 2.5×10^{-10} mol cm^{-2} surface concentration of the Ni catalyst was estimated from XPS and electrochemical characterisations. Under 33 mW cm^{-2} broadband LED irradiation, the onset HER potential was found at −0.06 V vs. normal hydrogen electrode (NHE), only 40 mV more negative than a control Pt-coated p-Si electrode. The covalently assembled semiconductor–molecular catalyst construct has an onset for light-driven HER shifted 200 mV more positive than a methylated p-Si photoelectrode in contact with a solution of the same catalyst.

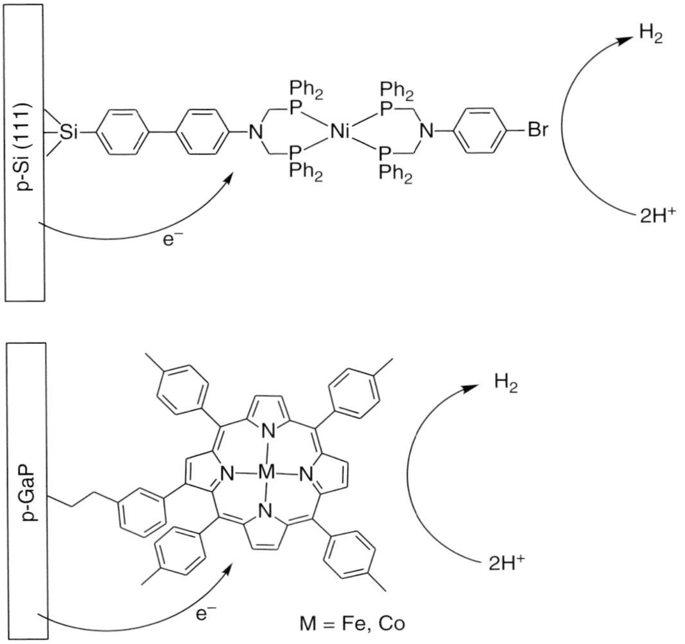

Figure 10.5 Structures of molecular-based electrodes based on p-Si [23] or p-GaP [24] semiconductor materials.

An alternative method consists in coating the surface of the semiconductor by a layer of a transparent conducting or semiconducting oxide, typically TiO_2. Then the catalyst can be attached onto the metal-oxide surface through the covalent binding of phosphonate or carboxylate groups. This stepwise functionalisation permits to take benefit of the three-dimensional structuration of the metal-oxide layer to increase catalyst loading. In addition, the deposition of a protection layer can be well controlled with such a stepwise strategy. A thin Ti layer or atomic layer deposition (ALD)-deposited TiO_2 layers have been used in this context. Actually, this strategy was first used by Chorkendorff and coworkers as an alternative of the drop-casting method presented above. Planar n^+p silicon electrode was first protected by a 7-nm-thick metallic Ti layer and then coated with a 100-nm-thick TiO_2 film. A $[Mo_3S_4]$ cluster bearing phosphonate functional groups was then attached onto the TiO_2 surface to yield an efficient photocathode developing 20 mA cm^{-2} current density at 0 V vs. RHE and proven stability over 1 h (Figure 10.6) [25].

Turner and coworkers used a similar strategy to graft a cobaloxime catalyst onto TiO_2-protected $GaInP_2$ semiconductor [26]. ALD was first used to protect the $GaInP_2$ surface against oxidative corrosion. A 35-nm-thick TiO_2 film proved optimal to obtain good photoelectrocatalytic performances after soaking the electrode in an ethanolic solution of a picolinic acid derivative of a cobaloxime. Further stabilisation of the electrode construct was demonstrated when additional layers of TiO_2 (10 cycles) were deposited using ALD (Figure 10.6). Current

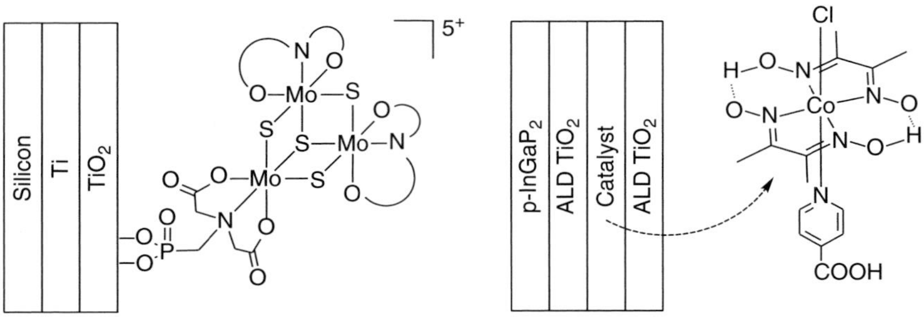

Figure 10.6 Structure of photoelectrodes based on H_2-evolving catalysts attached onto TiO_2 via phosphonate [25] and carboxylate linkages [26].

density up to 9 mA cm^{-2} was measured at 0 V vs. RHE under 1-Sun irradiation. The photocathode showed an initial decay of the photocurrent over the first 4 h and the current density was then sustained for 16 h at ~5 mA cm^{-2} corresponding to a turnover frequency (TOF) of 1.9 s^{-1} in 0.1 M sodium hydroxide (pH 13).

Similarly, a molecular nickel bidiphosphine catalyst was attached through phosphonate linkage onto a silicon photocathode protected by an organic dimethoxyphenyl monolayer embedded into an aluminium–zinc oxide layer capped with a 20-Å-thick TiO_2 layer [27]. The passivation of the silicon surface with organic and inorganic coatings improved the stability as well as charge extraction towards the interface with the electrolyte where the catalyst operates. The onset potential was ~200 mV more positive for such a semiconductor/organic coating/AZO/TiO_2 molecular catalyst system compared with semiconductor/organic coating/TiO_2 molecular catalyst system.

10.2.1.3 Covalent Attachment of the Catalyst Within an Oligomeric or Polymeric Material Coating the Semiconductor Surface

Although it is the last presented here, this strategy has been the first to be implemented to construct a H_2-evolving photocathode based on silicon. In 1984, Mueller-Westerhoff and Nazzal [28] covalently attached a 1-methylferrocenophane catalyst to polymethylstyrene and coated this polymer onto a p-type silicon photocathode (Figure 10.7). The polystyrene film was also used to protect the silicon surface from corrosion. H_2 evolution was observed with a saturating current density of 230 mA cm^{-2} under 9 suns for 5 days in neat boron trifluoride hydrate electrolyte, the high acidity of which probably prevents passivation of the silicon electrode. H_2 evolution was also observed in aqueous acidic electrolytes such as perchloric acid ($HClO_4$), fluoroboric acid and hydrochloric acid. This field experienced a renaissance 5 years ago with the work of Moore and coworkers [29]. These authors investigated cobaloxime catalysts on GaP semiconductor photocathodes for H_2 evolution. Two different polymeric frameworks, polyvinylpyridine (PVP) [30, 31] and polyvinylimidazole (PVI) [32, 33], were investigated, both of which are able to coordinate cobaloxime in axial positions thanks to their pyridine or imidazole residues. Table 10.2 gathers the results obtained for based on H-bridged cobaloximes [30, 31] as well as BF_2-annulated cobaloximes [34]. Parameters such as light intensity [31],

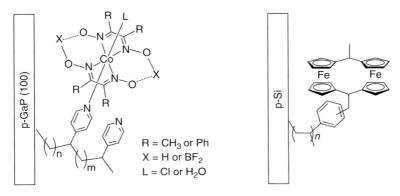

Figure 10.7 Structure of various photoelectrodes based on ferrocenophane [28] and cobaloxime [29] catalysts embedded or grafted in polymeric frameworks.

Table 10.2 The reports by Moore and coworkers based on GaP-modified cobaloxime photocathodes.

Cobaloxime	Grafting polymer	pH	Illumination (100 mW cm^{-2})	Orientation and crystal faces	j at 0 V vs RHE (mA cm^{-2})	References
H-bridged	PVP	7	Newport Oriel apex illuminator model 71228 light source	100	~2.80	[30]
BF$_2$-annulated	PVP	7	Solar light PV cell test simulator model 16S-300-005V4.0 xenon lamp	100	~0.90	[31]
H-bridged	PVP	7	PV cell testing 16S 300 W solar simulator	100	1.20	[34]
H-bridged	PVP	4.5	PV cell testing 16S 300 W solar simulator	100	~1.15	[34]
BF$_2$-annulated	PVP	7	PV cell testing 16S 300 W solar simulator	100	0.56	[34]
BF$_2$-annulated	PVP	4.5	PV cell testing 16S 300 W solar simulator	100	~1.15	[34]
H-bridged	PVI	7	100 W Oriel solar simulator	111A	~0.89	[33]
H-bridged	PVI	7	100 W Oriel solar simulator	111B	~0.89	[33]
H-bridged	PVP	7	100 W Oriel solar simulator	100	~1.3	[32]
H-bridged	PVI	7	100 W Oriel solar simulator	100	~1.2	[32]

pH [34], orientation and crystal faces of GaP [33] have been varied. Table 10.2 can be summarised that H-bridged cobaloxime shows greater efficiency when PVP was used a grafting polymer at pH 7, whereas BF_2-annulated cobaloxime shows greater efficiency with PVP grafting polymer at pH 4.5 (Figure 10.7). This observation is to be confirmed with PVI grafting polymer, but is in line with previous observations that BF_2-annulated cobaloximes require more acidic conditions to mediate electrocatalytic H_2 evolution [35, 36].

In conclusion, light-harvesting semiconductor-tethered molecular catalyst systems seem promising for H_2 production, especially as modern techniques allow to overcome surface passivation issues. However, the fabrication of bulk semiconductors often depends on rare earth elements such as In, Ga or Ge, which is not always economically viable; organic photovoltaic solar cells may be a cost-effective alternative with identical designs of semiconductor systems.

10.2.2 H_2-evolving Photocathodes Based on Organic Semiconductors

Organic semiconductor materials (OSCs) are conducting polymers, oligomers or self-assembled discrete molecules that use abundant materials and can be deposited using low-cost solution-based processes. Additionally, OSCs display high optical absorption in the visible region of the solar spectrum. They generally possess much more reducing excited states than inorganic semiconductors and their electronic properties can be easily tuned through structural modifications [37]. These materials allowed to develop organic solar cells reaching 10% of power conversion efficiency (PCE) [38]. Light-induced charge separation is achieved at the interface between the two OSCs. Upon absorption of a photon in one of the OSCs, an exciton (i.e. a pair of electrostatically bound electron and hole) forms and then migrates to the junction with the other OSC. Here, the exciton dissociates, resulting in the appearance of a hole in one OSC called the donor and an electron in the other OSC called the acceptor. In photovoltaic cells, electrons and holes then migrate to the electrodes where charges are collected. A key issue is the structuration of the two materials as excitons can only diffuse over a few tens of nanometres during their lifetime, that is, before spontaneous recombination takes place. Therefore, the two OSCs must be either deposited as two very thin layers, possibly with an interlayer containing both compounds, yielding a mixed interlayer junction, or as a blend of the two materials forming molecular p–n junctions all over the bulk layer (bulk heterojunction, BHJ). In general, BHJs are formed between a polymer or an array of small molecules, acting as the light absorber, and a fullerene derivative [39]. To use such organic junctions in the context of direct photoelectrochemical water splitting, different conditions should be met: (i) the organic layer in contact with a liquid electrolyte must maintain the capacity of light absorption and generating charges as well as the internal charge separation and transport to the active surface; (ii) an appropriate photovoltage must be reached to allow charge transfer between the organic layer and an acceptor located in the electrolyte or at the surface of the photoelectrode and (iii) the electrochemical reaction at the organic semiconductor/liquid electrolyte interface must be facilitated by an appropriate catalyst. Early reports described photocathodes made of conducting polymers, such as polyacetylene (CH) [40], polypyrrole [41], polyaniline [42], poly(3-methylthiophene) [43] or poly(3-hexylthiophene) (P3HT) [44, 45]. However, these photoelectrodes did not contain any junction between OSCs nor HER catalysts and charge separation arise from the Schottky junction between p-type OSC and electrolyte. Hence, only low photocurrents (few to tens of microamperes per square centimetre) were obtained in aqueous electrolytes, depending on the solution pH, irradiation intensity, excitation wavelength and film thickness. Another step in the photoactive cathode design was to blend the p-type P3HT-conjugated polymer, with low bandgap and a high degree of intermolecular order leading to high-charge carrier mobility [46], as the light-harvesting unit and a fullerene derivative, such as phenyl-C_{61}-butyric-acid methyl ester (PCBM), as electron acceptor in a BHJ structure in order to promote electron transfer. The resulting photocathode was used in aqueous saline solution [47]. Again, in the absence of any catalyst, very low photocurrents were observed and the faradaic yield for H_2 evolution was determined to be 0.3% [47].

Implementation of HER catalysts was found to be key to reach higher efficiencies. With Pt nanoparticles loaded onto the top of an organic p/n bilayer film of small molecules (metal-free phthalocyanine, H_2Pc/fullerene, C_{60}), photocurrent values reached several cents of microamperes per square centimetre and the faradic efficiency for hydrogen production in aqueous media approached or even surpassed 90% (Figure 10.8) [48]. H_2 production at Pt-coated C_{60}/water interface uses electrons from the conduction band of the C_{60} layer. Although the bottom of the conduction band of C_{60} is less negative than H_2 generation, the formation of the two-electron reduced C_{60}^{2-} species was evidenced spectroscopically after a 20-min induction period [48].

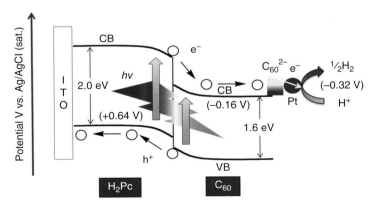

Figure 10.8 Mechanism of light-driven H_2 evolution in the ITO/H_2Pc/C_{60}-Pt system. H_2 evolves at Pt-loaded C_{60}, after a 20-min induction period during which C_{60}^{2-} is formed. *Source:* From Abe et al. [48]. Reproduced with permission of American Chemical Society.

Al-doped ZnO (40-nm thickness) was used to cover P3HT/PCBM and acted as electron-selective layer to a Pt/C catalyst [49]. The resulting photocathode delivers more than 1 mA cm^{-2} at 0 V vs. RHE under neutral pH conditions with 100% Faradic efficiency after equilibrium between the Zn^{2+}/Zn^0 concentration equilibrium in the AZO film is reached.

The photoelectrode shown in Figure 10.8 has been combined with photoanode based on OSC to oxidise thiol [50] or based on TiO_2 to oxidise water [51] (Figure 10.9). By the use of these photoanodes, H_2 could be generated without additional electric bias. Of note, it is positive to invert the coating order and obtain an (indium tin oxide) ITO/n-type OSC/p-type OSC architecture with photoelectrochemical activity for water oxidation [52] or hydrazine oxidation [53, 54]. In the latter case, concomitant H_2 generation occurs at the counter Pt electrode without any additional electric bias.

Amorphous molybdenum sulfide (a-MoS$_x$) can be used as an earth-abundant HER catalyst instead of Pt to develop photocathodes based on the same P3HT/PCBM BHJ [55]. It has been demonstrated recently that a-MoS$_x$ is a coordination polymer based on discrete trinuclear molybdenum disulfide clusters [56]. Photocathodes prepared by the direct spray deposition of mixtures of a-MoS$_x$ and TiO_2 on the organic BHJ-active layer (Figure 10.10) deliver a photocurrent of 200 μA cm^{-2} at 0 V vs. RHE and up to 300 μA cm^{-2} at −0.4 V vs. RHE with a 0.6 V anodic shift of the H_2 evolution reaction onset potential, a value close to the open-circuit potential of the P3HT:PCBM solar cell. The presence of TiO_2 nanoparticles in the MoS_3 film optimises both electron extraction at the PCBM interface and electron transport within the catalytic layer [55], as also observed in other architectures with compact TiO$_x$ layer separating the P3HT:PCBM BHJ layer and the catalytic film [57]. Different interfacial layers were investigated to improve the charge transfer between P3HT:PCBM and a-MoS$_x$. Metallic Al/Ti interfacial layers led to an increase in the photocurrent by up to 8 mA cm^{-2} at 0 V vs. RHE. A 50-nm-thick C_{60} layer also works as an interfacial layer, with a current density reaching 1 mA cm^{-2} at

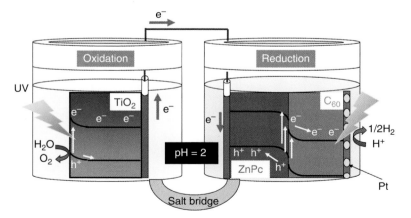

Figure 10.9 Schematic illustration of a two-compartment cell for unassisted solar water splitting using a TiO_2 photoanode and an OSC-based photocathode.

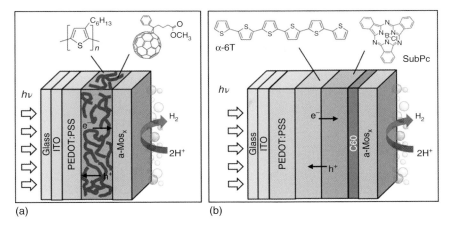

Figure 10.10 Architectures of (a) the P3HT:PCBM/a-MoS$_x$ [57] and (b) SubPc:C60/a-MoS$_x$ [58] H$_2$-evolving photocathodes. *Source*: From Haro et al. [57]. Reproduced with permission of American Chemical Society.

0 V vs. RHE [59]. A common issue is the long-term stability of the light-driven activity under operation. In a different architecture, the use of an insoluble cross-linked PEDOT:PSS hole selective underlayer between the transparent conducting oxide substrate and the P3HT:PCBM BHJ layer proved to be quite efficient to avoid delamination of the organic layer and to allow sustained PEC activity for more than 3 h with faradaic efficiency of 100% in pH 2 aqueous media [57]. Alternatively, reduced graphene oxide, molybdenum, tungsten and nickel oxide underlayers proved efficient for sustained photoelectrochemical activity in acidic aqueous electrolytes [60, 61]. However, photoelectrodes based on P3HT:PCBM BHJ core do not allow to store much energy as the onset potential for H$_2$ evolution is marginally positive with respect to the equilibrium potential of the H$^+$/H$_2$ couple. We then switched to planar p/n heterojunctions of small molecules and oligomers as this technology allows the use of a wider variety of donor and acceptor molecules. The use of α-sexithiophene (α-6T) and boron subphthalocyanine chloride (SubPc) as donor and acceptor OSC materials, respectively, in combination with C$_{60}$ as electron-extracting layer (Figure 10.10) indeed resulted in a significant increase of both the onset potential and the solar-to-hydrogen power saved [59, 60] for H$_2$ evolution of OPV-based photocathodes [58].

10.2.3 Dye-sensitised Photocathodes for H$_2$ Production

Dye-sensitised photocathodes for H$_2$ production are typically composed of three major parts: (i) a semiconductor (SC) layer for charge separation, (ii) a molecular dye for light harvesting and (iii) a catalyst for converting H$^+$ to H$_2$. Upon visible-light irradiation, two successive electron transfers take place (Figure 10.11): first, hole injection from the excited dye to the semiconductor valence band occurs and generates the reduced dye, which, in turn, transfers

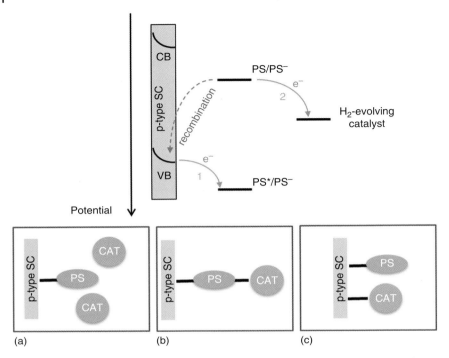

Figure 10.11 Schematic presentation of the electron transfers taking place upon irradiation of a dye-sensitised photocathode (top) and the three different H_2-evolving photocathode architectures (bottom): physisorbed or diffusing catalyst (a); covalent or supramolecular dye–catalyst assembly (b); co-grafted dye and catalyst (c).

the electron to the H_2-evolving catalyst. Obviously, two successive light-induced cycles are required for hydrogen production to occur.

The development of such H_2-evolving dye-sensitised photocathodes follows on from the knowledge acquired on p-type dye-sensitised solar cells (p-DSSCs) [62, 63]. Indeed, both systems rely on similar requirements: (i) a p-type semiconductor fulfilling precise specifications such as transparency in the visible-light domain, good carrier mobility, porosity to insure a high molecular loading, a valence band potential suitable for electron transfer to a molecular photosensitiser (Figure 10.11); (ii) the robust grafting of molecular photosensitisers onto the surface of the semiconductor and (iii) the choice of an efficient electron acceptor in order to avoid the detrimental recombination of the photoreduced dye with the holes generated in the semiconductor (Figure 10.11). However, two main differences also arise for the construction of H_2-evolving dye-sensitised photocathodes: first, the final electron acceptor is no more the electrolyte but the H_2-evolving catalyst; as a consequence, the efficiency of the process does not rely anymore on a single electron transfer but on multielectronic catalysis, thus introducing kinetic limitations. Second, the system should operate in aqueous electrolytes, which introduces new constraints both regarding stability and electronics because of the pH dependence of the surface state of the semiconductor.

Such additional complexity renders the development of efficient dye-sensitised H_2-evolving photocathodes more challenging. Most of the systems reported to date are assessed under a three-electrode configuration using a reference electrode and a Pt counter electrode and not in a complete device. Examples of complete devices, such as tandem-dye-sensitised photoelectrochemical cells, are scarce, owing to the difficulty to find suitable working conditions, matching the requirements of both the anode and the cathode activities.

Currently, the range of known transparent and stable p-type metal oxides that can be prepared from inexpensive materials and by techniques available to most laboratories is limited to NiO [64, 65], CuO [66], $CuCrO_2$ [67, 68], $CuAlO_2$ [68, 69] and $CuGaO_2$ [68, 70]. To date, the majority of reported photocathodes for H_2 production rely on NiO [5, 71], the preparation of which has been subject to optimisation for p-DSSCs [64]. Nonetheless, NiO remains sub-optimal as a p-type semiconductor due to low hole mobility, an absence of hole traps and low permittivity [72]. Alternatively, quasi-metallic oxides such as ITO also proved to be suitable porous materials for the construction of dye-sensitised photocathodes [73–75].

A wide range of dyes suitable for sensitisation of p-type semiconductors have been reported in recent years, and some structural prerequisites for good performance have been elucidated, thanks to intensive studies on p-DSSCs [63]. It is important that the dye offers a large driving force for hole injection and long-lived charge separation. The former can be optimised by modifying the electronic structure so that the oxidation potential of the dye (HOMO) is significantly more positive than the valence band of the NiO. The latter has been addressed by designing dyes based on a donor acceptor motif [72], also called "push–pull" dyes. The electron-rich donor moiety (HOMO) is anchored to the NiO, whereas the electron poor acceptor (LUMO) is located away from the NiO, thus disfavouring recombination. The bridge between the two moieties is as long as possible and rigid to avoid bending of the acceptor unit towards the NiO surface. A range of chemical groups are available for anchoring molecules to metal-oxide surfaces such as carboxylic acid, phosphonic acid, hydroxamic acid and silatranes [76]. Carboxylic and phosphonic acids are commonly utilised.

There is a wide range of molecular catalysts reported for proton reduction to hydrogen, which may be incorporated into a photocathode system. The catalyst may be dissolved in the electrolyte solution (Figure 10.11a) or attached to the electrode surface. In most cases where a catalyst is attached to the surface, it is functionalised with the same anchoring groups as the dyes used in DSSCs (Figure 10.11c). In some molecular systems, the catalyst is covalently attached to the dye and does not require a separate anchoring group (Figure 10.11b). It is desirable that the catalyst operates at a high rate (TOF), with minimal energy requirements (overpotential), and has high stability (turnover number, TON). One of the best performing catalysts for H_2 production is Pt. However, Pt is rare and there is not enough in the Earth's crust if all the vehicles in use today were to transition to a hydrogen economy [77], as such there is a large research effort to develop systems that use earth-abundant elements [9, 78, 79]. This chapter focuses on catalysts using relatively abundant metals and Pt will not be discussed.

10.2.3.1 Dye-sensitised Photocathodes with Physisorbed or Diffusing Catalysts

One of the earliest examples of a dye-sensitised photocathode for H_2 production was reported in 2012 [80]. An organic dye, denoted P1, well-established in the field of p-DSSCs, was used as the sensitiser (Figure 10.12). P1 consists of a triphenylamine (TPA) donor unit, linked to two dicyanovinylene acceptor units via thiophene bridges, and a carboxylate anchoring group. The H_2-evolving photocathode was built by simply dropping a cobaloxime catalyst solution onto the sensitised NiO surface and allowing it to dry. Cyclic voltammetry (CV) indicated that electron transfer from the reduced dye to the catalyst is energetically feasible. Chronoamperometry experiments (−0.4 V vs. Ag/AgCl) were performed under chopped light irradiation (>400 nm) and showed −4 µA photocurrent density with P1 alone compared with −15 µA in the presence of the cobaloxime. H_2 production was detected using a modified Clark-type electrode. In both cases, the photocurrent decays over tens of seconds, but the decay is much faster with the catalyst and this is most probably due to rapid leaching of the cobaloxime into the electrolyte.

Later in the same year, a photocathode allowing H_2 production without any applied bias was reported in conjunction with a photoanode [82]. It consisted of a NiO film sensitised with an organic donor–bridge–acceptor dye, relying on a TPA unit bis-functionalised with perylene monoimide (PMI) acceptors through sexithiophene bridges. This dye-sensitised photocathode allows H_2 production with unity Faradaic efficiency, without any additional catalyst, which is unprecedented in the field. It produced a stable 3.9 µA cm^{-2} photocurrent density and up to 0.6% incident photon to current efficiency (IPCE). In a tandem configuration with a $BiVO_4$ photoanode, H_2 was produced with 80% Faradaic efficiency without any applied bias. Furthermore, the dye-sensitised film is stable in HCl and NaOH solutions for days; the observed stability is likely to be the result of the hexyl chains protecting the surface from the solution.

This stability effect was also recently reported for a photocathode sensitised with a related TPA–quaterthiophene–PMI push–pull dye, inspired from the thylakoid membranes of photosystems I and II (Figure 10.12) [81]. The presence of hexyl chains on the organic dye creates a hydrophobic layer over the electrode surface protecting not only NiO from acid dissolution but also the carboxylate anchoring of the dye from surface desorption. As a consequence, the electrode was stable at pH 0, which is the optimal pH for the $[Mo_3S_4]^{4+}$ H_2-evolving catalyst present in solution. Under these conditions, the dye-sensitised photocathode maintained a photocurrent density of −183 µA cm^{-2} for 17 h at 0 V vs. NHE with a Faradaic efficiency for H_2 production of 49%.

An alternative strategy for improving stability is the application of a thin layer of Al_2O_3 by ALD to prevent dye desorption. In one example, a PMI-based dye-sensitised NiO film was protected by 30 ALD cycles of Al_2O_3 [83]. A further advantage of Al_2O_3 over layers lies in the insulating properties of Al_2O_3 that can slow down charge recombination between the reduced dye and the NiO surface (see another example in Ref. [84] described below), as demonstrated with femtosecond transient absorption spectroscopy. The resultant electrode was stable to air and light over 45 days and could be assessed in H_2SO_4 aqueous solutions.

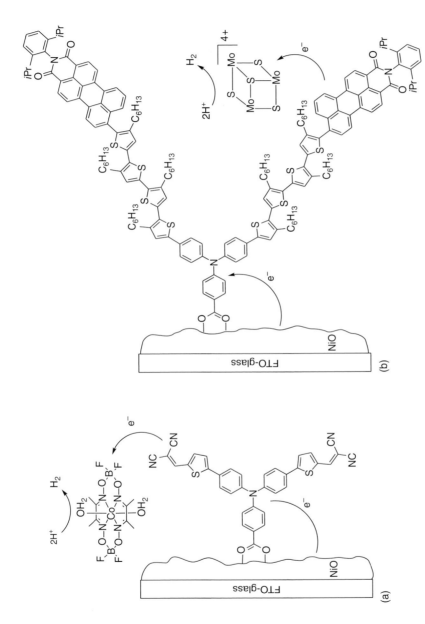

Figure 10.12 Structure of photocathodes based on dye-sensitised NiO with a physisorbed (a) [80] or a diffusing (b) [81] H$_2$-evolving catalyst. *Source:* From Click et al. [81]. Reproduced with permission of American Chemical Society.

Addition of either a nickel bis(diphosphine) or a cobaloxime H_2-evolving catalyst to the solution gave significant photocurrent enhancement and led to hydrogen production with Faradaic efficiencies of 98% and 60% for the Ni and Co catalysts, respectively. When films without ALD treatment were assessed during 2 h with the Ni catalyst in solution, a lower Faradaic efficiency (80%) was measured and complete loss of the dye was observed, whereas the dye remained intact on ALD-treated electrodes.

10.2.3.2 Dye-sensitised Photocathodes Based on Covalent or Supramolecular Dye–Catalyst Assemblies

Reports on supramolecular or covalent dye–catalyst dyad assemblies grafted onto NiO are scarce. In principle, this is advantageous because charge transfer is no longer diffusion-limited as for catalysts in solution, and the dye and catalyst are not competing for surface sites as occurs for co-adsorption strategies.

The first example of a H_2-evolving dye–catalyst assembly anchored onto NiO was reported in 2013 [84]. A cyclometalated tris-heteroleptic ruthenium photosensitiser was employed. The phenylpyridine ligand was anchored to NiO through a carboxylate group and pendant pyridines on the bipyridine ligands were coordinated to a cobaloxime catalyst (Figure 10.13). CV showed that hole injection from the excited Ru dye to NiO and electron transfer from the reduced Ru dye to the cobaloxime (see Figure 10.11) were both energetically favourable. The anchoring shows good stability in water and phosphate buffer (pH 7) attributed to the increased electron density from the phenylpyridine at the carboxylate group. Upon irradiation, the onset potential of the catalytic wave is shifted 40 mV more positive, indicating that solar energy can be stored by the system. The IPCE spectrum matches the absorption spectrum of the dye and is enhanced in the presence of the cobaloxime. A single ALD layer of Al_2O_3, coated onto NiO before grafting the dye–catalyst assembly, was shown to more than double the photocurrent density; this beneficial effect was attributed to reduced hole–electron recombination across the NiO surface. The photocathode produced 290 nmol H_2 with a Faradaic efficiency of 45%, at −0.2 V vs. NHE for 2.5 h (pH 7 phosphate buffer). Shifting the applied bias positive to 0.1 V reduced the unproductive dark current to give a Faradaic efficiency of 68%.

The first fully noble metal-free dyad was composed of a push–pull organic dye covalently attached to a cobalt diimine–dioxime catalyst by a copper-catalysed azide–alkyne cycloaddition ("click" chemistry) (Figure 10.13) [85]. It was demonstrated that the dye could transfer electrons to the Co catalyst both on thermodynamic grounds and by the observation of the photoaccumulation of Co(I) under visible-light irradiation in the presence of triethanolamine (TEOA) as electron donor. In MES buffer (pH 5.5) at 0.14 V vs. RHE, a photocurrent of −15 μA cm^{-2} was observed, and H_2 was produced with a Faradaic efficiency of 10%.

In 2016, a layer-by-layer strategy was reported to construct a H_2-evolving photocathode based on a [Ru(bpy)$_3$]$^{2+}$ derivative photosensitiser and a nickel bis(diphosphine) catalyst. Both of them are functionalised by phosphonic acid groups to allow a covalent anchoring of the Ru dye onto NiO and the chelation of Zr^{4+} cations (Figure 10.14a) [86]. The assembly was built through a step-by-step procedure, the NiO surface being first covered by a monolayer of phosphonated photosensitiser, before soaking in a $ZrOCl_2$ solution, and

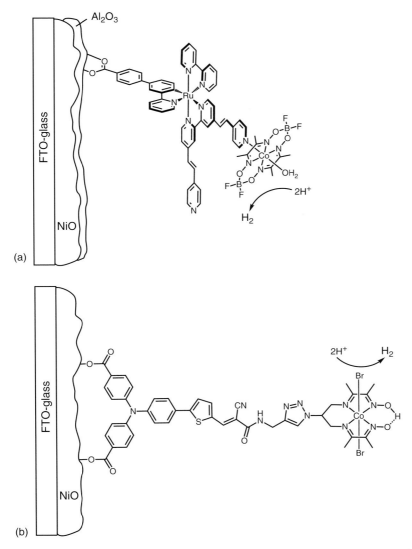

Figure 10.13 Structures of photocathodes based on supramolecular (a) [84] or covalent (b) [85] dye–catalyst assemblies. *Source*: (a) From Ji et al. [84]. Reproduced with permission of American Chemical Society. (b) From Kaeffer et al. [85]. Reproduced with permission of American Chemical Society.

finally in the phosphonic-acid-substituted catalyst solution. The best performing electrode was constructed with one layer of the phosphonic acid functionalised [Ru(bpy)$_3$]$^{2+}$ and two layers of Ni catalyst. It displayed a photocurrent of −9 µA cm$^{−2}$ at 0.3 V vs. RHE and a Faradaic efficiency for H$_2$ production of 10%. Post-photoelectrocatalysis characterisation of the electrode showed that no Ru was lost from the surface. The Ni catalyst was harder to detect due to the underlying NiO dominating XPS or inductively coupled plasma-optical emission spectrometry (ICP-OES) measurements. However, the CV of the electrode was

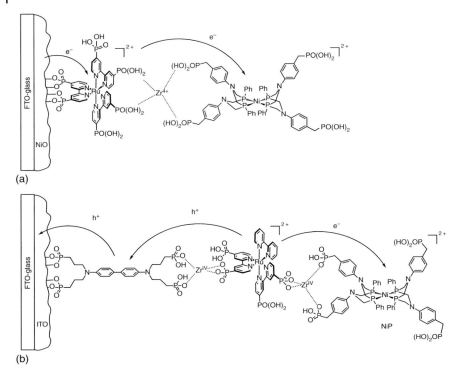

Figure 10.14 Structure of H_2-evolving dye-sensitised photocathodes based on layer-by-layer fabricated supramolecular dye–catalyst assemblies [75, 86]. *Source*: From Gross et al. [86]. Reproduced with permission of Royal Society of Chemistry.

unchanged after photoelectrolysis, suggesting little modification of the surface composition.

The following example is conceptually very similar to the previous one but relies on a conducting inverse opal nanostructured ITO electrode material (Figure 10.14b) [75]. Additionally, a phosphonic acid-substituted dianiline (DA) was inserted between NiO and the Ru photosensitiser to provide longer-lived charge separation, as confirmed with nanosecond transient absorption measurements: for the ITO-DA-Zr-RuP$_2^{2+}$ assembly in the absence of catalyst, a long-lived signal at 510 nm characteristic for the reduced dye is observed, in agreement with hole injection into ITO. For the assembly including NiP, this signal can still be observed but with a lower yield, indicative of electron transfer to the catalyst within the first 50 ns of the experiment. Under photoelectrocatalytic conditions, the photocathode produced H_2 with 53% Faradaic efficiency and with a quantum yield of 0.85%, in MES buffer pH 5.1 at an applied potential of −0.25 V vs. NHE.

10.2.3.3 Dye-sensitised Photocathodes Based on Co-grafted Dyes and Catalysts

Co-grafting dye and catalyst onto the electrode is an appealing alternative to covalent assembly of the two components, which often requires tedious synthetic steps. This strategy relies on electron hopping from the reduced dye to the catalyst on the surface for hydrogen production to occur. However, in sharp contrast with photoinduced intramolecular electron transfer processes within dyad assemblies, this process remained underexplored until the recent studies by the group of L. Hammarström. The first demonstration of electron hopping between molecules on NiO was reported in 2012 [87]. The dye used was Coumarin-343 due to its high reduction potential and fast hole injection into NiO. The selected catalyst was a derivative of a known Fe phosphine complex (**Fe1**), reported to produce H_2 with a TOF of 10 000 s^{-1} (Figure 10.14a). The radical anion of **Fe1** could be observed within the 50 ns time resolution of the equipment and the charge recombined with holes in NiO over 100 μs. Interestingly, these processes were unaffected by the dyeing order of the dye and catalyst. In the absence of **Fe1**, no signals were observed on the nanosecond timescale, in agreement with the previously reported femtosecond hole injection from dye to NiO followed by recombination on the picosecond timescale. To rationalise these data, it is postulated that electrons migrate across the surface via a sub-nanosecond self-exchange mechanism between dye molecules. Due to low mobility in NiO, holes are localised; this self-exchange mechanism may thus transport the electrons away from the holes to **Fe1**, explaining the nanosecond lifetime for the **Fe1** anion despite dye–NiO charge recombination occurring on the picosecond timescale.

Four years later, the photoinduced dynamics of **C343** on NiO was investigated with a different catalyst, a Fe carbonyl benzene dithiolate complex (**Fe2** in Figure 10.15a) [88]. For co-adsorbed **C343** and **Fe2**, hole injection is observed

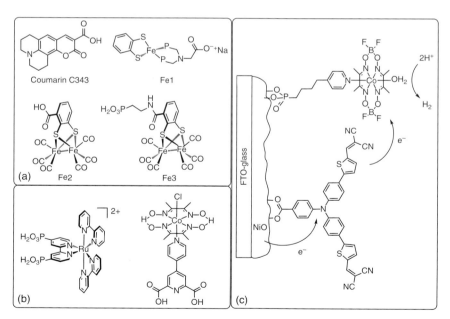

Figure 10.15 Dyes and catalysts employed for the construction of co-grafted H_2-evolving photocathodes. *Source:* (a) From Ref. [87–89]. (b) From Ref. [90]. (c) From Li et al. [92]. Reproduced with permission of American Chemical Society.

from **C343** to NiO on c. 200 fs timescale, as for **C343** alone. However, for **C343** alone, the reduced dye lifetime is hundreds of picoseconds, whereas with **Fe2** the reduced **C343** has a half-life of 10 ps. In this case, the neutral dye signal recovers and the reduced catalyst signature is observed and persists for up to 30 µs. The electron transfer from dye to catalyst is three orders of magnitude faster than 'hole-hopping' observed on TiO_2 [93, 94], and this may be due to the large driving force for electron transfer from dye to catalyst in this case. Unreactive **C343** anion cannot be observed, suggesting that there are no dye molecules unable to transfer electrons to **Fe2**, either directly or by rapid hopping from dye to dye.

Fe2 showed poor binding to NiO and to improve on this, a novel catalyst was prepared with a phosphonate anchoring group (**Fe3** in Figure 10.15a) [89]. Transient absorption spectroscopy of **C343** on NiO ($\lambda_{exc} = 440$ nm) shows hole injection in <150 fs, in agreement with the previous work, and charge recombination over 9 ps. A co-sensitised sample of **C343** and **Fe3** on NiO shows quite different kinetics. Hole injection is similar to **C343** alone, but the signal for the **C343** anion decays much faster and via several processes between 1 and 1000 ps. A signal corresponding to reduced **Fe3** can be observed from 1 ps and persists for >2 ns. The decay of the **C343** anion is assigned to the reduction of **Fe3**, and therefore, this reduction occurs via a combination of different processes. However, the reduction of **Fe3** mostly takes place between 1 and 10 ps and with an electron transfer yield of 70%. This electron transfer occurs more quickly and in greater yields than previously observed for **C343** with **Fe2**, and this may be related to the flexibility of the linking group. **C343** co-grafted with **Fe3** on NiO was measured on a longer timescale in order to elucidate the fate of the reduced **Fe3**. The signal assigned to reduced **Fe3** decays via multiple processes with time constants from ∼2 µs to >10 ms. The longer components, which outcompete the lifetime of reduced **Fe2**, are again attributed to the flexible linker in **Fe3**. Lifetimes of tens of milliseconds are important as protonation of the reduced **Fe3** is expected to occur on the millisecond timescale. Indeed, the system produced H_2 (acetate buffer pH 4.5, −0.3 V vs. Ag/AgCl) with a Faradaic efficiency of 50% and a photocurrent of −10 µA cm^{-2}. IR spectroscopy showed that deactivation is, at least in part, due to the loss of the CO ligands and desorption of **Fe3** from the surface, whereas the **C343** is quite stable.

A photoelectrode was assembled with a bis-phosphonate derivative of the well-known $[Ru(bpy)_3]^{2+}$ dye in combination with a cobaloxime catalyst possessing a pyridine dicarboxylate anchoring group (Figure 10.15b) [90]. Under a bias of −0.4 V vs. Ag/AgCl, a cathodic photocurrent of −13 µA cm^{-2} was observed. Hydrogen production was not quantified, but the photocathode was combined with a molecular photoanode under a two-electrode configuration. The tandem cell produced greater photocurrent than the photocathode alone but less than the photoanode alone, indicating that the photocathode activity is limiting the device performance.

Another example of a co-grafted photocathode was prepared with CdSe QDs as the light absorber combined with a cobaloxime catalyst [91]. The QDs were capped with thioglycolic acid, which provides a carboxylate function for grafting onto NiO, whereas the cobaloxime contained a phosphonate anchoring group. The electrode was constructed from an open porous NiO structure where nanoflakes were grown from a nickel nitrate precursor, in order to have a high surface area and promote high loadings of sensitiser and catalyst. Indeed, when co-grafted with QDs and cobaloxime, the photocurrent density was 1.7 times greater when compared with a classical doctor-bladed NiO preparation. The electrode displayed stable photocurrents up to 110 µA cm^{-2} at 0 V vs. NHE in neutral aqueous solution and produced hydrogen with 81% Faradaic efficiency. When a cobaloxime without a phosphonate anchoring group was used, the photocurrent rapidly decayed to 66 µA cm^{-2}, thus highlighting the benefit of co-grafting the catalyst. Other QDs/NiO-based photoelectrodes have been prepared with deposited MoS$_x$ catalyst [95] or molecular cobalt dithiolene catalyst in solution [96].

In 2015, a tandem water-splitting cell was reported based on molecular dyes and catalysts [92]. The photocathode consisted of the P1 dye and a cobaloxime catalyst, functionalised with an alkyl–phosphonate anchoring group, co-adsorbed onto NiO (Figure 10.15c). The photocathode displayed a photocurrent of −45 µA cm^{-2} on the timescale of the linear scan voltammetry (LSV) at applied potentials <−0.1 V vs NHE. This dropped to −20 µA cm^{-2} at longer times. Hydrogen production was confirmed with a Faradaic efficiency of 68%. The photocathode was integrated into a tandem cell with a photoanode composed of an organic dye and a Ru-based water oxidation catalyst co-adsorbed onto TiO$_2$. The tandem cell gave a Faradaic efficiency of 55% for H$_2$ without any applied potential and a solar to hydrogen efficiency of 0.05%. In one example of a molecular photocathode not using NiO, a [Ru(bpy)$_3$]$^{2+}$ derivative with pendant pyrrole groups was polymerised onto a carbon electrode, followed by electrodeposition of MoS$_x$ [97]. The electrode gave a photocurrent of −25 µA cm^{-2} with an applied bias of 0.1 V vs Ag/AgCl at pH 0.3. Under these conditions, H$_2$ was produced with 98% Faradaic efficiency. In terms of the quantity of MoS$_x$, a TON of 31 was obtained, equating to a TOF of 16 h^{-1}.

Proof of concept for dye-sensitised H$_2$-evolving photocathodes has been made in the last 5 years. However, low current density and poor durability remain as major challenges. A commercial device should perform efficiently for many months or years and yet most reports detail activity on the scale of hours.

10.3 Photocathodes for CO$_2$ Reduction Based on Molecular Catalysts

The reduction of CO$_2$ to obtain high-energy compounds has attracted much attention as one of the key technologies for addressing serious issues such as global warming and shortages in energy and carbon resources. Since the famous report from Lehn and coworkers about the highly selective reduction of CO$_2$ to CO using *fac*-Re(bpy)(CO)$_3$Cl as an electrocatalyst [98] as well as a photocatalyst [99], many types of molecular (photo)catalysts have been intensively developed [100]. For example, metal complexes of Re(I), Ru(II), Co(II), Ni(II), Mn(I) and Fe(II) and some pyridine derivatives have been proposed as potential catalysts for electrochemical and photochemical systems. In this section, some of the successful photoelectrochemical CO$_2$ reduction systems using molecular catalysts or molecular photocatalysts combined with electrodes are introduced.

10.3.1 Photocatalytic Systems Consisting of a Molecular Catalyst and a Semiconductor Photoelectrode

As already mentioned in the section about H_2 evolution, semiconductors with a narrow bandgap can absorb visible light, and it is expected that a combined system with molecular electrocatalysts will be useful even for selective CO_2 reduction (Figure 10.16).

In the initial stage of the studies, combinations of semiconductor photoelectrodes with solutions containing a molecular electrocatalyst were developed for CO_2 reduction. A current was obtained at potentials more positive than those of the theoretical redox potentials under irradiation of the semiconductor photoelectrodes. This indicated that the catalytic reduction reaction at the molecular catalyst progressed by utilisation of highly reducing conduction band electrons that were produced by photoexcitation of the semiconductor photoelectrodes. It was reported that macrocyclic Ni^{II} and Co^{II} complexes were found to be good electrocatalysts for CO_2 reduction to CO with p-Si (1.1 eV) [101, 102], p-GaAs (1.4 eV) [103] and p-GaP (2.2 eV) [104] photoelectrodes. Recently, Kubiak and coworkers examined in detail a photoelectrochemical system using p-Si as a semiconductor photoelectrode and $Re(bpy-Bu^t)(CO)_3Cl$ (bpy-Bu^t = 4,4′-di-*tert*-butyl-2,2′-bipyridine, Figure 10.17a) as a molecular catalyst for CO_2 reduction to CO [107]. Briefly, when the p-Si electrode was selectively photoexcited (λ_{ex} = 661 nm) in an Ar atmosphere in acetonitrile containing the Re complex and a supporting electrolyte, the potential of the reduction of the Re complex shifted in the positive direction by ~500 mV compared with that of the corresponding reaction using a platinum electrode. Under a CO_2 atmosphere, the system afforded CO with a Faraday efficiency of 97% ± 3%, and its quantum efficiency for light-to-chemical energy conversion reached 61%. An interesting feature of this system was that direct hydrogen production from the p-Si electrode surface could be controlled by the addition of water and variation of the concentration of the molecular catalyst, which afforded control of the CO/H_2 production ratio [108]. Fe [109] and Mn [110]

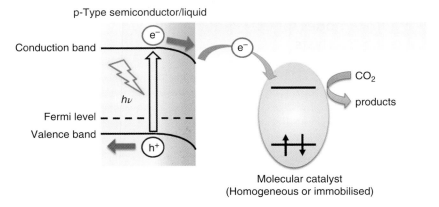

Figure 10.16 Schematic of CO_2 reduction using a combination of a molecular catalyst and a semiconductor photoelectrode.

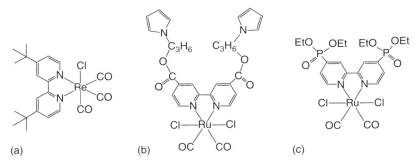

Figure 10.17 Structures of Re molecular catalyst (a) utilised in Ref. [104] and Ru catalysts with pyrrole (b) and phosphonate groups (c) used in Refs [105, 106]. *Source*: Schreier et al. [105]. https://www.nature.com/articles/ncomms8326. Licensed under CC BY 4.0.

complexes combined with silicon-based photoelectrodes were also successfully applied as molecular catalysts for CO_2 reduction.

Immobilisation of molecular electrocatalysts on semiconductor electrodes was examined with respect to efficient electron injection from the photoexcited electrode to the electrocatalyst. Grätzel and coworkers recently reported a CO_2 reduction system using polycrystalline Cu_2O in the form of a p-type semiconductor decorated with a Re(bpy)CO$_3$Cl molecular catalyst with a phosphonate anchoring unit (Figure 10.18) [111]. The photocathode was prepared by electrodeposition of Cu_2O followed by sequential modification with 20 nm of Al-doped ZnO (AZO) and 100 nm of TiO_2 by ALD [112]. Cu_2O is generally photoelectrochemically unstable; however, decomposition of the Cu_2O electrode was effectively suppressed by protecting the surface with a TiO_2 layer, whereas the AZO layer between these acted as an n-type buffer layer for constructing a p–n junction with the Cu_2O layer, thereby enhancing charge separation.

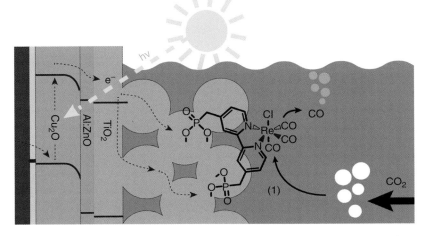

Figure 10.18 Protected Cu_2O photoelectrode with covalently bound Re(I) CO_2 reduction catalyst. *Source*: Schreier et al. [111]. Reprinted with permission of American Chemical Society.

Moreover, they synthesised a mesoporous scaffold with a thickness of 4.5–5 μm with 18 nm TiO_2 particles on the surface of the flat photoelectrode in order to support a sufficient amount of the Re catalyst. Photoelectrochemical measurements revealed that this photoelectrode generated substantial photocurrents exceeding 2.5 mA cm^{-2} under simulated sunlight in a CO_2-purged acetonitrile solution containing 0.1 M tetrabutylammonium hexafluorophosphate as an electrolyte. The mesoporous scaffold gave a 40-fold increase in current compared with that of the flat photocathode. Control experiments were performed in the absence of Cu_2O and light. Six hundred millivolt more negative potential was required to observe catalytic current when a mesoporous layer of TiO_2 on FTO was used as the electrode. This indicates that the photovoltage was produced by photoexcitation of the Cu_2O layer. Notably, both the photocurrent and obtained photovoltage showed an improvement over a previously reported system that used a TiO_2/AZO/Cu_2O photoelectrode in an acetonitrile solution containing a dissolved Re molecular catalyst (2.1 mA cm^{-2} and 560 mV) [105]; that is, the immobilisation of the molecular catalyst on the photoelectrode surface resulted in enhancement of the photoelectrochemical activity for CO_2 reduction. They suggested the following reasons for this: (i) electron transfer from the semiconductor electrode to the catalyst immobilised on the electrode surface should be more efficient compared to that in the combined system of the photocathode with the homogeneous Re catalyst in solution; (ii) immobilisation of the Re catalyst on the high surface area of the mesoporous TiO_2 scaffold could decrease light absorption by the catalyst, that is, the so-called *inner filter effect*, resulting in effective formation of photogenerated electrons on the photocathode. The system displayed >80% Faradaic efficiency for CO generation from CO_2 in the first hour, although the photocurrent gradually decreased. The authors noted the possibility of catalyst deactivation on the electrode, which was possibly caused by structural changes in the bipyridine ligand of the Re catalyst.

Aiming to couple CO_2 reduction with water oxidation, Morikawa, Sato and colleagues immobilised a Ru complex polymer ($[Ru(N^\wedge N)(CO)_2]^n$, RCP, $N^\wedge N$ = diimine ligand) as a catalyst layer on a zinc-doped indium phosphide photoelectrode (InP–Zn, $E_g = 1.35$ eV) [106]. This polymer was synthesised from the monomer shown in Figure 10.17b by photoelectrochemical polymerisation using the InP–Zn photoelectrode. Ru—Ru bonds were formed by a reductive reaction using electrons generated by photoexcitation of the InP–Zn photoelectrode, and then the pyrrole groups were oxidatively polymerised by anodic polarisation. This method was originally developed by Deronzier and coworkers [113]. The RCP-modified p-InP–Zn electrode generated a cathodic photocurrent under visible-light irradiation in a CO_2-purged aqueous solution, and the onset potential was 0.8 V more positive than that of the dark cathodic current. The formation of formic acid as a product of CO_2 reduction was confirmed using the RCP-modified InP–Zn photocathode at a potential of −0.6 V vs Ag/AgCl under visible light ($\lambda > 400$ nm), and the Faradaic efficiency was 62.3%. These results suggest that photoexcited electrons in the conduction band of InP–Zn transferred to the RCP, resulting in subsequent reduction of CO_2 by the reduced RCP. Notably, the photocathode with the RCP, which was prepared by cathodic polymerisation only, showed only 34.3% of the current efficiency for formate formation. The authors noted that the two-step polymerisation that formed the polypyrrole structure reinforced the contact between the RCP and p-InP–Zn, resulting in a higher Faradaic efficiency. They also pointed out the possibility of the formation of gaseous products, for example, CO and H_2, as the reason for the decrease in the Faradaic efficiency, whereas no detailed analysis of the gas phase component was performed. They subsequently reported a photocathode using Cu_2ZnSnS_4 (CZTS, E_g = 1.5 eV) as a p-type semiconductor, which consists of abundant and relatively cheap elements, with the same RCP catalyst [114]. CZTS modified with the RCP exhibited a much higher cathodic photocurrent than unmodified CZTS in CO_2-purged water, which suggests that the photoexcited electrons in CZTS were transferred to the RCP during the CO_2 reduction reaction. Furthermore, both a higher photocurrent and production of formic acid were obtained by adding an equimolar amount of a Ru complex with phosphonate groups as anchor (Figure 10.17c) in the polymerisation process on the photocathode. The authors suggested that stronger contact with the RCP on the photoelectrode could be obtained by chemisorption of the phosphonate anchors, and thus a higher activity for CO_2 reduction was achieved. A photo-electrode in which S was partially substituted by Se ($Cu_2ZnSn(S,Se)_4$ – CZTSSe) was combined with the RCP. The photocurrent and formic acid generation were improved due to higher hole conductivity obtained by Se substitution. In both systems using CZTS and CZTSSe-based photocathodes, the Faradaic efficiencies were ~80%.

Morikawa and coworkers successfully applied these photocathodes with an n-type semiconductor photoanode, which could use water as an electron source for the reduction of CO_2. The reaction proceeded via step-by-step excitation of both semiconductor electrodes, known as a Z-scheme-type electron transfer, using simulated solar light containing UV wavelengths. When the RCP-modified semiconductor photocathodes were used for CO_2 reduction, the valence bands of the semiconductors of the photocathodes were too 'shallow'; that is, they were more negative compared with the redox potential of water oxidation, and thus this reaction could not take place at the counter electrode without applying an external bias. To solve this problem, they utilised oxide-based n-type semiconductor electrodes (either TiO_2 or $SrTiO_3$) as the photoanode for water oxidation (Figure 10.19). First, a photoelectrochemical cell consisting of an RCP/InP–Zn photocathode and a Pt/TiO_2 photoanode was assembled [115]. The electrodes were partitioned by a Nafion film, and CO_2 and Ar were introduced into the cathodic and anodic sides, respectively. The cell was then irradiated with simulated sunlight (AM 1.5) from the anode side without bias. The TiO_2 absorbed only the UV light and the InP–Zn absorbed the remaining light transmitted through the photoanode. Formic acid was generated linearly from the cathodic side and, at the same time, oxygen was formed at the anodic side. Isotopic labelling experiments clearly indicated that formic acid and oxygen were derived from CO_2 and water, respectively. In this system, the conduction band edge potential of TiO_2 was 0.5 V more negative than the valence band edge potential of InP. Thus, photogenerated electrons in the conduction band of TiO_2 could move to the valence band of the excited InP through the outer circuit, resulting in Z-scheme-type

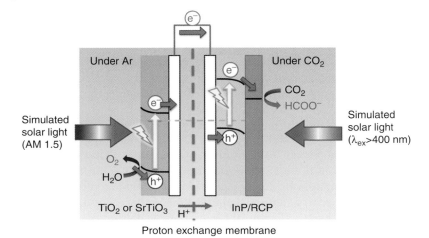

Figure 10.19 Schematic of the photoelectrochemical cell for tandem CO_2 reduction and water oxidation comprising a RCP/InP–Zn photocathode and metal-oxide photoanodes.

electron transfer. The efficiency of the conversion of sunlight energy to chemical energy was 0.03%.

The same group achieved much higher energy conversion efficiency by changing the photoanode from TiO_2 to $SrTiO_3$ [116]. As the bandgap of $SrTiO_3$ (3.2 V) is larger than that of TiO_2 (3.0 eV), shorter wavelengths of light would be absorbed. The $SrTiO_3$ photoanode generated an anodic photocurrent at a much more negative potential compared with that of the TiO_2 electrode, which indicated that the conduction band potential of $SrTiO_3$ was located at a more negative potential than that of TiO_2, and the driving force of electron transfer to the photocathode increased. The photocurrent generated from the photoelectrochemical cell using $SrTiO_3$ and RCP/InP–Zn was seven times higher than that from the cell using the TiO_2 photoanode, and the energy conversion efficiency reached 0.14%.

The same group also reported a monolithic stand-alone device composed of an amorphous Si/SiGe triple junction tandem photovoltaic cell and the RCP catalyst. This system was highly active for the reduction of CO_2 to HCOOH with 4.6% energy conversion efficiency, where the RCP catalyst was separated from the light absorber (the photovoltaic) with a carbon cloth substrate [117].

10.3.2 Dye-sensitised Photocathodes Based on Molecular Photocatalysts

The use of molecular photocatalysts immobilised on electrodes for CO_2 reduction is attractive because the molecular structure may be tuned in order to optimise redox properties. They can also display high catalytic efficiency and selectivity for CO_2 reduction. As is the case for the H_2-evolving photoelectrodes described in Section 10.2.2, molecular photocatalysts need to be fixed to a p-type semiconductor electrode (e.g. NiO) to drive CO_2 reduction with irradiation

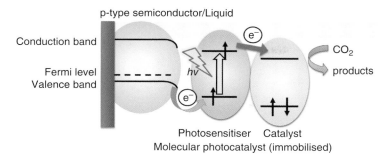

Figure 10.20 Schematic of CO_2 reduction using a molecular photocatalyst immobilised on a semiconductor electrode.

and using electrons injected from the valence band of the semiconductor (Figure 10.20).

In 2014, Inoue and colleagues reported a system in which a supramolecular photocatalyst composed of a zinc porphyrin sensitiser and a Re catalyst unit (Figure 10.21a) was immobilised on NiO with carboxylic groups as anchors [6]. This supramolecular photocatalyst strongly absorbs visible light, which is attributed to the Soret band of the Zn porphyrin at 430 nm. They first demonstrated the photocatalytic activity of this system under irradiation with visible light ($\lambda_{ex} = 430$ nm) in a DMF–triethylamine (TEA, 4 : 1, v/v) mixed solution. The TON of CO formation and the quantum yield were 14% and 0.12%, respectively. They then examined the immobilisation of the supramolecular complex on NiO in order to perform photoelectrochemical CO_2 reduction. The fluorescence lifetimes (τ) of both the Zn porphyrin mononuclear complex and the supramolecular complexes immobilised on NiO were shortened to about 30 ps from that in solution ($\tau < 2$ ns), which suggested efficient electron injection from NiO to the excited state of the Zn porphyrins. Nanosecond laser flash photolysis experiments showed the formation of a radical anion of the Zn porphyrins with a lifetime of 180 ns. CO_2 reduction was performed using the photoelectrode with the Zn porphyrin–Re supramolecular photocatalyst in CO_2-saturated DMF solution containing tetrabutylammonium hexafluorophosphate (0.1 M) as a supporting electrolyte under irradiation at $\lambda_{ex} = 430$ nm. This gave CO as a main product and a TON of 10 based on the immobilised complex, although the Faradaic efficiency was only 6.2%. An alternative photocathode, which was prepared by co-adsorption of the supramolecular complex with Zn porphyrin as an additional photosensitiser at a ratio of 24 : 1 (mononuclear sensitiser: supramolecular sensitiser), gave a TON of 122 based on the Re catalyst unit. This result was believed to imply that electrons are concentrated in the Re complex unit in the supramolecular photocatalyst by electron hopping between molecules.

Ishitani and colleagues employed a supramolecular photocatalyst, which consisted of a Ru(II) photosensitiser and a Re(I) catalyst with methylphosphonate groups as anchors (**RuRe**, Figure 10.21b), for photoelectrochemical CO_2 reduction by immobilisation on a NiO electrode [7]. This photocatalyst showed high activity for CO_2 reduction in an aqueous solution (quantum yield

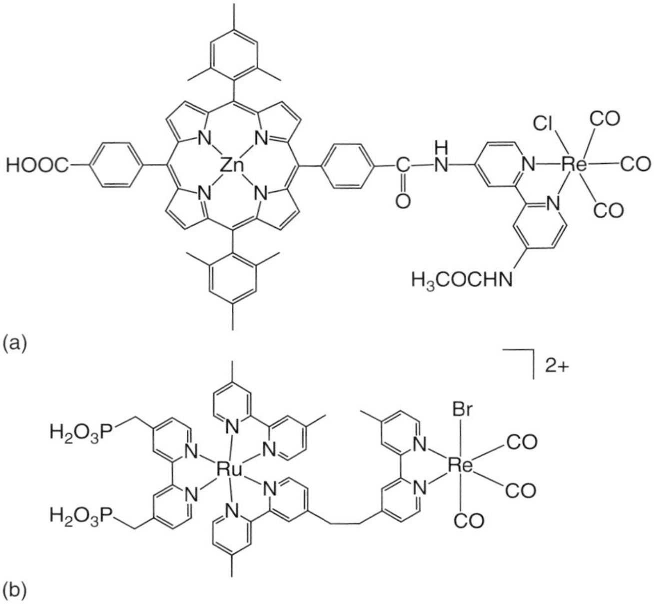

Figure 10.21 Structures of the supramolecular photocatalysts comprising a zinc porphyrin (a, [116]) or Ru(II) (b, [7, 117]) sensitiser and a Re(I) catalyst unit with carboxylic or methylphosphonate groups as anchors.

for CO formation = 13%, TON = 130) [118] as well as in organic solvent. As the valence band edge potential of NiO was more negative than the reduction potential of the excited state of the Ru photosensitiser unit, electron transfer from NiO to the light-absorbing Ru complex was expected to proceed. This process was clearly observed as a cathodic photocurrent when using a model mononuclear Ru complex on NiO in acetonitrile electrolyte (0.1 M Et$_4$NBF$_4$) under irradiation by visible light, which selectively excites the Ru complex. CO$_2$ reduction was examined using the photoelectrode modified with **RuRe** at an applied potential of −1.2 V vs. Ag/AgNO$_3$ in a DMF–TEOA (5 : 1, v/v) mixed solution. A cathodic photocurrent and corresponding selective CO generation were observed, and the TON of CO formation based on immobilised **RuRe** was 32 after 22 h irradiation. This photoelectrode was later successfully employed for CO$_2$ reduction in an aqueous solution [119]. In a CO$_2$-purged aqueous solution containing 50 mM NaHCO$_3$ (pH = 6.6), a photocathodic response was clearly observed under irradiation at λ_{ex} = 460 nm, where only **RuRe** was photoexcited. The photoelectrode was irradiated for 12 h with an applied bias of −0.7 V vs. Ag/AgCl, giving 361 nmol of CO (TON$_{CO}$ = 32 based on the amount of **RuRe** adsorbed on the NiO electrode). Although a small amount of H$_2$ was produced as a by-product, the total selectivity of CO formation was 91%. This photocathode was then successfully combined with a CoO$_x$/TaON photoanode, which was reported as an efficient photoanode for water oxidation under visible-light irradiation, in a functional photoelectrochemical cell (Figure 10.22). This cell produced 79 nmol of CO with a small amount of H$_2$ (6 nmol) and O$_2$ (77 nmol)

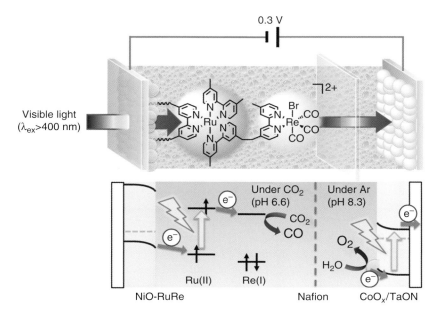

Figure 10.22 Schematic of hybrid photoelectrochemical cell comprising the NiO–**RuRe** photocathode and CoO$_x$/TaON photoanode [119].

from the photocathode and photoanode, respectively. The cell was operated under visible-light irradiation ($\lambda_{ex} > 400$ nm) with an applied electrical bias of 0.3 V and a chemical bias of 0.10 V originating from the pH difference between the electrolytes for each electrode. The light energy conversion efficiency of the reaction was 1.6×10^{-5}, considering the electrical and chemical bias. This is the first example of a visible-light-driven CO$_2$ reduction system using water as the reductant, based on molecular photocatalysts hybridised with semiconductor photocatalysts.

Acknowledgements

This work was supported by the French National Research Agency (Labex program, ARCANE, ANR-11-LABX-0003-01 and Joint JST-ANR project PhotoCAT, ANR 14 JTIC 0004 01).

References

1. Vesborg, P.C.K. and Jaramillo, T.F. (2012). *RSC Adv.* 2: 7933–7947.
2. Lewis, N.S. and Nocera, D.G. (2007). *Proc. Natl. Acad. Sci. U. S. A.* 104: 20142–20142.
3. Ashford, D.L., Gish, M.K., Vannucci, A.K. et al. (2015). *Chem. Rev.* 115: 13006–13049.

4 Yu, Z., Li, F., and Sun, L. (2015). *Energy Environ. Sci.* 8: 760–775.
5 Queyriaux, N., Kaeffer, N., Morozan, A. et al. (2015). *J. Photochem. Photobiol. C* 25: 90–105.
6 Kou, Y., Nakatani, S., Sunagawa, G. et al. (2014). *J. Catal.* 310: 57–66.
7 Sahara, G., Abe, R., Higashi, M. et al. (2015). *Chem. Commun.* 51: 10722–10725.
8 Ager, J.W., Shaner, M.R., Walczak, K.A. et al. (2015). *Energy Environ. Sci.* 8: 2811–2824.
9 Coutard, N., Kaeffer, N., and Artero, V. (2016). *Chem. Commun.* 52: 13728–13748.
10 Goetzberger, A., Hebling, C., and Schock, H.-W. (2003). *Mater.Sci. Eng. R Rep.* 40: 1–46.
11 Navarro, R.M., del Valle, F., Villoria de la Mano, J.A. et al. (2009). Photocatalytic water splitting under visible light: concept and catalysts development. In: *Advances in Chemical Engineering* (ed. I.d.L. Hugo and R. Benito Serrano), 111–143. Academic Press.
12 Nann, T., Ibrahim, S.K., Woi, P.-M. et al. (2010). *Angew. Chem. Int. Ed.* 49: 1574–1577.
13 Chandrasekaran, S., Nann, T., and Voelcker, N. (2016). *Nanomaterials* 6: 144.
14 Chandrasekaran, S., Macdonald, T.J., Mange, Y.J. et al. (2014). *J. Mater. Chem. A* 2: 9478–9481.
15 Chandrasekaran, S., Vijayakumar, S., Nann, T., and Voelcker, N.H. (2016). *Int. J. Hydrogen Energy* 41: 19915–19920.
16 Chandrasekaran, S., McInnes, S.J.P., Macdonald, T.J. et al. (2015). *RSC Adv.* 5: 85978–85982.
17 Chandrasekaran, S., Macdonald, T.J., Gerson, A.R. et al. (2015). *ACS Appl. Mater. Interfaces* 7: 17381–17387.
18 Hou, Y., Abrams, B.L., Vesborg, P.C.K. et al. (2012). *J. Photon. Energy* 2, 026001-026001-026001-026016.
19 Hou, Y.D., Abrams, B.L., Vesborg, P.C.K. et al. (2011). *Nat. Mater.* 10: 434–438.
20 Moore, G.F. and Sharp, I.D. (2013). *J. Phys. Chem. Lett.* 4: 568–572.
21 Le Goff, A., Artero, V., Jousselme, B. et al. (2009). *Science* 326: 1384–1387.
22 Huan, T.N., Jane, R.T., Benayad, A. et al. (2016). *Energy Environ. Sci.* 9: 940–947.
23 Seo, J., Pekarek, R.T., and Rose, M.J. (2015). *Chem. Commun.* 51: 13264–13267.
24 Khusnutdinova, D., Beiler, A.M., Wadsworth, B.L. et al. (2017). *Chem. Sci.* 8: 253–259.
25 Seger, B., Herbst, K., Pedersen, T. et al. (2014). *J. Electrochem. Soc.* 161: H722–H724.
26 Gu, J., Yan, Y., Young, J.L. et al. (2016). *Nat. Mater.* 15: 456–460.
27 Kim, H.J., Seo, J., and Rose, M.J. (2016). *ACS Appl. Mater. Interfaces* 8: 1061–1066.
28 Mueller-Westerhoff, U.T. and Nazzal, A. (1984). *J. Am. Chem. Soc.* 106: 5381–5382.

29 Cedeno, D., Krawicz, A., and Moore, G.F. (2015). *Interface Focus* 5: 20140085.
30 Krawicz, A., Yang, J., Anzenberg, E. et al. (2013). *J. Am. Chem. Soc.* 135: 11861–11868.
31 Krawicz, A., Cedeno, D., and Moore, G.F. (2014). *Phys. Chem. Chem. Phys.* 16: 15818–15824.
32 Beiler, A.M., Khusnutdinova, D., Jacob, S.I., and Moore, G.F. (2016). *Ind. Eng. Chem. Res.* 55: 5306–5314.
33 Beiler, A.M., Khusnutdinova, D., Jacob, S.I., and Moore, G.F. (2016). *ACS Appl. Mater. Interfaces* 8: 10038–10047.
34 Cedeno, D., Krawicz, A., Doak, P. et al. (2014). *J. Phys. Chem. Lett.* 5: 3222–3226.
35 Baffert, C., Artero, V., and Fontecave, M. (2007). *Inorg. Chem.* 46: 1817–1824.
36 Artero, V., Chavarot-Kerlidou, M., and Fontecave, M. (2011). *Angew. Chem. Int. Ed.* 50: 7238–7266.
37 Thompson, B.C. and Frechet, J.M.J. (2008). *Angew. Chem. Int. Ed.* 47: 58–77.
38 Green, M.A., Emery, K., Hishikawa, Y. et al. (2015). *Prog. Photovolt. Res. Appl.* 23: 1–9.
39 Sariciftci, N.S., Smilowitz, L., Heeger, A.J., and Wudl, F. (1992). *Science* 258: 1474–1476.
40 Chen, S.N., Heeger, A.J., Kiss, Z. et al. (1980). *Appl. Phys. Lett.* 36: 96–98.
41 Kaneko, M., Okuzumi, K., and Yamada, A. (1985). *J. Electroanal. Chem.* 183: 407–410.
42 Genies, E.M. and Lapkowski, M. (1988). *Synth. Met.* 24: 69–76.
43 Glenis, S., Tourillon, G., and Garnier, F. (1984). *Thin Solid Films* 122: 9–17.
44 El-Rashiedy, O.A. and Holdcroft, S. (1996). *J. Phys. Chem.* 100: 5481–5484.
45 Suppes, G., Ballard, E., and Holdcroft, S. (2013). *Polym. Chem.* 4: 5345–5350.
46 Li, C., Liu, M., Pschirer, N.G. et al. (2010). *Chem. Rev.* 110: 6817–6855.
47 Lanzarini, E., Antognazza, M.R., Biso, M. et al. (2012). *J. Phys. Chem. C* 116: 10944–10949.
48 Abe, T., Tobinai, S., Taira, N. et al. (2011). *J. Phys. Chem. C* 115: 7701–7705.
49 Haro, M., Solis, C., Blas-Ferrando, V.M. et al. (2016). *ChemSusChem* 9: 3062–3066.
50 Abe, T., Chiba, J., Ishidoya, M., and Nagai, K. (2012). *RSC Adv.* 2: 7992–7996.
51 Abe, T., Fukui, K., Kawai, Y. et al. (2016). *Chem. Commun.* 52: 7735–7737.
52 Abe, T., Nagai, K., Kabutomori, S. et al. (2006). *Angew. Chem. Int. Ed.* 45: 2778–2781.
53 Abe, T., Taira, N., Tanno, Y. et al. (2014). *Chem. Commun.* 50: 1950–1952.
54 Abe, T., Tanno, Y., Taira, N., and Nagai, K. (2015). *RSC Adv.* 5: 46325–46329.
55 Bourgeteau, T., Tondelier, D., Geffroy, B. et al. (2013). *Energy Environ. Sci.* 6: 2706–2713.
56 Tran, P.D., Tran, T.V., Orio, M. et al. (2016). *Nat. Mater.* 15: 640–646.
57 Haro, M., Solis, C., Molina, G. et al. (2015). *J. Phys. Chem. C* 119: 6488–6494.

58 Morozan, A., Bourgeteau, T., Tondelier, D. et al. (2016). *Nanotechnology* 27: 355401.
59 Bourgeteau, T., Tondelier, D., Geffroy, B. et al. (2015). *ACS Appl. Mater. Interfaces* 7: 16395–16403.
60 Bourgeteau, T., Tondelier, D., Geffroy, B. et al. (2016). *J. Mater. Chem. A* 4: 4831–4839.
61 Mezzetti, A., Fumagalli, F., Alfano, A. et al. (2017). *Faraday Discuss.* doi: 10.1039/C1036FD00216A.
62 Hagfeldt, A., Boschloo, G., Sun, L.C. et al. (2010). *Chem. Rev.* 110: 6595–6663.
63 Odobel, F. and Pellegrin, Y. (2013). *J. Phys. Chem. Lett.* 4: 2551–2564.
64 Dini, D., Halpin, Y., Vos, J.G., and Gibson, E.A. (2015). *Coord. Chem. Rev.* 304, 179–305, 201.
65 Wood, C.J., Summers, G.H., Clark, C.A. et al. (2016). *Phys. Chem. Chem. Phys.* 18: 10727–10738.
66 Al-Jawhari, H.A. (2015). *Mater. Sci. Semicond. Process.* 40: 241–252.
67 Powar, S., Xiong, D., Daeneke, T. et al. (2014). *J. Phys. Chem. C* 118: 16375–16379.
68 Yu, M., Draskovic, T.I., and Wu, Y. (2014). *Phys. Chem. Chem. Phys.* 16: 5026–5033.
69 Bandara, J. and Yasomanee, J.P. (2007). *Semicond. Sci. Technol.* 22: 20.
70 Srinivasan, R., Chavillon, B., Doussier-Brochard, C. et al. (2008). *J. Mater. Chem.* 18: 5647–5653.
71 Tian, H. (2015). *ChemSusChem* 8: 3746–3759.
72 Odobel, F., Pellegrin, Y., Gibson, E.A. et al. (2012). *Coord. Chem. Rev.* 256: 2414–2423.
73 Hamd, W., Chavarot-Kerlidou, M., Fize, J. et al. (2013). *J. Mater. Chem. A* 1: 8217–8225.
74 Huang, Z., He, M., Yu, M. et al. (2015). *Angew. Chem. Int. Ed.* 54: 6857–6861.
75 Shan, B., Das, A.K., Marquard, S. et al. (2016). *Energy Environ. Sci.* 9: 3693–3697.
76 Zhang, L. and Cole, J.M. (2015). *ACS Appl. Mater. Interfaces* 7: 3427–3455.
77 Gordon, R.B., Bertram, M., and Graedel, T.E. (2006). *Proc. Natl. Acad. Sci. U. S. A.* 103: 1209–1214.
78 Wang, M., Chen, L., and Sun, L. (2012). *Energy Environ. Sci.* 5: 6763–6778.
79 McKone, J.R., Marinescu, S.C., Brunschwig, B.S. et al. (2014). *Chem. Sci.* 5: 865–878.
80 Li, L., Duan, L.L., Wen, F.Y. et al. (2012). *Chem. Commun.* 48: 988–990.
81 Click, K.A., Beauchamp, D.R., Huang, Z. et al. (2016). *J. Am. Chem. Soc.* 138: 1174–1179.
82 Tong, L., Iwase, A., Nattestad, A. et al. (2012). *Energy Environ. Sci.* 5: 9472–9475.
83 Kamire, R.J., Majewski, M.B., Hoffeditz, W.L. et al. (2017). *Chem. Sci.* 8: 541–549.
84 Ji, Z., He, M., Huang, Z. et al. (2013). *J. Am. Chem. Soc.* 135: 11696–11699.

85 Kaeffer, N., Massin, J., Lebrun, C. et al. (2016). *J. Am. Chem. Soc.* 138: 12308–12311.
86 Gross, M.A., Creissen, C.E., Orchard, K.L., and Reisner, E. (2016). *Chem. Sci.* 7: 5537–5546.
87 Gardner, J.M., Beyler, M., Karnahl, M. et al. (2012). *J. Am. Chem. Soc.* 134: 19322–19325.
88 Brown, A.M., Antila, L.J., Mirmohades, M. et al. (2016). *J. Am. Chem. Soc.* 138: 8060–8063.
89 Antila, L.J., Ghamgosar, P., Maji, S. et al. (2016). *ACS Energy Lett.* 1: 1106–1111.
90 Fan, K., Li, F., Wang, L. et al. (2014). *Phys. Chem. Chem. Phys.* 16: 25234–25240.
91 Meng, P., Wang, M., Yang, Y. et al. (2015). *J. Mater. Chem. A.* 3: 18852–18859.
92 Li, F., Fan, K., Xu, B. et al. (2015). *J. Am. Chem. Soc.* 137: 9153–9159.
93 Hu, K., Robson, K.C.D., Beauvilliers, E.E. et al. (2014). *J. Am. Chem. Soc.* 136: 1034–1046.
94 Brennan, B.J., Durrell, A.C., Koepf, M. et al. (2015). *Phys. Chem. Chem. Phys.* 17: 12728–12734.
95 Dong, Y., Chen, Y., Jiang, P. et al. (2015). *Chem. Asian. J.* 1660–1667.
96 Ruberu, T.P.A., Dong, Y., Das, A., and Eisenberg, R. (2015). *ACS Catal.* 5: 2255–2259.
97 Lattach, Y., Fortage, J., Deronzier, A., and Moutet, J.-C. (2015). *ACS Appl. Mater. Interfaces* 7: 4476–4480.
98 Hawecker, J., Lehn, J.-M., and Ziessel, R. (1984). *J. Chem. Soc. Chem. Commun.* 328–330.
99 Hawecker, J., Lehn, J.-M., and Ziessel, R. (1986). *Helv. Chim. Acta* 69: 1990–2012.
100 Yamazaki, Y., Takeda, H., and Ishitani, O. (2015). *J. Photochem. Photobiol. C Photochem. Rev.* 25: 106–137.
101 Bradley, M.G. and Tysak, T. (1982). *J. Electroanal. Chem. Interfacial Electrochem.* 135: 153–157.
102 Bradley, M.G., Tysak, T., Graves, D.J., and Viachiopoulos, N.A. (1983). *J. Chem. Soc. Chem. Commun.* 349–350.
103 Beley, M., Collin, J.-P., Sauvage, J.-P. et al. (1986). *J. Electroanal. Chem. Interfacial Electrochem.* 206: 333–339.
104 Chartier, P., Beley, M., Sauvage, J.P., and Petit, J.P. (1987). *New J. Chem.* 11: 751–752.
105 Schreier, M., Curvat, L., Giordano, F. et al. (2015). *Nat. Commun.* 6: 7326.
106 Arai, T., Sato, S., Uemura, K. et al. (2010). *Chem. Commun.* 46: 6944–6946.
107 Kumar, B., Smieja, J.M., and Kubiak, C.P. (2010). *J. Phys. Chem. C* 114: 14220–14223.
108 Kumar, B., Smieja, J.M., Sasayama, A.F., and Kubiak, C.P. (2012). *Chem. Commun.* 48: 272–274.
109 Alenezi, K., Ibrahim, S.K., Li, P., and Pickett, C.J. (2013). *Chem. Eur. J.* 19: 13522–13527.

110 Torralba-Peñalver, E., Luo, Y., Compain, J.-D. et al. (2015). *ACS Catal.* 5: 6138–6147.
111 Schreier, M., Luo, J., Gao, P. et al. (2016). *J. Am. Chem. Soc.* 138: 1938–1946.
112 Paracchino, A., Laporte, V., Sivula, K. et al. (2011). *Nat. Mater.* 10: 456–461.
113 Chardon-Noblat, S., Deronzier, A., Ziessel, R., and Zsoldos, D. (1998). *J. Electroanal. Chem.* 444: 253–260.
114 Arai, T., Tajima, S., Sato, S. et al. (2011). *Chem. Commun.* 47: 12664–12666.
115 Sato, S., Arai, T., Morikawa, T. et al. (2011). *J. Am. Chem. Soc.* 133: 15240–15243.
116 Arai, T., Sato, S., Kajino, T., and Morikawa, T. (2013). *Energy Environ. Sci.* 6: 1274–1282.
117 Arai, T., Sato, S., and Morikawa, T. (2015). *Energy Environ. Sci.* 8: 1998–2002.
118 Nakada, A., Koike, K., Nakashima, T. et al. (2015). *Inorg. Chem.* 54: 1800–1807.
119 Sahara, G., Kumagai, H., Maeda, K. et al. (2016). *J. Am. Chem. Soc.* 138: 14152–14158.

11

Molecular Design of Glucose Biofuel Cell Electrodes

Michael Holzinger[1,], Yuta Nishina[2,*], Alan Le Goff[1], Masato Tominaga[3], Serge Cosnier[1], and Seiya Tsujimura[4]*

[1] *University of Grenoble Alpes – CNRS, Department of Molecular Chemistry (DCM, UMR 5250), rue de la Chimie, F 38000, Grenoble, France*
[2] *Okayama University, Graduate School of Natural Science and Technology, Research Core for Interdisciplinary Sciences, Tsushimanaka, Kita-ku, Okayama-shi, Okayama, 700-8530, Japan*
[3] *Saga University, Graduate School of Science and Engineering, Department of Chemistry and Applied Chemistry, Honjyo-machi, Saga-shi, Saga, 840-8502, Japan*
[4] *University of Tsukuba, Division of Materials Science, Faculty of Pure and Applied Sciences, Tennodai, Tsukuba, Ibaraki, 305-8573, Japan*

11.1 Introduction

The first example of an enzymatic biofuel cell (EBFC) using enzymes as catalysts was proposed in 1964 [1]. This concept just slowly evolved over two decades until new achievements in enzyme wiring and new nanotechnological approaches, initially developed for biosensing, led to an impressively growing interest in biological energy production in the year 2000 [2].

The principle of EBFCs is similar to that of classic fuel cells, which is based on a catalytic fuel-oxidizing anode and a catalytic oxidizer-reducing cathode. The difference lies in the nature of the catalysts used, which are of biological origin in the case of biofuel cells [3], contrary to abiotic fuel cells where principally noble-metal-based catalysts or alloys are used. Compared with conventional fuel cells, EBFCs are safe due to the enzyme reactions that can operate under mild conditions such as room temperature, atmospheric pressure, and neutral pH. Additionally, EBFCs can be used for several biologically related reductants (fuels) as electron donors such as sugars, alcohols, amines, organic acids, and hydrogen at the anode side. On the cathode side, O_2 or H_2O_2 is mostly used as an electron acceptor. Such advantages of EBFCs lead to a large variety of potential applications. One promising application of glucose EBFCs is the power supply of implanted medical devices such as pacemakers, sensors, or actuators as actually glucose is the "fuel" and oxygen is the oxidizer in living organisms. There are many reviews about implantable power generators using biological and abiotic catalysts [4–8]. In the highly complex media of body fluids, abiotic catalysts have

---

* Contact information to whom correspondence should be addressed: Michael Holzinger (michael.holzinger@univ-grenoble-alpes.fr) and Yuta Nishina (nisina-y@cc.okayama-u.ac.jp)

Molecular Technology: Energy Innovation, Volume 1, First Edition.
Edited by Hisashi Yamamoto and Takashi Kato.
© 2018 Wiley-VCH Verlag GmbH & Co. KGaA. Published 2018 by Wiley-VCH Verlag GmbH & Co. KGaA.

the disadvantage of insufficient selectivity toward glucose oxidation and oxygen reduction and its inhibition by various compounds, and these catalysts generally show low efficiency at neutral pH. In contrast, such conditions are ideal for optimal operation of enzymes, but the drawback here is the insufficient lifetime of these catalysts. Up to date, no competitive setup exists that can replace the currently used lithium batteries.

Another promising sector for glucose fuel cell application is the power supply of low-power-consuming portable devices. Taking into account the limited time of operational (bio)catalysis, one-time-use devices such as sensors can be focused [9, 10].

Despite the remaining issues keeping this research field at the academic level, tremendous progresses were achieved in terms of power output that increased by factor 1000 from microwatts to milliwatts for individual glucose biofuel cells [11–13]. The challenges of enzyme wiring are now better understood leading to these improvements. In fact, dependent on the nature and catalytic center of each enzyme used either for the electrocatalytic oxidation of glucose or the electrocatalytic reduction of oxygen, the concept of efficient electron transfers from or to the electrode has to be adjusted. There are two common ways to transfer electrons to or from the catalytic centers of enzymes. The first and generally envisioned type is the direct regeneration of the enzyme by the electrode. This so-called direct electron transfer (DET) allows achieving optimal cell voltages and transfer rate kinetics but strongly depends on the position of the redox center in or on the enzyme.

DET reactions between proteins (enzymes) and electrodes have been extensively studied from the viewpoints of both understanding the fundamental features and for applications as biosensors and biofuel cells. The first DET reactions of proteins were achieved by Eddowes and Hill [14], Yeh and Kuwana [15], and Niki et al. [16], in which the reversible cyclic voltammetric responses of cytochrome c on gold electrode coated with 4,4'-bipyridyl and tin-doped indium oxide electrode, and cytochrome c_3 on mercury electrode, were reported. Before the achievement of the first DET reactions, it was impossible to observe a voltammetric response based on DET reaction of a protein because of extremely slow DET reaction of a protein even though an electron transfer protein in respiratory chain in mitochondrion was used. During the period of late 1970s to early 1980s, key concept for successful DET was based on specific electrode surface structures [17]. The electrode surface was functionalized to inhibit adsorptive surface denaturation of proteins and adsorption of passivating impurities. Additionally, the functionalized electrode may control other factors such as orientation.

Among functional electrodes, so-called promoter-modified electrodes were very convenient to use within several years [18, 19]. Promoters or electron-transfer promoters stand for functions that accelerate DET kinetics but are themselves electrochemically inactive at the potentials of interest [17]. Especially chemisorption of organothiols on gold or silver is a powerful approach to functionalize electrode surfaces. Using bis(4-pyridyl)disulfide-modified gold as a promoter-functionalized electrode, the first "promoted" DET reactions of cytochrome c were performed by Taniguchi et al. [20]. Such chemisorption functionalization of organothiols on gold or silver has been extensively developed using SAM (self-assembled monolayer) techniques. During late 1980s to early 2000s, the surface structure of SAM-modified single-crystal gold with atomically flat surface has been well analyzed and understood at molecular scale using, for example, electrochemical reductive desorption method [21], *in situ* surface-enhanced IR adsorption spectroscopy [22], X-ray photoelectron spectroscopy [23], or *in situ* electrochemical scanning tunneling microscopy [24]. At this stage, the understanding of the design of the effective electrode surface structure for fast DET rates with enzymes resulted from comprehensive investigations at molecular scale [19, 25, 26].

Lipid-membrane-modified electrodes are also efficient examples for promoted DET with enzymes. In this regard, lipids are not only biological surfactants that are usually components of biomembranes but also artificial surfactants. A lipid membrane structure can be formed artificially on solid surfaces. Enzyme can be adsorbed on the lipid membrane surface or be embedded within, which has a stabilizing effect for enzymes [27]. A typical achievement for promoted DET using lipid-membrane-modified electrodes was demonstrated in the case of cytochrome *c* oxidase. Cytochrome *c* oxidase (or complex IV) has a huge molecular weight (204 kDa) and catalyzes the final step in the mitochondrial electron transfer chain and is regarded as one of the major regulation sites for oxidative phosphorylation. Immobilized cytochrome *c* oxidase on lipid- membrane-modified electrodes showed promoted DET with reduced cytochrome *c*. With this setup, the final step of the model in the mitochondrion could be artificially reproduced [28].

These approaches were developed for the DET reactions of redox proteins at the early stage, but these achievements remained at a fundamental level.

Redox enzymes are generally large molecules of 20–800 kDa in mass. The average hydrodynamic diameters are 50 to several hundred Ångstrom [29]. Therefore, DET is difficult to achieve, as in many cases, the redox center of enzymes is deeply buried within the protein shell. The electron transfer rate is exponentially dependent on the distance between the redox-active centers as predicted by the Marcus theory [30]. Therefore, in order to shorten the electron transfer distance between the active center of the enzyme and the electrode, surface functionalization of an electrode to control the orientation might be an essential or suitable approach.

When no DET can be achieved, redox molecules with appropriate redox kinetics and diffusion capacities to shuttle the electrons via mediated electron transfer (MET) are necessary. A mediator can be organic molecules or metal complexes, which may have access to a redox-active center located deeply inside the protein molecule. MET is still a powerful method and is used in biosensors employing enzymes that do not satisfyingly exchange electrons with the electrodes. The cornerstone of commercial glucose biosensors is based on MET. In general, MET-based EBFCs show higher electric power density in comparison with DET-based EBFCs even when a reduced cell voltage has to be accepted [31]. This reduced cell voltage accounts for the electrocatalytic reaction occurring at the redox potential of the redox mediator that requires additional overpotential in order to provide a sufficient driving force for the establishment of a fast MET. The principles of these electron transfer types are depicted in Figure 11.1.

A further advantage of MET is the possibility to establish electronic communication in solution, whereas DET is a purely surface-dependent phenomenon.

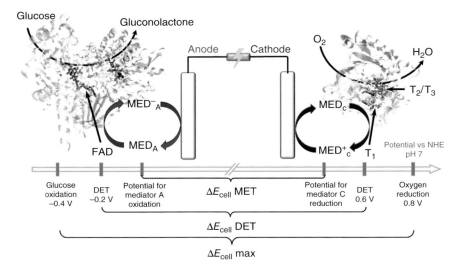

Figure 11.1 Schematic presentation of the influence of the electron transfer type to the cell voltage of EBFCs related to the standard redox potential of glucose and oxygen vs NHE. FAD-dependent glucose-oxidizing enzymes and oxygen-reducing multicopper enzymes are chosen as examples.

In this context, highly porous conductive nanostructures became the material of choice for efficient wiring of enzymes. Carbon nanotubes (CNTs) have very favored properties as electrode material for electron transfer with enzymes due to their nanowire structures enabling naturally close contact to the redox centers of some enzymes [12]. CNTs can further be shaped to give freestanding electrodes, which facilitates the integration on electronic devices [32]. Another promising approach is to use nanostructured carbon materials with controlled pores as electrodes. Mesoporous carbons with narrow pore size distributions could be tailored to the target enzymes while providing high specific surface area. These carbons were investigated as enzyme supports with an increase in the total amount of electrochemically active enzymes. The enzymes could be stabilized by encapsulating them in the pores of the support, thus preventing their loss from the support and their aggregation, or degradation of their molecular structure [33]. The enzyme-support interactions can be adjusted by the pore characteristics, including the pore structure and morphology, and by surface chemical characteristics, such as hydrophobic/hydrophilic interactions, electrostatic interactions, and hydrogen bonding. Among these factors, the pore size is an important parameter affecting enzyme immobilization [34].

Regarding the constant improvements in glucose biofuel cell performances, molecular technology became an important tool not only in terms of mediator design but also for oriented immobilization of enzymes, thus enabling ET. The different principles using molecular technology for improved performances of biocatalytic anodes and cathodes in glucose biofuel cell setups are described and illustrated with relevant examples.

11.2 Molecular Approaches for Enzymatic Electrocatalytic Oxidation of Glucose

The performance of EBFCs can be limited by mass transport of fuels, catalytic activity, or electron transfer rate. If the electron transfer is the rate-limiting process, molecular design of mediators can improve the performance of the cell and is of particular interest for bioanodes. The choice of the mediators can increase electron transfer rate; therefore, MET is generally faster by several orders of magnitude than DET. In contrast, the resulting cell potential (ΔE_{cell}) is reduced by the electron-mediating process (Figure 11.1). This implies that redox potential of a mediator is essentially important to obtain high power densities. Furthermore, the aim of the design and conception of glucose-oxidizing bioanodes is the fast and repeatable regeneration (oxidation) of the enzyme by the electrode material, which is often in competition with its natural process.

In the case of glucose oxidase (GOx), the enzyme is regenerated by oxygen in a two-electron reduction producing hydrogen peroxide [35]. GOx is a flavoprotein that catalyzes oxidation of β-D-glucose utilizing molecular oxygen as the electron acceptor. GOx is a homodimeric enzyme, with a noncovalently but tightly bound FAD molecule at the active site. This catalytic center is surrounded by a vast glycosylated protein shell (total mass ~160 kDa), which is responsible for its outstanding stabilities and activity over several pH values. GOx resists furthermore to chemical functionalization that even tagged enzymes could be commercialized. The disadvantage of such protected catalytic sites is that DET with electrodes is very hard to achieve as the tunneling distance is too high [36]. Several approaches were described to achieve DET with GOx. One strategy consisted in the deglycosylation of the enzyme, which presumably decreases the distance between the active site and the electrode surface [37]. Another biotechnological pathway was the production of FAD less GOx. This nonactive apo-enzyme could be successfully reactivated and wired on FAD-functionalized nanostructured conductors. Xiao et al. first proposed in 2003 this approach by the reconstitution of an apo-GOx using FAD-modified gold nanoparticles. The thus reactivated enzyme showed a sevenfold improved electron transfer turnover rate compared with the natural electron transfer rate to oxygen [38]. One year later, Patolsky et al. attached FAD to CNTs enabling the wiring of apo-GOx and showed a sixfold improved electron transfer turnover rate [39]. The challenge to obtain DET with native commercially available GOx was successfully overcome by Zebda et al. [40]. Bioanodes were obtained by the compression of a CNT/GOx mixture. However, the low yield of wired enzymes had to be compensated by their quantity and the enzyme catalase had to be added to reduce the high amount of produced hydrogen peroxide by unwired GOx.

MET might be a more appropriate way to wire GOx even when a related reduction of the cell voltage has to be taken into account. The Heller group first described a MET with GOx using an osmium-based redox polymer [41]. As the redox potential of osmium can be adjusted by the ligand, more appropriate osmium redox hydrogels were synthesized targeting glucose biofuel cell applications [42–44]. These hydrogels were successfully combined with CNT microfibers [45] or CNT-based textiles [46], reaching power output of 2 mW cm^{-2}.

Besides metal complexes, pure organic redox molecules showed to be promising alternatives due to clearly enhanced stabilities of the biocatalytic electrodes. Reuillard et al. used naphthoquinone in GOx/CNT pellets and obtained a 14-fold increase of catalytic current compared with the DET setup mentioned earlier [31] while accepting an increase in the open-circuit potential and consequently a cell voltage loss of ∼200 mV of the constructed glucose biofuel cell. The most common strategy to wire glucose-oxidizing enzymes with quinones or Os hydrogels is depicted in Figure 11.2.

Quinones are becoming more popular as mediators in bioanodes. Therefore, quinones are excellent mediators for other glucose-oxidizing enzymes such as glucose dehydrogenases (GDH) as the redox potential of quinones can be tailored to suit the desirable value by changing the substituents [47, 48]. In addition, heteroatoms or π-conjugated systems connected to quinone frameworks can enhance the interaction with carbon-based electrode surfaces [47]. The advantage of GDHs over GOx is that no oxygen is reduced to hydrogen peroxide, which is generally considered a negative effect. The regeneration of these enzymes is assured by mediators, which can especially be designed for biofuel cell applications [49]. There are three main classes of GDHs that are related to their respective cofactors. These cofactors are pyrroloquinoline quinone (PQQ), nicotine adenine dinucleotide (NAD) with or without phosphate (P), and flavin–adenine–dinucleotide (FAD).

The soluble group of pyrroloquinoline quinone–glucose dehydrogenase (PQQ-GDH) is composed of two identic subunits each with strongly bound PQQ and three calcium ions [50]. Tanne et al. functionalized CNT layers on gold electrodes with PQQ and coupled covalently a reconstituted apo-GDH. The authors could thus achieve MET via the intermittent PQQ layer [51]. Few years ago, the same research team realized DET with this reconstituted apo-GDH by using poly(3-aminobenzoic acid-co-2-methoxyaniline-5-sulfonic acid) – PABMSA, a sulfonated polyaniline that seems to form favored interactions with PQQ-GDH for DET [52, 53]. DET could also be achieved via simple amide coupling using 1-pyrenebutanoic acid succinimidyl ester with PQQ-GDH π-stacked on a CNT buckypaper [54]. The authors explained this electron transfer phenomenon by the porous 3D structure of buckypaper where the distance of embedded enzymes to the CNT matrix is on average short enough for DET. This phenomenon was also observed by Ivnitski et al. [55]. An operational glucose biofuel cell based on bioelectrocatalytic buckypaper electrodes could successfully be inserted in a snail [54], in clams [56], onto exposed rat cremaster tissue [57], and in lobsters [58].

NAD-dependent GDH is composed of four identic subunits where the cofactor NAD is not confined in the protein structure [59]. In order to achieve electron transfer from the enzyme, this cofactor has to be included together with a catalyst that re-oxidizes the generated NADH and a redox mediator. Even when many compounds are necessary for this setup, the overpotential of NAD for glucose oxidation seems to be interesting enough to face the issues of multimolecular

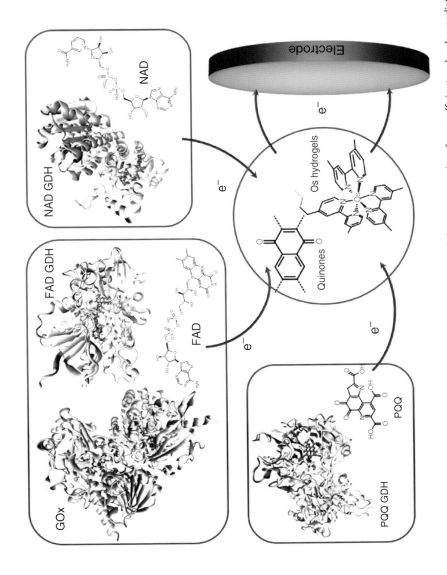

Figure 11.2 Illustrations of the different glucose-oxidizing enzymes, their cofactors, and the two categories of most efficient molecular mediators.

engineering. Furthermore, there is a wide series of NAD-dependent enzymes that can be used as anodic catalyst of the oxidation of biofuels such as glucose [60], alcohols [61], and L-lactate [62]. Therefore, once the optimal strategy is determined for the regeneration of NAD, it can serve for many different biofuel cell applications. Lalaoui et al. studied different setups for the electrocatalytic oxidation of NADH using multiwalled carbon nanotube (MWCNT) defects, immobilized diaphorases, and immobilized ruthenium complexes [63] where the ruthenium-based compound provided the best compromise between current density and overpotential. Another original example for a molecular approach to oxidize NADH was proposed by Giroud et al. They used a dithiobis (nitrobenzoic acid)-modified pyrene derivative for this purpose and used this molecule at the same time for oriented immobilization of multicopper enzymes at the cathode [64]. However, for these and other examples [65], NAD has to be dissolved in solution to obtain the reported performances, which is quite inconvenient for practical applications. To circumvent this issue, Sakai et al. proposed subsequent deposition of poly-L-lysine as the cationic ground layer followed by the enzyme NAD-GDH, NAD, diaphorase for efficient NADH oxidation, the mediator 2-methyl-1,4-naphthoquinone, and finally polyacrylic acid to form a polyion complex with the ground layer tightly fixing the different components in between [66]. With this design, a power density of 1.4 ± 0.24 mW cm^{-2} at 0.3 V with an open-circuit voltage of 0.8 V could be obtained. Based on this principle, an original glucose biofuel cell design was proposed combining two fuel cells that were connected in parallel and could provide doubled power outputs. Another strategy involved the immobilization of NADH on CNT-based electrodes using modified pyrene derivatives [67] or polymers [68].

The complexity to incorporate cofactors in a biofuel cell design guided the research to another, here, FAD-dependent GDH. The active site for glucose oxidation of this enzyme is similar to that of GOx, but the enzyme is independent of oxygen and therefore needs artificial electron mediators [69]. Zafar et al. tested several FAD-dependent GDH for glucose biofuel cell applications [60]. FAD GHDs from *Glomerella cingulate*, a recombinant form expressed in *Pichia pastoris*, and the commercially available glycosylated enzyme from *Aspergillus* sp. were wired with the osmium redox polymer [Os(4,4'-dimethyl-2,2'-bipyridine)$_2$ (PVI)$_{10}$ Cl]$^+$ on graphite electrodes. All of them showed excellent performances for electrocatalytic oxidation of glucose in terms of current density, selectivity, and turnover. Nonetheless, the deglycosylated form of FAD-GDH from *G. cingulate* provided higher catalytic currents due to the reduced size of the enzyme, which enabled higher densities of immobilized enzymes. FAD-GDH from *Aspergillus terreus* was co-immobilized with a PVI-Os(2,2-bipyridine)$_2$Cl derivative on a glassy carbon (GC) electrodes. The steady-state catalytic current for glucose oxidation was 2.6 mA cm^{-2} at pH 7 and 25 °C. This value increased 1.6-fold after oxidative deglycosylation of the enzyme [70]. MgO-templated porous carbon electrode further coated with the deglycosylated FAD-GDH and PVI-Os(bipyridine)$_2$Cl showed a 33-fold increase in glucose oxidation current density (ca. 100 mA cm^{-2} at pH 7 phosphate buffer containing 0.5 M glucose, at 25 °C) compared with that of the flat electrode [71]. FAD-GDH-hydrogel-modified porous carbon electrode showed exceptionally long-term stability by caging effect of porous carbon scaffold and shrinking effect by increasing the phosphate buffer concentration [72]. Hexacyanoferrate ($[Fe(CN)_6]^{3-}$; ferricyanide) was used as a redox mediator for commercially available FAD-GDH-based blood glucose sensor strips because of its high solubility in water, low cost, and high stability. The reactivity between FAD-GDH and ferricyanide is quite low, leading to a bimolecular rate constant to be as low as $10^3 \, M^{-1} \, s^{-1}$ in phosphate buffer (pH 7.0) at room temperature [73]. The rate constant for FAD-GDH toward quinones and organic redox dyes, such as phenothiazines, was approximately 2.5 orders of magnitude higher than that for GOx [74]. The difference suggests that the electron transfer kinetics is determined by the potential difference (the driving force of electron transfer), as well as the electron transfer distance between the redox-active site of the mediator and the FAD, affected by steric or chemical interactions. Naphthoquinone-based hydrogels have also been successfully used for FAD-GDH entrapment and MET wiring, reaching catalytic glucose oxidation currents of $2 \, mA \, cm^{-2}$, accompanied with onset potentials of $-0.13 \, V$ vs Ag/AgCl [48].

11.3 Molecular Designs for Enhanced Electron Transfers with Oxygen-Reducing Enzymes

Most enzymes used for the bioelectrocatalytic reduction of oxygen into water are from the multicopper enzyme family where laccases and bilirubin oxidases (BOD) can be considered enzymes of choice. They are composed of two distinct redox centers where a trinuclear 2 and 3 type (T_2/T_3) center reduces oxygen to water in a four-electron process and a mononuclear 1 type (T_1) center usually oxidizes its natural substrate (in general phenolic compounds) in a one-electron process and supplies the T_2/T_3 center with the harvested electrons [75]. These multicopper enzymes are generally smaller than the ones for glucose oxidation and the electron transfer is more evident to achieve as the active sites are more accessible.

Several strategies were proposed to wire laccases via MET and DET [76]. Os-based redox polymers are again famous examples for efficient electron shuttling from the electrode to the enzyme [77], but the molecule 2,2'-azino-bis(3-ethylbenzothiazoline-6-sulphonic acid) (ABTS) became the mediator for laccases of choice due to its appropriate redox potential, electron transfer rates, and stability [78]. Furthermore, ABTS retains entirely these beneficial properties after chemical modifications for its immobilization [79]. By modifying ABTS with two pyrene groups, this mediator could also act as a crosslinking agent and reinforce buckypapers leading to freestanding redox-active electrodes [80]. Recently, an original approach has involved the design of chimeric protein based on a prion domain and a rubredoxin domain. These proteins are able to form self-assembled amyloid nanofibers [81]. Thanks to the rubredoxin domain, these protein nanofibers were not only able to entrap enzymes but were also able to trigger MET with entrapped laccases. This type of bioassembly represents a promising alternative in the design of versatile "all-protein" bioelectrodes.

An original alternative to "all immobilized" component setup was proposed with these mediator-functionalized buckypapers as higher current densities were obtained when laccase was dissolved in solution. With this configuration, the biological catalysts can easily be replaced when the catalytic activity decreases [82].

Besides the highly efficient approaches for MET for laccases, the fact that the active centers are close to the protein surface motivates to achieve the generally preferred DET. Furthermore, laccase from *Trametes versicolor* is particularly suitable for optimized DET wiring as the protein structure provides a hydrophobic domain close to the T_1 center. The F. A. Armstrong group first discovered the possibility of oriented immobilization of this laccase via hydrophobic interactions on anthracene-modified surfaces, thus enabling DET for the electrocatalytic reduction of oxygen [83]. Several examples followed reporting the immobilization, orientation, and wiring of laccase using polyaromatic hydrocarbons such as anthracene [84], naphthalene [85], pyrene [86], or anthraquinone [87] derivatives attached to carbon surfaces and therein mainly CNTs. Lalaoui et al. studied this phenomenon in more detail and calculated the binding energy of the anthraquinone representing the polyaromatic compounds and adamantane, a saturated hydrocarbon [88]. The calculations were accompanied by electrochemical and quartz crystal microbalance (QCM) experiments. The modeling revealed a higher binding energy (-15.4 ± 1.8 kcal mol^{-1}) for anthraquinone than for adamantane (-7.8 ± 1.5 kcal mol^{-1}), but the experiments showed higher catalytic currents for adamantane (2.13 mA cm^{-2} at 0.55 V vs SCE) than for anthraquinone (0.74 mA cm^{-2} at 0.52 V vs SCE) both, immobilized of CNTs. It was concluded that anthraquinone tends to form π-stacking interactions with the CNT walls and this leads to a lower amount of available anchor groups and thus to a lower surface coverage of oriented enzymes. This π-stacking issue for polyaromatic substances was already evoked 2 years before and qualitatively studied by Bourourou et al. [87]. The same group also adapted this strategy for the covalent modification of reduced graphene oxide. This nanomaterial showed lower DET properties compared with CNTs but exhibited strong π-stacking interactions with CNT films [89]. Another strategy also involved the specific modification of laccase mutants with pyrene groups, localized at the vicinity of the T_1 center [90]. These enzymes were successfully immobilized on CNT electrodes and gold nanoparticles modified with β-cyclodextrin groups, showing efficient DET at both nanomaterials. In addition to this latter work, several works have underlined the excellent properties of gold nanoparticles in terms of DET toward oriented laccases, either via supramolecular [90] or via covalent interactions [91].

BOD from *Myrothecium verrucaria* is a promising alternative to laccase as its highest catalytic activity is in the neutral range [92–94]. Similar to other multicopper enzymes, BOD consists of a single subunit with four redox-active Cu atoms (T_1 and T_2/T_3), but the substrate binding site is hydrophilic and is incompatible with the oriented immobilization and wiring strategies for laccase [95]. Nonetheless, the strong interaction with its substrate bilirubin allows site-specific immobilization of this enzyme, enabling efficient DET [95–97]. More efficient performances were obtained using the protoporphyrin IX, a mimic of bilirubin [98]. It was concluded via several parallel experiments that

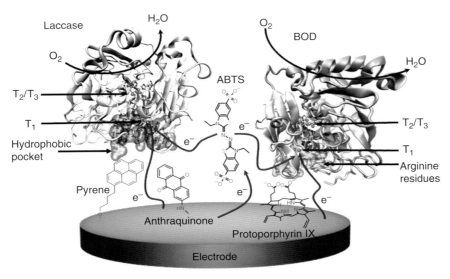

Figure 11.3 Sketch of the mostly used oxygen-reducing enzymes (laccase and BOD) for biofuel cell applications and appropriate molecular functions for oriented immobilization and promoted DET. In the center, the structure of ABTS used for MET modes is displayed.

the presence of carboxylates mainly contributes to the oriented immobilization of BOD via both electrostatic interactions and favorable dipolar moment of the enzyme. Based on these results, Lalaoui et al. proposed a simplified and less expensive approach by functionalization of CNTs via covalent diazonium grafting with carboxynaphthyls [99]. A scheme of oriented wiring of BOD and laccase is presented in Figure 11.3. Tsujimura's group previously reported DET reaction of BOD using pore-size-controlled mesoporous carbons, including carbon gel and MgO-templated carbon [100, 101]. By using MgO template with pore diameters of 38 nm, the DET catalytic current density was found to be 6 mA cm^{-2} with an electrode rotation rate of 8000 rpm at pH 5, 25 °C, with O_2 saturation. To further improve the current production efficiency, a three-dimensional (3D) hierarchical pore structure was fabricated using a MgO-templated porous carbon produced from two MgO templates with sizes of 40 and 150 nm [102]. The macropores improve mass transfer inside the carbon material, and the mesopores improve the electron transfer efficiency of the enzyme by surrounding the enzyme with carbon. The electrode showed 13 mA cm^{-2} of oxygen reduction current at pH 5 without any further surface modification.

11.4 Conclusion and Future Perspectives

EBFCs have unique features compared with other energy harvesters and batteries, including the potential for miniaturization, high theoretical power densities, and high biocompatibility. Considerable efforts have been made to develop EBFCs as novel power sources that are cost–effective, environmentally friendly,

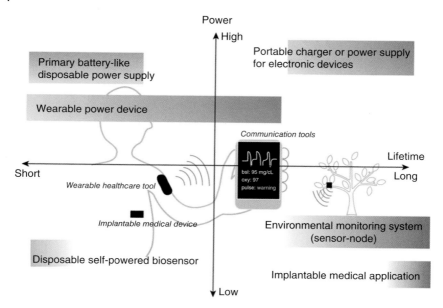

Figure 11.4 Potential application of enzymatic biofuel cell.

and readily available in order to drive implantable, epidermal, or wearable (bio)electronic devices, as well as the ubiquitous sensor-node systems required for technologies related to the Internet of things (IoT). Potential applications of EBFCs are illustrated in Figure 11.4. However, EBFC technology is still at an early stage of development, with many fundamental scientific and engineering problems that have still to be resolved. Two critical issues related to EBFCs are their short lifetimes and poor power densities; the number of electroactive enzymes available for a reaction is limited because enzymes on electrode surfaces are generally unstable, and there is a large barrier to electron transfer between enzyme-active sites and electrode surface. Usually, a trade-off is attempted between the output power density and lifetime, as it is highly challenging to improve both simultaneously. Considering these inherent drawbacks, there are two possibilities for the application of EBFCs, with the first focusing on disposable products and the second on longer-lasting devices that require less power.

The advantage of EBFCs for primary battery-like single-use disposable applications is the easy disposal of used EBFCs, as they are basically composed of enzymes and carbon electrodes, both of which can be made using ecofriendly, biodegradable materials. Primary batteries are usually used as portable chargers, which need to be recycled. The present requirement of powering electronic devices such as wireless communication tools anytime and anywhere will increase considerably in the future. To compare with existing primary batteries, the output performance as well as the price competitiveness of EBFCs should be improved. EBFCs can also meet the demands of powering wearable or epidermal electronic devices because of their high safety, as they lack strongly basic electrolytes and metal packages. Additionally, EBFCs that operate using sweat, urine, tear, saliva, or blood are better suited to disposable use. Such disposable energy devices would open up new routes for applications such as epidermal healthcare electronics, contact lens or mouth guard-type sensing devices, self-powered urine/blood glucose monitoring devices, and communication tools for emergencies [103].

Low-power but longer-lasting devices could power implantable medical devices such as pacemakers or neurostimulators using glucose and oxygen in the body as fuels. The simple structure of EBFCs would allow for straightforward miniaturization, reducing the burden imposed on a patient. However, improving the stabilities of enzymes that have been immobilized on electrode surfaces still presents a significant challenge for the development of EBFCs with lifetimes greater than 1 year.

It should be noted that another possible application that makes the best use of EBFCs could be single-use disposable self-powered biosensors, which require low amounts of electricity for transmitting and amplifying sensor signals. To achieve this, it will be necessary to design cell configurations that allow the output power to depend on the fuel concentration.

The most feasible fuel for EBFCs is glucose due to its chemical stability, safety, and accessibility. Although the number of electrons available from glucose oxidation is limited to two in the present technology, this should be increased by developing novel biomimetic (bioinspired), enzymatic, or abiotic electrocatalyst cascade systems. The capacities of EBFCs could exceed those of lithium ion batteries or other secondary batteries if 24 electrons could be taken from one glucose molecule. Although the output power density depends on the kinetics of the enzymes, exceptionally high capacities (energy densities) would open up new applications for EBFCs not only as fuel cells but also as primary batteries (Figure 11.4).

Much research still focuses on monosaccharides such as glucose, but stable and abundant polysaccharides, including paper, cotton, leaves and stems of plants, and starch, can be utilized in combination with specific hydrolases. On the other hand, alcohols, carbohydrates, or organic acids of low molecular weight can be considered as alternative anodic fuels to increase energy density. Recently, lactic acid from sweat has gathered considerable attention as an energy source for epidermal or wearable power devices [62]. Oxygen is most widely used as a final electron acceptor in enzymatic biocathodes. Although it is the most abundant and accessible oxidant, the delivery of oxygen to enzymes can be the rate-limiting factor in some systems, especially in implantable EBFCs, due to its low solubility in solution. Alternative oxidants that have high formal potentials as well as that are stable, of low cost, and safe should be developed.

A promising approach to overcoming both the short lifetimes and low power densities of EBFCs and allowing them to reach practical applications is to use nanostructure-controlled porous carbon materials as an electrode. However, enzymes in confined nanospaces have yet to be elucidated with respect to electrochemical and biological enzymatic reactions, as well as the 3D structural changes in the enzymes. A combination of more efficient electron transfer technology by modification of the microscopic interface between the enzyme and the porous carbon materials with macro–meso hierarchical structures would be helpful for achieving higher and more stable current outputs with lower amounts of enzymes, contributing to a practical advancement in fuel cell technology. Eliminating diffusional mediators by facilitating efficient electron transfer between conductive nanostructured materials and dehydrogenases could improve EBFC stability, which would be beneficial for the development of implantable EBFCs.

Molecular technology, protein engineering, nanostructured materials, hydrogels, and polymers have all been used to enhance the kinetics of electron transfer between enzyme-active sites and electrode surfaces, as well as the stability of the three-dimensional structures of enzymes inside electrode nanospaces, thereby improving the performance, stability, and durability of EBFCs. The ultimate goal for an EBFC with both high output power and stability is application to a ubiquitous wireless power supply for any electronic devices that require electricity.

References

1 Yahiro, A.T., Lee, S.M., and Kimble, D.O. (1964). Bioelectrochemistry: I. Enzyme utilizing bio-fuel cell studies. *Biochim. Biophys. Acta* 88 (2): 375–383.
2 Rasmussen, M., Abdellaoui, S., and Minteer, S.D. (2016). Enzymatic biofuel cells: 30 years of critical advancements. *Biosens. Bioelectron.* 76: 91–102.
3 Minteer, S.D. (2017). Methods in biological fuel cells. In: *Springer Handbook of Electrochemical Energy* (ed. C. Breitkopf and K. Swider-Lyons), 743–755. Berlin, Heidelberg: Springer Berlin Heidelberg.
4 Cosnier, S., Le Goff, A., and Holzinger, M. (2014). Towards glucose biofuel cells implanted in human body for powering artificial organs: review. *Electrochem. Commun.* 38: 19–23.
5 Katz, E. and MacVittie, K. (2013). Implanted biofuel cells operating *in vivo* – methods, applications and perspectives – feature article. *Energy Environ. Sci.* 6 (10): 2791–2803.
6 Healey, M. and Lee, J. (2016). Meta-study focusing on abiotic cells for human implants. *PAM Rev.* 3 (subject 68412): 100–112.
7 Kerzenmacher, S., Ducrée, J., Zengerle, R., and von Stetten, F. (2008). Energy harvesting by implantable abiotically catalyzed glucose fuel cells. *J. Power Sources* 182 (1): 1–17.
8 Aghahosseini, H. et al. (2016). Glucose-based biofuel cells: nanotechnology as a vital science in biofuel cells performance. *Nano Res.* 1 (2): 183–204.
9 Shitanda, I., Kato, S., Hoshi, Y. et al. (2013). Flexible and high-performance paper-based biofuel cells using printed porous carbon electrodes. *Chem. Commun.* 49 (94): 11110–11112.
10 Narváez Villarrubia, C.W. et al. (2014). Practical electricity generation from a paper based biofuel cell powered by glucose in ubiquitous liquids. *Electrochem. Commun.* 45: 44–47.
11 Cosnier, S., Holzinger, M., and Le Goff, A. (2014). Recent advances in carbon nanotube based enzymatic fuel cells. *Front. Bioeng. Biotechnol.* 2 (45).

12 Holzinger, M., Le Goff, A., and Cosnier, S. (2012). Carbon nanotube/enzyme biofuel cells. *Electrochim. Acta* 82: 179–190.
13 Cosnier, S., Gross, A.J., Le Goff, A., and Holzinger, M. (2016). Recent advances on enzymatic glucose/oxygen and hydrogen/oxygen biofuel cells: achievements and limitations. *J. Power Sources* 325: 252–263.
14 Eddowes, M.J. and Hill, H.A.O. (1977). Novel method for the investigation of the electrochemistry of metalloproteins: cytochrome c. *J. Chem. Soc. Chem. Commun.* (21): 771b–772b.
15 Yeh, P. and Kuwana, T. (1977). Reversible electrode reaction of cytochrome c. *Chem. Lett.* 6 (10): 1145–1148.
16 Niki, K., Yagi, T., Inokuchi, H., and Kimura, K. (1979). Electrochemical behavior of cytochrome c3 of *Desulfovibrio vulgaris*, strain Miyazaki, on the mercury electrode. *J. Am. Chem. Soc.* 101 (12): 3335–3340.
17 Armstrong, F.A., Hill, H.A.O., and Walton, N.J. (1988). Direct electrochemistry of redox proteins. *Acc. Chem. Res.* 21 (11): 407–413.
18 Nuzzo, R.G. and Allara, D.L. (1983). Adsorption of bifunctional organic disulfides on gold surfaces. *J. Am. Chem. Soc.* 105 (13): 4481–4483.
19 Gooding, J.J., Mearns, F., Yang, W., and Liu, J. (2003). Self-assembled monolayers into the 21st century: recent advances and applications. *Electroanalysis* 15 (2): 81–96.
20 Taniguchi, I., Toyosawa, K., Yamaguchi, H., and Yasukouchi, K. (1982). Reversible electrochemical reduction and oxidation of cytochrome c at a bis(4-pyridyl) disulphide-modified gold electrode. *J. Chem. Soc. Chem. Commun.* (18): 1032–1033.
21 Weisshaar, D.E., Lamp, B.D., and Porter, M.D. (1992). Thermodynamically controlled electrochemical formation of thiolate monolayers at gold: characterization and comparison to self-assembled analogs. *J. Am. Chem. Soc.* 114 (14): 5860–5862.
22 Stole, S.M. and Porter, M.D. (1990). *In situ* infrared external reflection spectroscopy as a probe of the interactions at the liquid–solid interface of long-chain alkanethiol monolayers at gold. *Langmuir* 6 (6): 1199–1202.
23 Bard, A.J. et al. (1993). The electrode/electrolyte interface – a status report. *J. Phys. Chem.* 97 (28): 7147–7173.
24 Nishiyama, K. et al. (2008). Conformational change in 4-pyridineethanethiolate self-assembled monolayers on Au(111) driven by protonation/deprotonation in electrolyte solutions. *Phys. Chem. Chem. Phys.* 10 (46): 6935–6939.
25 Willner, I., Willner, B., and Katz, E. (2007). Biomolecule–nanoparticle hybrid systems for bioelectronic applications. *Bioelectrochemistry* 70 (1): 2–11.
26 Armstrong, F.A. and Wilson, G.S. (2000). Recent developments in faradaic bioelectrochemistry. *Electrochim. Acta* 45 (15–16): 2623–2645.
27 Salamon, Z., Hazzard, J.T., and Tollin, G. (1993). Direct measurement of cyclic current-voltage responses of integral membrane proteins at a self-assembled lipid-bilayer-modified electrode: cytochrome f and cytochrome c oxidase. *Proc. Natl. Acad. Sci. U. S. A.* 90 (14): 6420–6423.

28 Cullison, J.K., Hawkridge, F.M., Nakashima, N., and Yoshikawa, S. (1994). A study of cytochrome c oxidase in lipid bilayer membranes on electrode surfaces. *Langmuir* 10 (3): 877–882.

29 Rusling, J.F., Wang, B., and Yun, S.E. (2008). Electrochemistry of redox enzymes. In: *Bioelectrochemistry* (ed. P.N. Bartlett), 39–85. John Wiley & Sons, Ltd.

30 Marcus, R.A. and Sutin, N. (1985). Electron transfers in chemistry and biology. *Biochim. Biophys. Acta Rev. Bioenerg.* 811 (3): 265–322.

31 Reuillard, B. et al. (2013). High power enzymatic biofuel cell based on naphthoquinone-mediated oxidation of glucose by glucose oxidase in a carbon nanotube 3D matrix. *Phys. Chem. Chem. Phys.* 15 (14): 4892–4896.

32 Holzinger, M., Haddad, R., Le Goff, A., and Cosnier, S. (2016). Enzymatic glucose biofuel cells: shapes and growth of carbon nanotube matrices. In: *Dekker Encyclopedia of Nanoscience and Nanotechnology*, Thirde (ed. S.E. Lyshevski), 1–10. CRC Press.

33 Inagaki, M., Toyoda, M., Soneda, Y. et al. (2016). Templated mesoporous carbons: synthesis and applications. *Carbon* 107: 448–473.

34 Suzuki, A., Mano, N., and Tsujimura, S. (2017). Lowering the potential of electroenzymatic glucose oxidation on redox hydrogel-modified porous carbon electrode. *Electrochim. Acta* 232: 581–585.

35 Ferri, S., Kojima, K., and Sode, K. (2011). Review of glucose oxidases and glucose dehydrogenases: a Bird's eye view of glucose sensing enzymes. *J. Diabetes Sci. Technol.* 5 (5): 1068–1076.

36 Wohlfahrt, G. et al. (1999). 1.8 and 1.9 A resolution structures of the Penicillium amagasakiense and *Aspergillus niger* glucose oxidases as a basis for modelling substrate complexes. *Acta Crystallogr. Sect. D* 55 (5): 969–977.

37 Courjean, O., Gao, F., and Mano, N. (2009). Deglycosylation of glucose oxidase for direct and efficient glucose Electrooxidation on a glassy carbon electrode. *Angew. Chem. Int. Ed.* 48 (32): 5897–5899.

38 Xiao, Y., Patolsky, F., Katz, E. et al. (2003). "Plugging into enzymes": Nanowiring of redox enzymes by a gold nanoparticle. *Science* 299 (5614): 1877–1881.

39 Patolsky, F., Weizmann, Y., and Willner, I. (2004). Long-range electrical contacting of redox enzymes by SWCNT connectors. *Angew. Chem. Int. Ed.* 43 (14): 2113–2117.

40 Zebda, A. et al. (2011). Mediatorless high-power glucose biofuel cells based on compressed carbon nanotube-enzyme electrodes. *Nat. Commun.* 2: 370.

41 Gregg, B.A. and Heller, A. (1991). Redox polymer films containing enzymes. 2. Glucose oxidase containing enzyme electrodes. *J. Phys. Chem.* 95 (15): 5976–5980.

42 Mano, N., Mao, F., and Heller, A. (2002). A miniature biofuel cell operating in a physiological buffer. *J. Am. Chem. Soc.* 124 (44): 12962–12963.

43 Heller, A. (1992). Electrical connection of enzyme redox centers to electrodes. *J. Phys. Chem.* 96 (9): 3579–3587.

44 Pinyou, P. et al. (2016). Design of an Os complex-modified hydrogel with optimized redox potential for biosensors and biofuel cells. *Chem. Eur. J.* 22 (15): 5319–5326.

45 Gao, F., Viry, L., Maugey, M. et al. (2010). Engineering hybrid nanotube wires for high-power biofuel cells. *Nat. Commun.* 1 (1): 2.

46 Kwon, C.H. et al. (2014). High-power biofuel cell textiles from woven biscrolled carbon nanotube yarns. *Nat. Commun.* 5 (3928).

47 Milton, R.D. et al. (2015). Rational design of quinones for high power density biofuel cells. *Chem. Sci.* 6 (8): 4867–4875.

48 Hou, C., Lang, Q., and Liu, A. (2016). Tailoring 1,4-naphthoquinone with electron-withdrawing group: toward developing redox polymer and FAD-GDH based hydrogel bioanode for efficient electrocatalytic glucose oxidation. *Electrochim. Acta* 211: 663–670.

49 Kavanagh, P. and Leech, D. (2013). Mediated electron transfer in glucose oxidising enzyme electrodes for application to biofuel cells: recent progress and perspectives. *Phys. Chem. Chem. Phys.* 15 (14): 4859–4869.

50 Oubrie, A. et al. (1999). Structure and mechanism of soluble quinoprotein glucose dehydrogenase. *EMBO J.* 18 (19): 5187.

51 Tanne, C., Göbel, G., and Lisdat, F. (2010). Development of a (PQQ)-GDH-anode based on MWCNT-modified gold and its application in a glucose/O_2-biofuel cell. *Biosens. Bioelectron.* 26 (2): 530–535.

52 Göbel, G., Schubart, I.W., Scherbahn, V., and Lisdat, F. (2011). Direct electron transfer of PQQ-glucose dehydrogenase at modified carbon nanotubes electrodes. *Electrochem. Commun.* 13 (11): 1240–1243.

53 Scherbahn, V. et al. (2014). Biofuel cells based on direct enzyme–electrode contacts using PQQ-dependent glucose dehydrogenase/bilirubin oxidase and modified carbon nanotube materials. *Biosens. Bioelectron.* 61 (0): 631–638.

54 Halámková, L. et al. (2012). Implanted biofuel cell operating in a living snail. *J. Am. Chem. Soc.* 134 (11): 5040–5043.

55 Ivnitski, D., Atanassov, P., and Apblett, C. (2007). Direct Bioelectrocatalysis of PQQ-dependent glucose dehydrogenase. *Electroanalysis* 19 (15): 1562–1568.

56 Szczupak, A. et al. (2012). Living battery – biofuel cells operating in vivo in clams. *Energy Environ. Sci.* 5 (10): 8891–8895.

57 Castorena-Gonzalez, J.A. et al. (2013). Biofuel cell operating *in vivo* in rat. *Electroanalysis* 25 (7): 1579–1584.

58 MacVittie, K. et al. (2013). From "cyborg" lobsters to a pacemaker powered by implantable biofuel cells. *Energy Environ. Sci.* 6 (1): 81–86.

59 Yamamoto, K. et al. (2001). Crystal structure of glucose dehydrogenase from Bacillus megaterium IWG3 at 1.7 Å Resolution 1. *J. Biochem.* 129 (2): 303–312.

60 Zafar, M.N. et al. (2012). Characterization of different FAD-dependent glucose dehydrogenases for possible use in glucose-based biosensors and biofuel cells. *Anal. Bioanal. Chem.* 402 (6): 2069–2077.

61 Kim, Y.H., Campbell, E., Yu, J. et al. (2013). Complete oxidation of methanol in biobattery devices using a hydrogel created from three modified dehydrogenases. *Angew. Chem. Int. Ed.* 52 (5): 1437–1440.

62 Jia, W., Valdés-Ramírez, G., Bandodkar, A.J. et al. (2013). Epidermal biofuel cells: energy harvesting from human perspiration. *Angew. Chem. Int. Ed.* 52 (28): 7233–7236.

63 Lalaoui, N. et al. (2016). Enzymatic versus Electrocatalytic oxidation of NADH at carbon-nanotube electrodes modified with glucose dehydrogenases: application in a Bucky-paper-based glucose enzymatic fuel cell. *ChemElectroChem* 3 (12): 2058–2062.

64 Giroud, F., Sawada, K., Taya, M., and Cosnier, S. (2017). 5,5-Dithiobis(2-nitrobenzoic acid) pyrene derivative-carbon nanotube electrodes for NADH electrooxidation and oriented immobilization of multicopper oxidases for the development of glucose/O_2 biofuel cells. *Biosens. Bioelectron.* 87: 957–963.

65 Gao, F., Yan, Y., Su, L. et al. (2007). An enzymatic glucose/O_2 biofuel cell: preparation, characterization and performance in serum. *Electrochem. Commun.* 9 (5): 989–996.

66 Sakai, H. et al. (2009). A high-power glucose/oxygen biofuel cell operating under quiescent conditions. *Energy Environ. Sci.* 2 (1): 133–138.

67 Narváez Villarrubia, C.W., Artyushkova, K., Garcia, S.O., and Atanassov, P. (2014). NAD^+/NADH tethering on MWNTs-Bucky papers for glucose dehydrogenase-based anodes. *J. Electrochem. Soc.* 161 (13): H3020–H3028.

68 Zhang, M., Mullens, C., and Gorski, W. (2007). Coimmobilization of dehydrogenases and their cofactors in electrochemical biosensors. *Anal. Chem.* 79 (6): 2446–2450.

69 Yoshida, H. et al. (2015). Structural analysis of fungus-derived FAD glucose dehydrogenase. *Sci. Rep.* 5 (13498).

70 Murata, K., Akatsuka, W., Sadakane, T. et al. (2014). Glucose oxidation catalyzed by FAD-dependent glucose dehydrogenase within Os complex-tethered redox polymer hydrogel. *Electrochim. Acta* 136: 537–541.

71 Tsujimura, S., Murata, K., and Akatsuka, W. (2014). Exceptionally high glucose current on a hierarchically structured porous carbon electrode with "wired" Flavin adenine dinucleotide-dependent glucose dehydrogenase. *J. Am. Chem. Soc.* 136 (41): 14432–14437.

72 Suzuki, A. and Tsujimura, S. (2016). Long-term continuous operation of FAD-dependent glucose dehydrogenase hydrogel-modified electrode at 37 °C. *Chem. Lett.* 45 (4): 484–486.

73 Tsujimura, S. et al. (2006). Novel FAD-dependent glucose dehydrogenase for a dioxygen-insensitive glucose biosensor. *Biosci. Biotechnol. Biochem.* 70 (3): 654–659.

74 Tsuruoka, N., Sadakane, T., Hayashi, R., and Tsujimura, S. (2017). Bimolecular rate constants for FAD-dependent glucose dehydrogenase from Aspergillus terreus and organic electron acceptors. *Int. J. Mol. Sci.* 18 (3): 604.

75 Solomon, E.I., Sundaram, U.M., and Machonkin, T.E. (1996). Multicopper oxidases and Oxygenases. *Chem. Rev.* 96 (7): 2563–2606.

76 Le Goff, A., Holzinger, M., and Cosnier, S. (2015). Recent progress in oxygen-reducing laccase biocathodes for enzymatic biofuel cells. *Cell. Mol. Life Sci.* 72 (5): 941–952.

77 Barrière, F., Ferry, Y., Rochefort, D., and Leech, D. (2004). Targeting redox polymers as mediators for laccase oxygen reduction in a membrane-less biofuel cell. *Electrochem. Commun.* 6 (3): 237–241.

78 Kenzom, T., Srivastava, P., and Mishra, S. (2014). Structural insights into 2,2′-azino-bis(3-ethylbenzothiazoline-6-sulfonic acid) (ABTS)-mediated degradation of reactive blue 21 by engineered *Cyathus bulleri* laccase and characterization of degradation products. *Appl. Environ. Microbiol.* 80 (24): 7484–7495.

79 Bilewicz, R., Stolarczyk, K., Sadowska, K. et al. (2009). Carbon nanotubes derivatized with mediators for laccase catalyzed oxygen reduction. *ECS Trans.* 19 (6): 27–36.

80 Bourourou, M. et al. (2014). Freestanding redox buckypaper electrodes from multi-wall carbon nanotubes for bioelectrocatalytic oxygen reduction via mediated electron transfer. *Chem. Sci.* 5 (7): 2885–2888.

81 Altamura, L. et al. (2017). A synthetic redox biofilm made from metalloprotein–prion domain chimera nanowires. *Nat. Chem.* 9 (2): 157–163.

82 Rubenwolf, S. et al. (2012). Prolongation of electrode lifetime in biofuel cells by periodic enzyme renewal. *Appl. Microbiol. Biotechnol.* 96 (3): 841–849.

83 Blanford, C.F., Heath, R.S., and Armstrong, F.A. (2007). A stable electrode for high-potential, electrocatalytic O_2 reduction based on rational attachment of a blue copper oxidase to a graphite surface. *Chem. Commun.* 0 (17): 1710–1712.

84 Meredith, M.T. et al. (2011). Anthracene-modified multi-walled carbon nanotubes as direct electron transfer scaffolds for enzymatic oxygen reduction. *ACS Catal.* 1 (12): 1683–1690.

85 Karaśkiewicz, M. et al. (2012). Fully enzymatic mediatorless fuel cell with efficient naphthylated carbon nanotube–laccase composite cathodes. *Electrochem. Commun.* 20 (0): 124–127.

86 Lalaoui, N., Elouarzaki, K., Le Goff, A. et al. (2013). Efficient direct oxygen reduction by laccases attached and oriented on pyrene-functionalized polypyrrole/carbon nanotube electrodes. *Chem. Commun.* 49 (81): 9281–9283.

87 Bourourou, M. et al. (2013). Supramolecular immobilization of laccases on carbon nanotube electrodes functionalized with (methylpyrenylaminomethyl)anthraquinone for direct electron reduction of oxygen. *Chemistry* 19 (28): 9371–9375.

88 Lalaoui, N. et al. (2016). Hosting adamantane in the substrate pocket of Laccase: direct bioelectrocatalytic reduction of O_2 on functionalized carbon nanotubes. *ACS Catal.* 6 (7): 4259–4264.

89 Lalaoui, N., Le Goff, A., Holzinger, M. et al. (2015). Wiring Laccase on covalently modified Graphene: carbon nanotube assemblies for the direct bio-electrocatalytic reduction of oxygen. *Chem. Eur. J.* 21 (8): 3198–3201.

90 Lalaoui, N. et al. (2016). Direct electron transfer between a site-specific pyrene-modified laccase and carbon nanotube/gold nanoparticle supramolecular assemblies for bioelectrocatalytic dioxygen reduction. *ACS Catal.* 6 (3): 1894–1900.

91 Gutiérrez-Sánchez, C., Pita, M., Vaz-Domínguez, C. et al. (2012). Gold nanoparticles as electronic bridges for Laccase-based biocathodes. *J. Am. Chem. Soc.* 134 (41): 17212–17220.

92 Jian, G., Xl-xian, L., Pei-sheng, M., and Gao-xiang, L. (1991). Purification and properties of bilirubin oxidase from *Myrothecium verrucaria*. *Appl. Biochem. Biotechnol.* 31 (2): 135–143.

93 Tsujimura, S. et al. (2001). Bioelectrocatalytic reduction of dioxygen to water at neutral pH using bilirubin oxidase as an enzyme and 2,2′-azinobis (3-ethylbenzothiazolin-6-sulfonate) as an electron transfer mediator. *J. Electroanal. Chem.* 496 (1–2): 69–75.

94 Mano, N. and Edembe, L. (2013). Bilirubin oxidases in bioelectrochemistry: features and recent findings. *Biosens. Bioelectron.* 50: 478–485.

95 Cracknell, J.A., McNamara, T.P., Lowe, E.D., and Blanford, C.F. (2011). Bilirubin oxidase from *Myrothecium verrucaria*: X-ray determination of the complete crystal structure and a rational surface modification for enhanced electrocatalytic O_2 reduction. *Dalton Trans.* 40 (25): 6668–6675.

96 Tominaga, M., Ohtani, M., and Taniguchi, I. (2008). Gold single-crystal electrode surface modified with self-assembled monolayers for electron tunneling with bilirubin oxidase. *Phys. Chem. Chem. Phys.* 10 (46): 6928–6934.

97 Murata, K., Kajiya, K., Nakamura, N., and Ohno, H. (2009). Direct electrochemistry of bilirubin oxidase on three-dimensional gold nanoparticle electrodes and its application in a biofuel cell. *Energy Environ. Sci.* 2: 1280–1285.

98 Lalaoui, N., Le Goff, A., Holzinger, M., and Cosnier, S. (2015). Fully oriented bilirubin oxidase on porphyrin-functionalized carbon nanotube electrodes for Electrocatalytic oxygen reduction. *Chem. Eur. J.* 21 (47): 16868–16873.

99 Lalaoui, N., Holzinger, M., Le Goff, A., and Cosnier, S. (2016). Diazonium functionalisation of carbon nanotubes for specific orientation of multicopper oxidases: controlling electron entry points and oxygen diffusion to the enzyme. *Chem. Eur. J.* 22 (30): 10494–10500.

100 Tsujimura, S., Kamitaka, Y., and Kano, K. (2007). Diffusion-controlled oxygen reduction on multi-copper oxidase-adsorbed carbon aerogel electrodes without mediator. *Fuel Cells* 7 (6): 463–469.

101 Tsujimura, S. and Murata, K. (2015). Electrochemical oxygen reduction catalyzed by bilirubin oxidase with the aid of 2,2′-Azinobis(3-ethylbenzothiazolin-6-sulfonate) on a MgO-template carbon electrode. *Electrochim. Acta* 180: 555–559.

102 Funabashi, H., Takeuchi, S., and Tsujimura, S. (2017). Hierarchical meso/macro-porous carbon fabricated from dual MgO templates for direct electron transfer enzymatic electrodes. *Sci. Rep.* (in press doi: 10.1038/srep45147).

103 Bandodkar, A.J. and Wang, J. (2016). Wearable biofuel cells: a review. *Electroanalysis* 28: 1188–1200.

Index

a
air/liquid interface 32
  grazing incidence (GI) mode 8
α-diimine complexes 221
α-sexithiophene (α-6T) 263
amorphous molybdenum sulfide (a-MoS$_x$) 262
amsterdam density functional (ADF) program 63
Anderson-type static disorder potentials 16, 19
antenna effect 132
artificial carbon-cycle system 210
artificial photosynthesis 209, 210, 242, 251
artificial Z-Scheme mechanism 243
atomic force microscope (AFM) measurements 141
atomic layer deposition (ALD) 257, 266, 268
2,2′-azino-bis(3-ethylbenzothiazoline-6-sulphonic acid) (ABTS) 295
azobenzene, photoisomerization of 201

b
band narrowing 9, 10
band transport 4, 7–9, 16, 19
Belousov–Zhabotinsky (BZ) reaction 191, 203
bent-shaped heteroacenes 66–71
benzobis(thiazole) (BBT) 74
benzothieno[1]benzothiophene (BTBT) 63

1-benzyl-1,4-dihydronicotinamide (BNAH) 219
bioanodes 291, 292
bioelectrocatalytic buckypaper electrodes 292
biomolecular machines 204
biosensing 287
bipyridine ligand 215, 268, 276
bis(2-hydroxypyridine) chelate complexes 89–93
block copolymer (BCP) 108
Boltzmann constant 5, 158
boron subphthalocyanine chloride (SubPc) 263
Brewster angle microscopy (BAM) 32–34, 39
brickwork-packing (2D) 59
bulky-type substituents 64–65

c
carbon nanotubes (CNTs) 256, 290
C$_8$–BTBT 66
C$_{10}$–DNBDT–NW 68
C$_{10}$–DNT–VW 68
C$_{10}$–DNTT 65, 66
Chebyshev polynomials 14
chemical actuators 187–188, 191
chemical energy 188, 243, 274, 278
chirality control, molecular technology for
  chiropitcal properties and optical activity
    amphiphilic barbituric acid derivative 141

Molecular Technology: Energy Innovation, Volume 1, First Edition.
Edited by Hisashi Yamamoto and Takashi Kato.
© 2018 Wiley-VCH Verlag GmbH & Co. KGaA. Published 2018 by Wiley-VCH Verlag GmbH & Co. KGaA.

chirality control, molecular technology for (contd.)
    high-sensitive detection technique 141
    light--matter interaction 141
    macroscopic polarization 142
    microscopic CD measurements measurements via far-field field detection 142
    PEM technique 143
    transition dipole moments 141
chiral lanthanide(III) complexes
  circularly polarized luminescence (CPL) 130
  electric and magnetic dipole transitions 132
  Frontier applications 135
  luminescence dissymmetry factor 131
  optical activity of 132–135
circularly polarized luminescence 129
MCD and MCPL
  ferromagnetic materials 135
  magnetic fields break, light absorption and emission 136–137
  molecular materials and applications 137–138
molecular self-Aassembled helical structures:
  chiral liquid crystalline phases 139
  CLC laser action 139–140
  electromagnetic wave reflection 138
  natural photonic crystals 138
near-field optics 145
chiral lanthanide(III) complexes 146
  circularly polarized luminescence (CPL) 130
  electric and magnetic dipole transitions 132
  Frontier applications 135
  luminescence dissymmetry factor 131
  optical activity of 132–135
cinnamic acid (CA) 110
circularly polarized light 129, 130, 135, 138–140
cobaloxime 259, 260, 268, 272, 273
cobalt diimine-dioxime catalyst 268
coherent band transport model 7–9
coherent polaron transport model 9–10
coherent transport approach 19
composite films 110
concavo-convex pattern 201
conventional band and hopping models 18, 19
Coulomb coupling 157
coupled-cluster theory (CC2) 160
cyclic voltammetry (CV) 64, 74, 266
cytochrome c oxidase 289

d

DBTDT 62, 63
density functional theory (DFT) 7, 11–15, 159
depolarization 145
diarylethenes 196, 197
dielectric constant 139, 192
1,8-dihydroxyanthraquinone (DHA) 144
dimer approach 14
N,N-dimethylformamide (DMF) 221
1,3-dimethyl-2-phenyl-2,3-dihydro-1H-benzo[d]imidazole (BIH) 231
dinaphtho[2,3-b:2′,3′-f]thieno[3,2-b]thiophene (DNTT) 4, 63
direct electron transfer (DET) 288
distributed feedback 139
dye-free system 107, 108, 116, 117
dye-sensitized photocathodes 278–281
  covalent/supramolecular dye-catalyst assemblies 268–270
  multielectronic catalysis 264
  p-type dye-sensitized solar cells (p-DSSCs) 264
  p-type semiconductors 265
  physisorbed/diffusing catalysts 266–268
  quasi-metallic oxides 265

e

electronic-couplings 156, 157, 159, 160
  for triplet energy transfer 161

electron–phonon couplings 5, 6, 8–12, 19
electron–phonon scatterings 12, 16
emitters, chemical structures of 168
enzymatic biofuel cells (EBFCs) 287, 297, 299
external quantum efficiency (EQE) 172, 176

f
Faradaic efficiency 254, 266, 277, 279
Fe porphyrin 220, 221
ferrocenophane 258, 259
ferromagnetic materials 135
field-effect transistor (FETs) 2, 28, 57, 61
finite-difference time domain method 139
Fischer–Tropsch process 210
flashing ratchet 191, 192
flavin–adenine–dinucleotide (FAD) 292
fluorescence resonance energy transfer (FRET) 156–157
flushing ratchet 193
formic acid 91–93, 210, 277
Förster energy transfer model 156
fossil resources 209
freeze-pump-thaw degassing 168
full-width at half maximum (FWHM) 45
functional organic molecules
    with long alkyl chains 27
    without long alkyl chains 27

g
global warming 209, 273
glucose biofuel cell electrodes
    carbon nanotubes (CNTs) 290
    cytochrome c oxidase 289
    direct electron transfer (DET) 288
    electrochemical and quartz crystal microbalance (QCM) experiments 296
    enzymatic electrocatalytic oxidation
        bioelectrocatalytic buckypaper electrodes 292
    FAD dependent glucose dehydrogenases 294
    glucose oxidase (GOx) 291
    NAD dependent enzymes 294
    osmium redox hydrogels 291
    PABMSA 292
    quinones 292
laccases and bilirubin oxidases (BOD) 295
lipid-membrane-modified electrodes 289
low power consumptions 288
mediated electron transfer (MET) 289
molecular technology 290
self-assembled monolayer techniques 288
graphite 1, 294

h
H-bridged cobaloximes 258, 260
Herringbone-packing (2D) 59
heteroacene 58, 61–63, 65, 74, 175
hexacene 65, 175
hydrogen, photocathode materials
    low band gap semiconductors
        covalent attachment 258, 260
        physisorb, semiconductor surface 253, 255, 257
    organic semiconductor materials
        α-sexithiophene (α-6T) 263
        bulk heterojunction 260
        HER catalyst implementation 261
        solar spectrum 260
    p-n junction 253
hydrogenases 82, 89, 251, 292, 300

i
incoherent hopping transport model 6–7
inner filter effect 276
intersystem crossing (ISC) 163, 224
Ir(III)-Re(I) and Os(II)-Re(I) systems 234

k
Kasha's rule 198
kinetic energy 9, 188, 189, 193, 197

l

Langmuir–Blodgett (LB) film 26, 141
Langmuir film 26–28
layer-by-Layer (LbL) deposition 38, 41
ligand-centered electric dipole transitions 134
light energy conversion efficiency 281
linear alkyl chain substituents 65–66
linear fused acene-type molecules 59–61, 75
linear heteroacenes 61, 64
linear polarization anisotropy 142–145
linear polarization component 142, 143
lipid-membrane-modified electrodes 289
liquid crystals (LC)
   photo-contraction of 106
   polymer networks 119
liquid-phase interfacial synthesis
   air/liquid interfaces 27
   Langmuir-Blodgett (LB) film 26
   metal-organic frameworks (MOFs) 29
   porphyrins on metal ion solutions 28, 29
   with long alkyl chains 27
   without long alkyl chains 27, 28

m

magnetic field 129, 135–137, 147, 172, 175, 178
magneto-chiral dichroism 138
Marcus theory 3, 6, 7, 10, 15, 19, 158, 289
mediated electron transfer (MET) 289
merocyanine dye films 2
mesoporous TiO_2 scaffold 276
mesoscale motion
   autonomous motion 203, 204
   in gradient fields 199, 201
   movement triggered by mobile molecules 201, 202
mesoscopic motion 188–191, 193, 201, 203
metal and functional groups
   hydrogenases 82
   nitrogenase 83
   peroxidases 82, 83
   proton-responsive metal complexes 84, 93
metal-organic frameworks (MOFs) 28
   formation process 32, 34
   in-plane molecular 46, 49
   layer stacking motif 41, 46
   LbL deposition 38, 41
   nanosheet creation 29, 31
   post-injection method 35, 37
1-methyl-ferrocenophane catalyst 258
molar absorption coefficients 196, 198
molecular design
   for long diffusion length 155, 158
   of SF compounds 172
molecular machine 189, 193, 197, 201–204
molecular power 189
   gliding motion 189, 191
   motion generated by 189
   photochromic molecules 196, 197
   photochromic reaction 197, 199
   photoisomerization and actuation 199
   stimuli-responsive compounds 193, 199
   supramolecular structure 193, 196
monosaccharides 299
multiexciton generating process (MEG) 179
multi-walled carbon nanotube (MWCNT) defects 294

n

nafion film 277
nanofabrication techniques 199
nanosheet domain size 34–38
naphthalene diimide (NDI) 72
   and perylene diimide 72, 74
naphthalenes 4, 17, 18, 59, 60, 63, 72–74
naphthoquinone 292, 294, 295
naphthoquinone-based hydrogels 295
near-field optics 145
nematic liquid crystals (NLCs) 112

nicotine adenine dinucleotide (NAD) 292
nitrogenase 83–84, 86, 87
non-dye LC materials 107
non-polycyclic π-conjugated compounds 177, 178
non-polycyclic SF compounds 179
nonlinear optical (NLO)
  photo-functional materials 106
  systems 112
n-type semiconductor photoanode 242, 277

o

oil-shock 209
organic conductors 2
organic light-emitting diode 57, 135
organic semiconductors
  anthracenes 2
  coherent band transport model 7–9
  coherent polaron transport model 9–10
  computational molecular technology:
    power-law temperature dependence 4
    rubrene single-crystal FETs 4
  dimensionless electron-phonon coupling constant 5
  effective charge transport of 58, 59
  electroluminescent (EL) devices 2
  electron-donor-acceptor complex 1
  electronic device applications 2
  external electric field 2
  field-effect transistor (FET) 2
  graphite 1
  incoherent hopping transport model 6–7
  insulating materials 1
  intrinsic mobilities evaluations 17–18
  linear fused acene-type molecules 59, 75
  low-cost printing processes 1
  low-mobility regime 3
  multi-component systems 2
  n-type 71, 74
  π-conjugated cores for 61, 64, 66, 71
  π-conjugated polymers 3
  polaron formation energy vs. transfer integrals 15–16
  requirements of 58
  single-component organic compounds 2
  substituents for 64, 65
  temperature dependence of mobility 17–18
  tetracyano-p-quinodimethane (TCNQ) 2
  tetrathiafulvalene (TTF) 2
  time-of-flight technique 2
  trap potentials 10
  wave-packet dynamics approach 11–15
organic synthesis techniques 177
organothiols 288
osmium redox hydrogels 291

p

PEM technique 143
pentacene 3, 5, 7, 16, 61, 64, 172–174
peroxidases 82, 83, 87
perylene 1, 72, 74, 165, 175–176, 266
perylene diimide (PDI) 72–74, 165
phenyl-C_{61}-butyric-acid methyl ester (PCBM) 261
photoactuators 118–122
photo-alignable polymer film 110
photo-alignment techniques 107
photocatalytic CO_2 reduction
  catalyst (CAT) 212
  CO_2 low concentrations 236–241
  CuI complexes 222
  earth-abundant metals 220–223
  hybrid systems, supramolecular photocatalyst and semiconductors 241–245
  initiation mechanism 213
  multicomponent Systems 218
  one-photon excitation 212
  reaction mechanism
    anionic ligands 216
    one-electron-reduced species (OERS) 216
  redox photosensitizer 212

photocatalytic CO$_2$ reduction (*contd.*)
  redox reactions   211
  rhenium(I) complexes
    bipyridine ligand   215
    cis, trans-[Re(4,4′-X$_2$-2,2′-bpy)(CO)$_2$(PR$_3$)(PR′$_3$)]$^+$ 215
    π–π interaction   215
    [Ru(dmb)$_3$]$^{2+}$ and functions   223
  supramolecular photocatalysts
    heterogeneous materials   223
    hybrid systems   224
    Ir(III)-Re(I) and Os(II)-Re(I) systems   234–236
    Ru(II)-Re(I) systems   224–233
    Ru(II)-Ru(II) systems   233–234
photocatalytic reactions   162, 212, 213, 216, 218, 220, 222, 229, 233, 239–243
photocathode materials
  CO$_2$ reduction
    dye-sensitised photocathodes   278–281
    molecular catalyst and semiconductor photoelectrode   274–278
  H$_2$ evolution
    dye-sensitized photocathodes   263–265
    low band gap semiconductors   253–260
    organic semiconductor materials   260–263
    p-n junction   253
photo-chemical alignment   107–112
photo-chemical reactions   108
photo-crosslinkable polymer surface   107
photo-elastic modulator (PEM)   142
photo-electrochemical cells (PECs)   252
photoisomerization   196
  and actuation   199
photonic crystals   138
photon upconversion   158, 162–171
photo-physical alignment   112–115, 123

photo-physico-chemical alignment   107, 115–118, 123
photo-polymerization-induced molecular   118
photoreaction kinetics   197
photostationary state (PSS)   198, 231
photo-switching effect   109, 123
pincer-type bis(azole) complexes   84, 89
pincer-type ruthenium   89, 93
π-stacking (1D)   59
poly(3-aminobenzoic acid-co-2-methoxyaniline-5-sulfonic acid) (PABMSA)   292
polycyclic π-conjugated compounds
  heteroacene   175
  hexacene   175
  pentacene   172, 174
  perylene   175
  terrylene   175
  tetracene   174, 175
polymer-stabilized LCs   114
polystyrene film   258
polyvinylimidazole (PVI)   258
polyvinylpyridine (PVP)   258
porous coordination polymers (PCPs)   28
power conversion efficiency (PCE)   171, 174, 180, 260
proton-coupled electron transfer (PCET)   82
proton-responsive metal complexes
  bis(2-hydroxypyridine) chelate complexes   89, 93
  pincer-type bis(azole) complexes   84, 89
  with three appended protic groups on tripodal scaffolds   94, 98
p-type dye-sensitized solar cells (p-DSSCs)   264
pyrazolato-imidazolyl carbonyl complex   89
pyrazolato-pyrazole complex   86
pyrene derivatives   294
PyrroloQuinoline Quinone (PQQ)   292

Pyrroloquinoline Quinone Glucose Dehydrogenase (PQQ GDH) 292

q
quinones 292, 295

R
Raman scattering 147
reversible photo-switching 106
rhenium(I) complexes, bipyridine ligand 215
robots 187, 188
rubredoxin domain 295
Ru(II)-Ru(II) systems 233
rylene-type SF compounds 177

s
scanning electron microscopy 170
Schrödinger equation 158
second-order nonlinear optical (NLO) materials 109
self-assembled monolayer techniques 288
semiclassical Marcus theory 6
sensitizers
   chemical structures of 166
   conventional triplet 164
Shockley–Queisser limit 159, 162, 169, 171
single-crystal field-effect transistors 3, 61, 74
solar batteries 210
solar energy conversion 210, 253
solvothermal reactions 29
spectroscopic analyses 147
spiropyrans 196, 197, 201
supramolecular photocatalysts 279
   Ir(III)-Re(I) and Os(II)-Re(I) systems 234–236
   Ru(II)-Re(I) systems 224–233
   Ru(II)-Ru(II) systems 233–234
supramolecular polymerization 135
surface photochemistry 107

t
temperature-dependent behavior 7
terrylene 165, 175, 177
tetrabutylammonium hexafluorophosphate 276, 279
tetracene 59, 61, 165, 174, 175
tetracyano-p-quinodimethane (TCNQ) 2
thermal fluctuation 187, 188
threshold excitation intensity 164
time-dependent density functional theory (TD-DFT) 160
time-of-flight technique 2
time-resolved resonance-Raman spectroscopy 177
TIPS-pentacene 173–175
trap potentials 10–12, 16, 19
triethanolamine (TEOA) 213, 236, 238, 241, 268
triisopropylsilylethynyl (TIPSE) 64, 66
triplet energy migration-based photon upconversion (TEM-UC) 164
triplet excitons
   diffusion coefficient (D) and length 156
   electronic transition processes 158, 162
triplet-triplet annihilation (TTA) process 158, 162
triplet-triplet annihilation-based photon upconversion (TTA-UC) 162
   NIR-to-visible 165, 169
   properties of 164
   quantum yield 164
   sensitized by metal complexes with S-T absorption 169, 170
triplet-triplet energy transfer (TTET) 163
turnover number for CO formation (TON_{CO}) 239

v
van der Waals and π-π interactions 28
van der Waals force 58
van der Waals interactions 4, 15, 27, 41, 65, 70, 192
visualization 145, 147

x

X-ray photoelectron spectroscopy   289

y

Young's modulus   106

z

zinc porphyrin sensitiser   279
Z-scheme-type electron transfer   277